Csaba Varga

# THE HISTORY OF
# NUMERALS
# AND
# NUMBER-WRITING

Original title

A SZÁMJELEK ÉS A SZÁMÍRÁS TÖRTÉNETE

by Csaba Varga
Pilisszentiván, Hungary, 2012

Translated by
Dr. L.Kontur

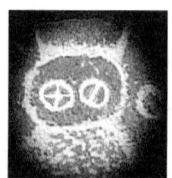

FRIG Publisher

Varga Csaba

# THE HISTORY OF NUMERALS
# AND
# NUMBER-WRITING

FRIG Publisher

# CONTENT

The serial numbers of the parts and tables are **to be read from right to left** as it was used by the Etruscans.

**PART Λ.**     Reckoning

**PART IΛ.**  Our present numerals

**PART IIΛ.**  The calculator

**PART IIIΛ.**  Well known ancient number-writing methods

**PART IIIΛ.** Subsequent number-writing methods

**PART X.** Letters as numerals

**PART IX.** Appendix

# PART I

## THE SPIRITUAL INFLUENCE OF THE PAST

## I. PREFACE

This book does not stand on its own. It accompanies my very success-ful book JEL JEL JEL (2001, five editions), in English: Signs Letters Al-phabets, the 30,000-year history of the alphabet (2009) and in German: Zeichen Buchstaben Zahlen (2010). Their messages are interdependent, because writing and number-writing are like twins, or like two sides of a coin.

I am convinced that this new book of mine: The history of numerals and number-writing will wake the interest of just as many readers and in-vestigators.

The pronunciation of the Hungarian alphabet for English speaking peo-ple as it is used in this script:

| | | | | | | | | | |
|---|---|---|---|---|---|---|---|---|---|
| A | <a> | („flop") | I | <i> | (ink) | P | <p> | (**pair**) |
| Á | <aa> | (**arm**) | Í | <ii> | (**evening**) | R | <r> | (**run**) |
| B | <b> | (**boat**) | J | <y> | (**young**) | S | <sh> | (di**sh**) |
| C | <cz> | (**cz**ar) | K | <k> | (mil**k**) | SZ | <s> | (li**s**t) |
| CS | <ch> | (**church**) | L | <l> | (**play**) | T | <t> | (**time**) |
| D | <d> | (**wand**) | Ly | <y> | (**young**) | Ty | <tj> | (**Katja**) |
| E | <ɛ> | (**gender**) | M | <m> | (**milk**) | U | <u> | (**book**) |
| É | <e> | (illu**strate**) | N | <n> | (**no**) | Ú | <uu> | (**ooze**) |
| F | <f> | (**friend**) | Ny | <nj> | (el ni**nj**o) | Ü | <ue> | (**Munich**) |
| G | <g> | (**go**) | O | <o> | (h**o**ld) | V | <v> | (**have**) |
| GY | <dj> | (**George**) | Ó | <ó> | (**open**) | Z | <z> | (**zoo**) |
| H | <h> | (**have**) | Ö | <œ> | (**her**) | Zs | <ž> | (**treasure**) |

# 2. INTRODUCTION

It was recognized long ago that our knowledge about the history of writing numbers was very sketchy. The threads leading into the past have regularly been lost in the fog. Similarly, we haven't found the start of the old civilizations and even the oldest known culture must have originated from a previous, older one. There are no signs of insecure starting. The high cultures always appeared out of the earth in full armor for the archaeologists. Nature must have destroyed all the traces going further back.

It is not easy to recognize gaps in knowledge and this is probably the reason why we handle the numeral-writing of different cultures as developments independent from each other. However, there is no proof for that. Number-writing became something of a stepchild because of this uncertainty and most books about the history of writing don't even mention it. For this reason, therefore, it did not attract attention that the few humble-looking signs which do exist give a very clear track of the ancient movement of intellect through history and the written-down numbers represent clearly measurable knowledge of high intellectual worth.

Let's put aside the different doctrines and follow with attention the outward forms of the numerals and the basic thoughts about how they were written. We are going to do this on the basis of the rapidly growing number of finds. We will analyse carefully written numbers and we will see that the seemingly rootless threads are all connected in the number-writing method of "dots+lines" which is with us still. If we look from the far past into the present – and do not follow the usual bad method of trying to understand the past from the present – then we can see that there were only partial and small changes in writing numbers since ancient times. Nowhere and never was there a substantial change apart from a few recent stumbling and inventive tryouts. However, the same can generally be said about the history of writing-signs, writing methods, numerals, writing numerals, languages and music. The strong influence of one special ancient culture is still alive today. One example of this: did the reader ever think about the age of the "abacus"? It can be bought in a local store, but nevertheless it has been around for several tens of thousands of years.

Finally, I state and will prove that writing numbers and the history of our numerals has been a continuous process for at least 30,000 years. This statement may startle some readers for its uniqueness. However, I have already proved this statement for the history of the writing-signs in the book "Jel Jel Jel" (Signs Letters Alphabets) and it is equally valid for the similar age of the numerals as well. It is so startling because of our distorted view of the past. Let me give an example of this.

Somebody reading scientific articles regularly may often find statements such as: "Writing was already known at the beginning of the third millennium BC. The writing of numbers, respectively, the development of the numerals happened parallel to writing"

The second half of the statement is acceptable. Everybody who writes certainly writes numbers as well. But the first part, saying that "writing was known in 3000BC", suggests that it started around 5000 years ago. It does not say directly, but anyone who reads it would probably think that writing would have started shortly before that. However, many findings show us that writing and number-writing were present 30,000 years ago. It did not start then, it was already present in different places. When did it start? [1] We don't know. It could be much-much older. Some inexplicable additional matters tell me that both, writing text and numbers including the so-called "proto-nostratic" ancient language, are the inheritance of an ancient high culture. Should somebody in 10,000 years' time evaluate our history, they certainly will see it as one long-lasting civilization of at least 30,000 years including our present (if nothing else happens until then).

Let's look at some data from the very early history of humanity in order to be able to imagine how things really might have happened. We should not be afraid of thinking in a large time-scale about human cultures.

---

1 What we really think of, when we say „start"? We may wipe important questions under the carpet while connecting to this inconceivable idea.

# 3 ABOUT THE INTELLECTUAL CONTINUITY

Chinese people live unaffected by the knowledge about their ten thousand years of continuity. Archaeological findings prove that they were growing different kinds of millet seeds and rice. They had dogs, domesticated swine and ceramics, etc. The oldest known find does not mean that agriculture and livestock-breeding started then. These are the oldest known finds found in China. The traces point to a much earlier time. Twelve thousand-year-old grain and grinding stones were found along the Nile, 15,000-year-old cooked rice in Japan and 30-35,000-year-old granaries in the area of the Caucasus. Even the Aboriginals of Australia have had plantations probably for thousands of years and, even while moving around, they regularly returned to those caring for them.

Intellectual continuity may also have stayed unbroken for a very long time. Examples for this are the three pictures below. We see the continuity of a kind of Mother-Goddess representation:

a. Çatal Hüyük 9,000 years

b. Kebele M-goddess, Minos, 4-5.000 years

c. Symbol of England 19th cent. with a "French sauce"..

Picture 1. Everything is the same, only the English changed her into a martial lady, but they could not know 200 years ago that the shield made out of the coat of arms will look like a plate on a car's wheel.

We see on picture 'c' that the English, no longer aware of the essence, even put a harpoon, the emblem of Neptune into the hand of the life-giving Mother-Goddess and changed one lion for the disc. Despite all of that, the starting point stayed clear. We can see clearly here that this thought and its representation stayed continuously in people's mind for nine thousand years, even during the time between these pictures. There was no break of memory in people's minds. This continuity even leads us into much earlier times. Thus, the people of Çatal Hüyük were already taking care of a custom inherited from an endless past before them.

The two pictures below prove to us a much longer-lasting intellectual continuity.

Picture 2, A 'shaman' (medicine man) 14,000 years ago (Les Trois Freres). On the right is a Siberian shaman (drawn by Nicholas Witsten 1705). Despite the 14,000 years difference in time the essence remained the same.

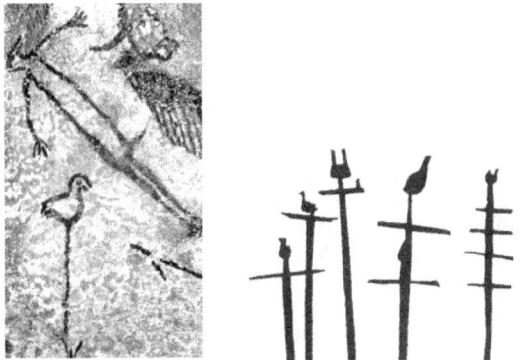

Picture 3. On the left is a wall-picture in the cave of Lascaux (17,000 years old), on the right are head-boards of shamans in Siberia. There we have 17,000 years of continuity.

14

Picture 4. Above is the Hittite winged Sun, below is the winged Sun pictured in the embroidery of Kalotaszeg, Hungary.

Such concordances tell us about intellectual movements of world-historical importance. (These are very useful to unmask false theories.)

The game below is a proof of continuity:

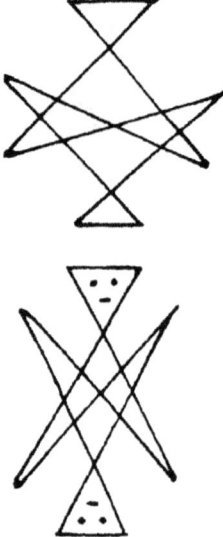

Picture 5. Above: Magic knotting from Ancient Iran. Below: exactly the same today in the middle of Europe

The pictures are from the book Kabay Lizett: "Kulcskérdésekhez kulcsszavak" (Keywords for pivotal questions). One produces this line-structure by holding a loop of thread stretched between the fingers of both hands and putting the line crosswise from one to the other fingers. We all played this game as children.

Picture 6. Left are the "busók" <bushook> from Mohács (Hungary) and right the Basque traditionalist procession in Navarra. In Mohács they carry two cattle-bells in the front or on their back. We see the bells in Navarra only on peoples' backs. It would be interesting to know why the two bells have to be carried as regulated by the ancient tradition.

One of our other childhood games had many names: bige, ige, pike, brige, brincka, peca or dole. It was even played by Sumerian children. The 'bige' (a short stick sharply carved at both ends) is laid into a small hole in the earth. If you hit its outstanding point, then it will jump and you hit it again and again in the air until it goes down. The winner is the one whose 'bige' flies the farthest.[2]

A king of Pentecost remains a living custom in many villages. Young fellows play different games (running horses, fighting a bull) and the winner will be crowned as King of the village for one year, earning many privileges for the period (like a Prince of Carnival in Cologne, Germany). We must look into the very early past for the origin of this custom. There we find that the destiny of this king was originally not a game. He was given everything he wanted for one day, but after this day he was sacrificed to God, probably to ask for a plentiful harvest. Pentecost has been the day when the fulfilment and spreading of springtime's promise happened. The date of Pentecost (7x7 = 49 days after Easter) was determined in ancient times, when 7 was a holy number and life was regulated by the Moon-cycle. Christianity adopted it with a new, but very similar meaning.

*Let's quote Katona Sándor from the book "The History of Humanity" by Barabás László:* [3] *"A group of international scientists belonging to YALE*

---

2 Gönczi Ferenc: „Somogyi gyermekjátékok" (Games for Children in Somogy) (Kaposvár, 1949)

3 Barabási László unfortunately he did not tell the title of the book.

*University left in 1932 to do geological, palaeontological and archaeological investigations in the Himalayas. H. de Terra - the known American archaeologist and a member of the group – wrote later that they particularly examined, in the middle of Kashmir, a dried-out lake-bottom of 10,000 square kilometres at 1830 metres and surrounded by large mountains. Surprisingly, they found in the Lama cloister of Srinagar in Tibet - where Körösi Csoma Sándor also once worked - a folk legend in a very old Tibetan handwritten book called "Radzsatarangini" (Circlet of the Kings). This legend tells in every detail, how the lake once suddenly moved down to the valley following earthquakes and landslides The story is still alive among the population of today's Kashmir.*

*The investigations of the scientists proved step by step every detail of the legend and they were able to find the former water level along the slope of the mountains. Countless bones of animals and humans made a later carbon-14 determination possible and revealed the probably date of the catastrophic event to be 400,000 years ago. This means that the accurate description of that event had survived, through the story-telling tradition, over 8,000 generations."*

We must see this continuous process of grandmothers' story telling as a fire fighters' chain, where water-filled buckets are passed from hand to hand. This is the way in which knowledge spreads. Should we invite all those people to dinner who have been involved in bringing us the 10,000-year-old knowledge of baking bread, then we may count on around 200 guests. Or, dear reader, the good news about Jesus' birth needed only 40 couriers to reach us. It seems very few, compared to the large time-span. Thus it was possible that the very important knowledge of number-writing might stay unchanged for such a long time. Why change a time-honoured method, and one which is the best? You will doubtless agree with this statement after reading this book.

See in the following one more example of continuity in history. There was a tool used for stripping off and degreasing animal-skins. ".. we have countless finds built from flint from the time of the middle Palaeolithic, in the so-called "Mousterian culture", 35,000 years ago. Two thousand years ago this tool was still chiselled from stone here and 1000 years later in the Alaskan 'proto-Eskimo' culture. Today you can buy an exact copy of this tool in a shoemaker's supply store, made of iron as a 'meat-processing knife'. Its form hasn't really changed."[4]

The fragmentation of history is only an illusion.

---

4 Gáboriné Csánk Vera: *Az ősember Magyarországon.* (Prehistoric Man in Hungary) Page 52, Gondolat 1980

# 4. GEOMETRY'S KNOWN HISTORY

We rightly look at geometry as at one of the highest intellectual activities. Geometry is the inseparable twin of mathematics.

The same conditions are needed for both. We must practise both by drawing them. We draw the numbers too.

The traces of doing geometry go back a long way. It is wrong to say that the ancient Greeks developed it, because they only learned it from somewhere else. They were industrious in writing down what they had previously learned abroad. Their best teachers were the Scythian magis and the Egyptians. However their own scientific efforts were also great.

Let's take a brief look into the past:

Picture 7. A pupil practised Geometry 3,800 years ago in Babylon.

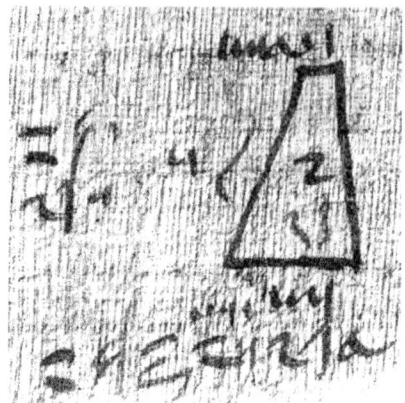

Picture 8 Calculating the volume of a truncated cone. (Part of the MMP, the Moscow Mathematical Papyrus, created in Egypt 4,000 years ago).

We see below some 30,000-year-old geometrical drawings. The pentagon drawn with one line is particularly interesting.

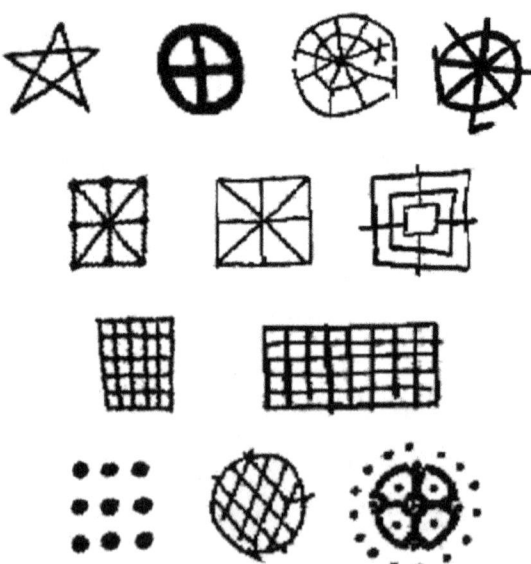

Picture 9 These 30,000-year-old drawings are proof of the very long use of planimetry. The nine dots in the left lower corner signal certainly the number nine (Fontainebleau, Dordogne).
http://www.dinosoria.com/fontainebleau.html

The engraved stones of Tászoktető are well-known. Their age has been estimated by some people as 30,000 years, but no official dating is known. In any case, the stones and the condition of the drawings are witnesses to a great age. I know of 4 "written" stones. Two of them, saved by Barabási László, are now in the Museum of Gyergyó. People speak about many more stones, but those were collected and used for building houses in the neighbouring villages.

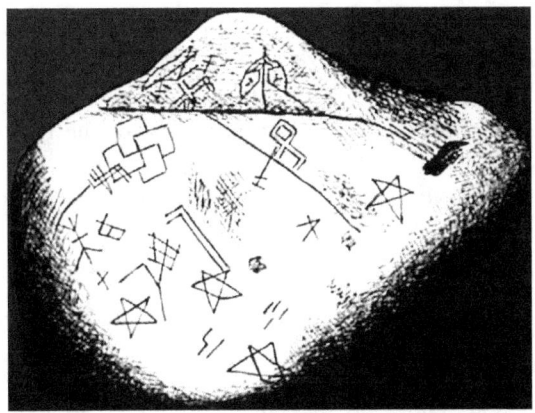

Picture 10. One side of stone No 1 from Tászoktető.

Picture 11 A very interesting geometrical figure on stone No 1.

Picture 12 One side of stone No 3.

There are many stars drawn on the stones, all of them (except one) drawn with one continuous line.

Picture 13. Geometrical polygons: star formations on the Tászoktető stones

Older than 17,000-year-old geometrical forms in the cave of Lascaux

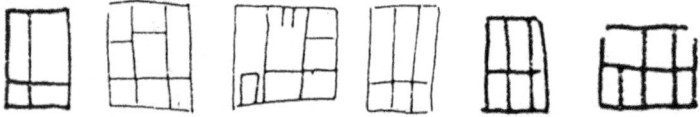

Picture 14 We don't know for what purpose, but they drew the above pictures, which means they were thinking in geometrical forms. We know this for certain, because we can see it.

The find below is around 77,000 years old. It was found in the Blombos cave, South Africa, a few hundred kilometres from Cape Town.

Picture 15. Even spearheads and picking tools had been found

Picture 16. Hunting tools in the Blombos cave

Look at the sword-like tool (indicated by the white arrow). It has a perfect geometrical design and its carving is flawlessly regular. Its proportions look rather attractive. The net of quadrates scratched onto the stone supports the evidence that those people applied their creative activity to the abstract world of forms.

Let's compare picture 14 with the following two figures. Both carry the geometrically identical pattern, but while the first is 77,000 years old, the two following figures are only 30-35,000.

?

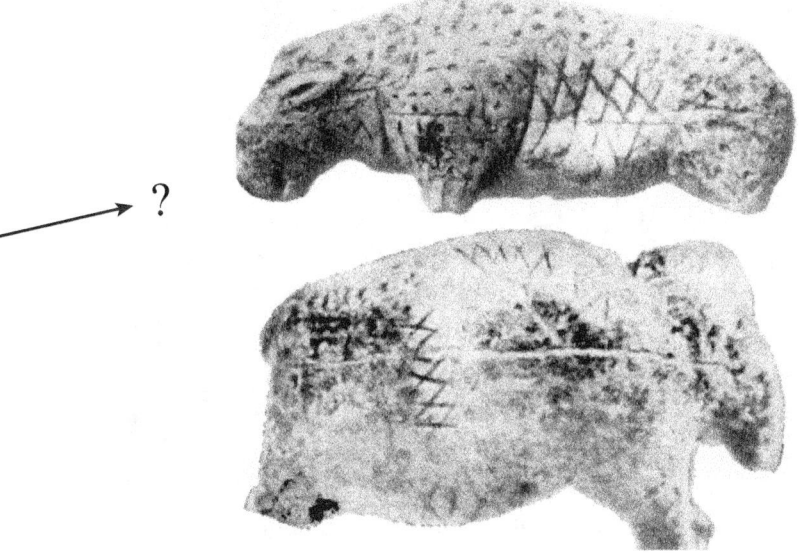

Picture 17 Found in Vogelherde cave close to Stetten ob Lontal, Germany

Vilmos Zolnai wrote the following about these rows of quadratic patterns found on mammoth and bear(?) in his book "Művészetek eredete"[5] (Origin of the Arts): *"It is unthinkable that the scribbles on the animal-flanks were meant just as decorations. Did the hunters of the Ice Age make these scribbles because they liked it better that way? The serial pattern has a definite meaning. Its meaning appears from the sacrificial practices of the Voguls, Bedouins and other folk-groups.[6] Otherwise, similar patterns of zigzag lines decorate the field marshal's batons and many statues of the mother goddess as well."*

Was this pattern so widespread that we should assume a common origin, now lost in the mists of time? We can't imagine a fast dissemination process over large distances with occasional possibilities for building up intellectual connections. Therefore, we may rightly suggest that an intellectually-based custom presented in the Blombos cave 77,000 years ago, turning up again 30-35,000 years later in the Vogelherede cave and still being used in our times, must have spread from a common root.

I think this quadratic net meant the texture of life or proliferation, the chain of multiplication. In reality, both mean the same. The sense of the quadrate standing on its corner meant, in known antique cultures: many, multiplies. It is even used twice on a 12,000-year-old drawing as a writing-sign. See the following beautiful drawing.

Picture 18 Musée les Eyzies, Les Eyzies de Tayac. (Find Lorteti)

---

5   Gondolat Publisher, 2001.

6   Unfortunately, the author of the book does not tell us any more about the meaning of these signs in his book

Looking up what the oldest suitable Sumerian signs tell us: the quadrate means much, plenty and if doubled: very much. The perpendicular line means *water*. Both together: *flooding, irrigate*, and doubled this meant *big flood*. The swimming animals and the large fishes between their legs support this interpretation.[7]

The conceptual deepness of *flooding, much, multiplies and plenty* is the same, we can spread its general meaning and its writing-sign forwards and backwards in time. Are we still not familiar with this really ancient sign?

---

7 See the detailed explanation in the book "The Living Language of the Stone Age" page 123 by Csaba Varga, Frig Publisher.

# 5. THE OLDEST KNOWN WRITING SIGNS[8]

Picture 19. Signs engraved into an aurochs bone around 200,000 years ago, found in a cave close to Dordogne in Southern France by Francois Bordes 1968). (See more in the book "Mysterious Writings" by Mandics György 1987)

The following find turned up in 1830 in a stone pit not far from Philadelphia around 22 meters under the surface. The workers cut out and raised a big block of marble after cleaning the area from mica- and clay- slate and gneiss layers. By cutting the block they found these sharply engraved signs. There is a regular rectangular niche with smooth flat bottom and steep walls, 4x1.6 cm. The two regular signs stand out from the smooth surface:

Picture 20 The engraved signs from the marble pit close to Philadelphia. It was first publicized in the Journal of Science 1831, 19, page 361. See further Michael A. Cremo – Richard L. Thompson: Forbidden Archaeology. *The hidden History of the Human Race*. Bhaktivedanta Publishing P. 797.

The two signs have certainly been made by humans. The age of the find is without doubt greater than several hundred thousand years, from the over 20 metres of deposits above it. Some people estimate it at several million years old.

From 30,000 years on, the finds of written signs  get more numerous and it is easier to demonstrate the continuity.

---

8 I demonstrated earlier in my book "Signs Letters Alphabets" that the 30,000-year-old history of writing can be perfectly proved through archaeological finds.

# 6. THE EVEN TODAY CLEARLY CONCEIVABLE PAST OF WRITING AND NUMBER-WRITING

Looking at number writing since earliest times – this means according to the finds, at least 30,000 years ago – there were two kinds of number-signs: dot and stroke. These remained in use in different places on the continents and are even hidden within the Arabic and Hindu numerals. The custom became looser in ancient times, when they started to write letters for numerals. We will deal with those later and first go through the history of writing numbers with dots and strokes.

Writing numbers with dots and strokes is still alive today and this proves that our culture is the inheritor of a much older culture. Our statement is further supported by the fact that the use of the dot-stroke numerals runs parallel with the history of using linear signs for writing. As we will see later, both definitely determined our literacy.

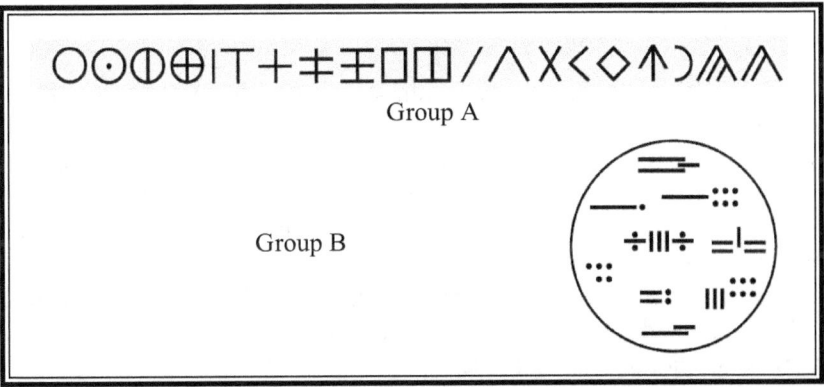

Picture 21. It's obvious that those groups contain two quite differently built signs.

The signs in the groups A and B have been used continuously for 30,000 years and even now form the basis of our literacy. Group B can only be connected with calculation and must be numerals because they build a system. Those are relics of a fully efficient number writing system. Recognizing this fact was the reason for starting to investigate the history of number-writing and this book presents my explorations up to the present time.

The following picture shows a selection of number and writing signs. This much already makes the aspect and mentality of archaic literacy clearly perceptible. Even if the finds are far apart in time – we embrace 30,000 years – the overall picture is real, because we can look at it as the result of random sampling, a method used often in other sciences. Further support for the authenticity and exclusiveness of this aspect is the fact that no counter-examples, no other sign-group was found:

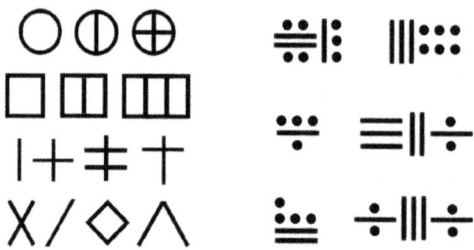

Picture 22 Aspect of archaic literacy. We see typical writing-signs on the left and a few numerals on the right. One can sense the perfect harmony of the conceptual cleanness, deep spirituality and empirical simplicity.

In this book, we will prove only our statement about the history of number-writing.

We must accept large time-spans as natural. In early times, some cultures were extremely durable, like China, which can be proud of its 10,000 years. Moreover, nothing proves that it first started 10,000 years ago nor are there any signs that it will end soon.

This will be supported by the sign-collection of Kate Ravilious (Picture 22). She collected all recognizable signs around the world. There are even dot-stroke numerals among those.

I ask the reader not to form an opinion too early.

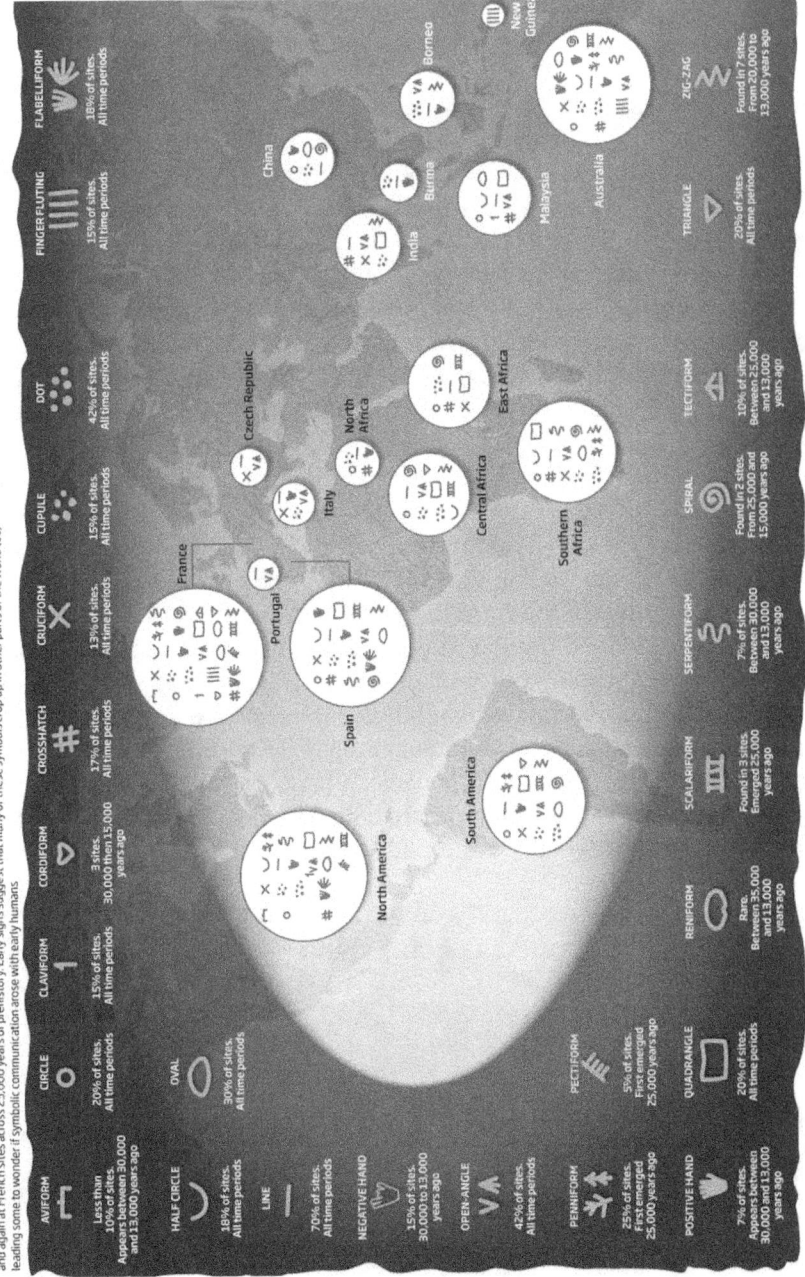

Picture 23. The sign collection of Kate Ravilious. We can see dot-stroke numerals here as well.

# 7. WHY ARE WE BLIND?

Why do we resist recognizing the intellectual accomplishments of our ancestors so hard? One possible answer is easy to find. Speaking about any folk groups living on the lowest level (living in the Amazonian jungle or somewhere in Asia) we hear immediately: "they live at the Palaeolithic level" This means then for everybody - we were brainwashed this way - that earlier, at all times everywhere everybody was living at this level. However, "we" managed to get out of the continuous, dull nothing during the last 8-10,000 years.

But it couldn't happen this way, and this can be proved easily with an experiment.

If we can pick any randomly-chosen moment in the 500,000 years of human history, then we can pick the day we are reading this book as well. Since, if we eradicate the false thought from our mind that our moment is an exceptional second in the infinity of time, then we must accept that this moment is equal to any other randomly-selected second of history. What are we seeing in this randomly-chosen moment? We see that, just as the lamellas between both ends of an open fan fill the spread - all the different cultural stages, high and low, that we can imagine fill the space even in our own time. This is in perfect harmony with nature's function, as living creatures fill out every possible habitat using appropriate body forms.

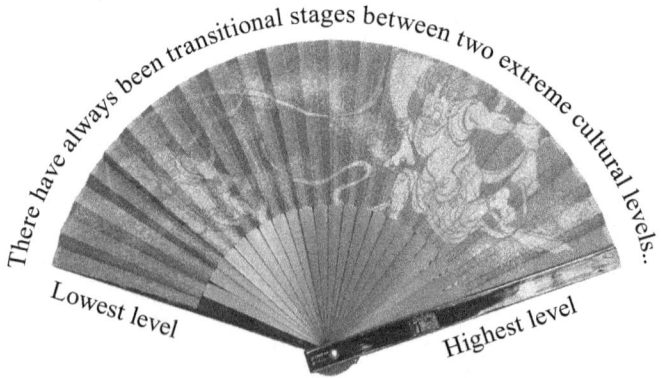

Picture 24. The cultural stages could never be uniformly the lowest.

We may only argue about the question: how much was the fan open: how much was the difference between the highest and lowest culture at a given time?

It is not possible that the fan was always closed during the million (?) years' history of humanity and opened first suddenly 8-10,000 years ago. Neither is it possible that the fan opened slowly, at the same speed continuously to become as open as possible for today.

We can only say, knowing nature's rules, that there was always an up and down of the highest cultural levels throughout the endless time of past history. Naturally, the nadirs always followed great catastrophic events of our earth. By no means can we state with certainty that there was no higher culture than ours on earth before.

However, one can certainly state:

1) there has always been a high stage of culture and parallel to it
2) there were always many people living on the lowest level as well.
3) and all the stages between these two extreme cultural levels
    were necessarily filled out at all the times as happens today.

Naturally, this doesn't mean that the same culture dominated humanity at the highest level or that the same population group always lived at the highest level. Neither do we have to prove that the same culture must be connected to one special folk group. Nothing, however, will exclude the possibility of any of the above. One thing we can say is that the highest culture of today has remained the same for 40,000 years and we don't see its end[9] yet nor that it is about to give its leading role to a quite different culture.

The big fractures and pauses in the cultural continuity of the last great ages exist only in the history books. Most historians describe the cultures unearthed in different places as starting at level zero and describe their origin rather as parthenogenesis out of nothing.

However, they can't do anything else. Every empire starts world history with itself and expects: historians, archaeologists and linguists approve that theory. This way, it breaks into pieces what has been in reality unbroken, since human history has been necessarily continuous. I write this reluctantly, but it can't be left unsaid, that the history books even of today bristle

---

9 There are several prophecies about the near death of our culture, but these point merely at our modern European branch of the ancient culture, and in which there are so many lies collected about the past, language and anything else that it must crash pretty soon.

with big frauds[10] Well, our history-writing stands on such bases and for these it needs to be sharply separated into prehistoric and historic time. This practice is wrong for two reasons.

First: 6-7,000 years is the time when written material naturally disappears and older material is gone long before that.

Second: the division into sections. In reality, there were never stages or breaks, because human history is necessarily continuous.

This way, the past looks unavoidably foggy. The past of our culture, its origin, has more or less disappeared. But the fact is that the earliest stages of all archaeologically found old cultures were already fully developed. It disproves the possibility that the ancient or even earlier cultures could have developed from nothing. Nowhere do we see a start. For example, although this is not generally known, it is becoming more and more clearly seen that literacy, number-writing and folk-music of the cultures in Old Egypt, Mesopotamia and China have the same very old cultural root.

Our thinking about the past is influenced by the false teaching that everything before us happened only as a preparation for us to reach the peak. But who can tell what the highest possible peak is?[11] We live with the wrong idea that creating machines is the highest goal of human intellect. However, to reach the peak through spiritual erudition needs a much higher level of culture than fabricating machines. There were certainly several cultures in the past which reached higher spiritual peaks than we do. Several million years are a long time.

A building or anything else becomes more ruined the older it is, but we don't conclude from this that people built ruins in times past.

---

10 For example: There is no "Indo-European" folk, neither is there a group of Indo-European languages. Those are the results of the large scale history fabrication, or rather, falsification of history in the 19th century. We like to say: "a word disappears but a writing stays". This is true, but it matters very much what stands written in a text. However, we know now that 90 % of the old documents written in Western Europe are falsifications.

11 I once heard a man say, at a party 'We live in the best possible world' and somebody answered: 'That's what I'm afraid of'.

# PART II.

INTRODUCING THE DOT-LINE NUMBER-WRITING

## 1. THE FOUR MAIN STATEMENTS OF THIS BOOK

This book proves that the principle of dot-line number-writing and dot-line numerals as well – with a few not very old interesting exceptions – is still around. With some "marring", this is what we use now all over the world.

We can establish the following four statements based on our investigation of the dot-line number-writing practice:

1) We must break away from the supposition that number-writing and general literacy first started around 5-6,000 years ago.

2) We have to stop thinking that using the place-value in number-writing is an invention of our Antique cultures. We will see that dot-line number-writing can only exist by using place-value. Therefore, since there is dot-line number-writing, people write numbers with place-value. This has been happening, according to our finds of today, for at least 30,000 years.

3) We must stop believing that marking '0' was an intellectual invention of the known Antique cultures as well. We will explain this misunderstanding in the chapter dealing with the abacus (soroban).

4) For as long as there has been number-writing – it doesn't matter by which method – there has been a numeral system extensible to an optional size. Naturally, it was used only in cultures where the people played with or had to deal with large numbers.

*

I am asking the reader not to put any of the presented material about dot and line or number-building with those in any "evolutionary" order. This wouldn't make any sense. Nothing is sequential there. The main point is that everything is always the same. If somebody lets her hair grow and curls it, her hair still stays the same. For the last 150 years, we have had to see development in everything. Forcing this way of thinking we may tend to see development even in pebble-stones if we put them in sequence by their size. This is very conspicuous even in the different books written about the history of number-writing. The authors put the finds obsessively in 'developmental' sequence and they don't care about the fact that the different versions were used in the same time or even that there is an earlier example of the "more developed" variant. There is merely a timely succession in the creation and in the discovery of archaeological findings, but the principle of number-writing has stayed unchanged. However, no scientific study or book can become successful without presentation of evolution. This is how our general view of the past became so much distorted.

# 2. THE LINE (rod)

The topic of this book is therefore the dot and line as number and the history of numeral building using these signs. I will state and prove that this kind of number-writing has been continuously in use for at least 30,000 years. I won't say that this was the time when somebody started using it. I state that the principle of place value has existed as long as dot-line number-writing has existed and the numbers were written according to this principle. The major essence of dot-line writing is the place-value.

Let's look at the line first.

The easiest way to remember the numeric value of things is to make a stroke (write or scratch) for each one. See this waiter's note:

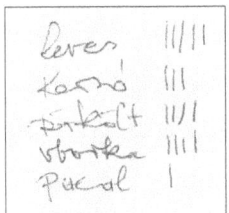

Picture 25, Isn't this a frequently seen picture?

he "Ishango" bone was found in 1950 in the then Belgian Congo and its age was supposed to be 25,000 years. There are strokes on the bone in three rows ordered into well separated groups.

Picture 26 The Ishango bone. Taken from the Internet.

We receive an interesting result by counting the strokes in the separate groups and comparing them row by row.

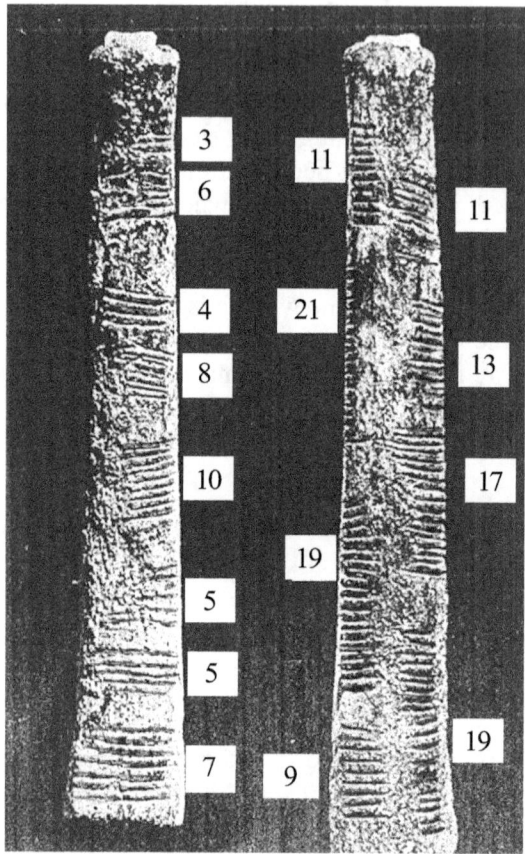

Picture 27. The strokes are grouped, as many groups as numbers.
(picture from Internet)

This must mean much more than just recording numbers randomly. The sums of the row in the middle and on the right are equal. The numbers on the right are (11, 13, 17, 19), which are prime numbers. In the middle: 9 and 19 are one short of 10 and 20, but 11 and 21 are one more than 10 and 20. The sum of both is 60. These can't all be by chance. Chance appears randomly and not multiplied at the same location and time.

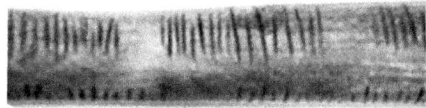

Picture 28. Not all photos of the Ishango bone are of good quality, but on this one we can see that the lines are carefully carved.

One can see a periodicity on the find below. The groups are divided rather with a stroke made carefully longer than the rest. The values are 25 and 30:

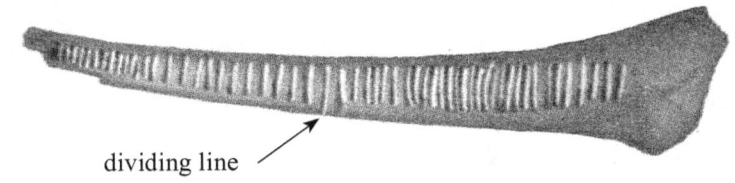

dividing line

Picture 29. These scores were scratched 25-29,000 years ago onto this bone. It was found in Dolni Vestonice (Filep László-Bereznai Gyula:The history of number-writing, Gondolat 1982)

Picture 30 The grouping of the numbers is visible. Found in Saint Marcel, Indre, France. A piece of the Aurignacian culture 26-30,000 years ago. Musee des Antiquités, St-Germain-en-Laye

Picture 31. Made in the Magdalene Period, 14-21,000 years ago. In Musee des Antiquités Nationale, St.Germain-en-Laye

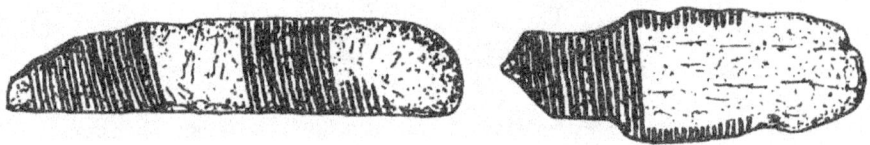

Picture 32. Both are from the Aurignacian culture 29-37,000 years old, found in Pekarna-cave, Czech Republic

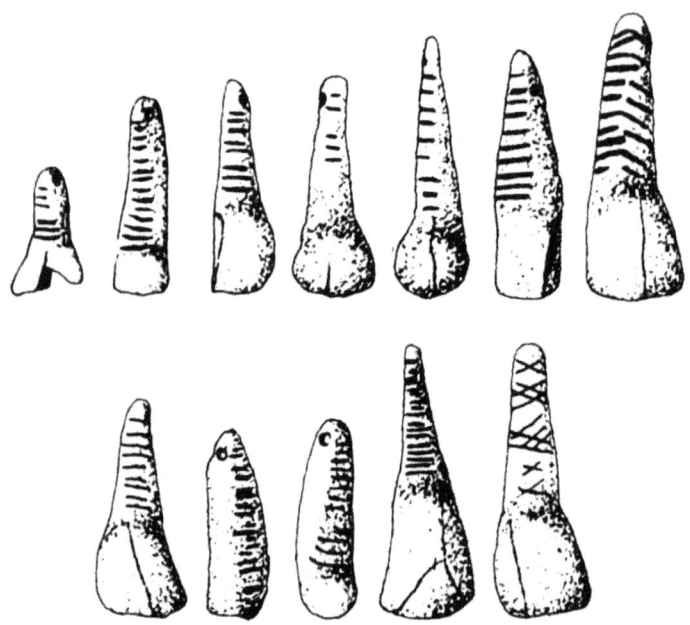

Picture 33. Teeth with numbers on them from the Magdalene culture. Goyet, Belgium

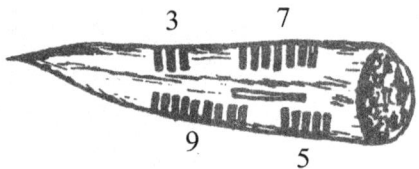

Picture 34. Numerals engraved into a tusk: 3, 7 and below 9 and 5. 22-29,000 years old. Found in Brassenpouy, France. Bordeaux Museum of Aquitanie.

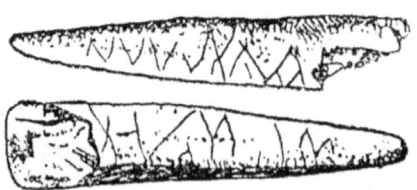

Picture 35. There is a text on both sides of this tusk. Found in Jankovich cave. It is around 15-41,000 years old, from between the Seleta and Magdalene Culture.(Lambresht Kálmán: Ősember1931)

The next find is unique, it carries text and numbers.

Picture 36. Around 13,000-year-old drawing from Rouffignac.

The method of marking numeric values with strokes has remained unchanged for ten thousand years. I think, however, this number could be multiplied by several hundreds or thousands as well.

The simplest method is to draw a line if one sees one, two while seeing two or three lines when seeing three of something. Why did they draw the strokes? There is only one answer for this question: to be able to look up later the sum of the pieces, the numbers.

Well, this is what we call number-reading.

How can the numbers of the strokes be "read"?[12] Reading the strokes we will receive a summing up number with a **name** to be remembered. Numbers can be different and only then is it worth marking numbers if every number has a name independent from the time of marking it with an adequate number of strokes.

See one more find from the very distant past:

Picture 37. 400,000-year-old notes (late Acheulian culture)

---

12 It is not a mistake to call it reading (megolvasni). In Hungarian, this meant counting in earlier times.

# 3. NUMBER AND NUMERAL

Marking numbers only with strokes we must draw as many strokes as the number of things we look at. Once we have drawn the necessary amount of strokes then it is not really of any further interest what has been counted. Because of this, we can more easily ponder upon the number itself, as the people certainly did 20-25,000 years ago as they engraved the groups of strokes on the Ishango bone.

Well, how can we count those belonging into one group enabling us to give their exact numbers by using strokes? It is only possible with numerals. We can count only with numerals. We can establish this rule: If there are no numerals than there is no counting either. However, number-writing has existed since ancient times; therefore, there have been numerals for such a long time that we can't even comprehend it.

If you want to check the above statement, please try to count a box of matches without naming any numeral. Take away one by one the pieces from a heap on the table and say only an 'r' for every piece, avoiding naming the numeral. At the end you will just look at the matches without knowing how many there are. Therefore, counting an unknown amount is not possible without numerals.

For the same reason, it is not possible to mark a certain number using strokes without naming out loud or silently the appropriate numeral. Take a piece of paper and a pencil to write down 21 with strokes. Make the strokes but say again only 'r' for every one of them. After a certain time you will no longer know how many strokes you drew.

It is so simple to prove that one can't count anything, or draw a certain number of lines without using numerals. I ask the reader to try at least one of the above experiments. It is a delightful experience to become aware of this interesting barrier in our brain.

We state again: the language of the people who wrote the numbers on the bones - found at Ishango and Dolno Vestonice - certainly contained names for numbers.

Think about this again: if the person who draws the lines does not know the numerals than his or her engraving makes no sense at all, because he or she can't tell the result. What would be remembered after the job is done, if not the name of the number?

We can even go one step further.

The person who drew 19, 17, 21, or 55 strokes, as seen on the previous 20-30,000 years old finds, must have had not just names for every number, but even a system of numerals. Because then - as today – when counting up to 85 there couldn't have been 85 individual 'custom made' names for every number not connected to any system. One can't apply so many names instantly. If we count up to 85, we are certainly able to continue to 86, 98 or even to 198.[13] But so many individual number-names are certainly not useful in the practice without a system. Only the numeral system can serve as a system of number-names.

Let us put it this way: There is no number-writing without numerals and the numerals must build a kind of system. However, this system is necessarily the mirror-picture of the then used numeral system.

*

I didn't want to say that the above statement was valid for everybody living in ancient times. There were certainly groups of people, as there are even today, who are not interested at all in larger numbers. These people name often just 1 and 2 separately and apply a repetition system using these names for higher numbers. (In this case we may speak about a base 2 number-system). We shouldn't be fooled by many school or scientific books, where it often appears that this was the 'stone-age' level and **everybody** - just a 'short' time ago - was only able to do that much[14] **everywhere** on our earth.[15] We will simply say, as we have proved as well and as the whole book will demonstrate, as far as numbers (mathematics) are concerned the interests of different population groups have always been very different, as is the case today.

---

13 There can be large differences between cultures. There are certainly some, who do not deal with numbers larger than 1000. However, if somebody can say a thousand, they can say 10 or 1000 thousand as well.

14 Moreover, this is not a question of ability; it is rather that of interest.

15 One has to force this foolishness for being able to speak about a compulsorily demonstrated "evolution". It is not possible here, more precisely neither possible here. One has to demonstrate an evolution every times seeing comparable things. This is a misuse of the in reality unexplained sense of "evolution". For example, the bureaucracy "evolves", and the communism was steadily "evolving". In these cases and in many others we really call deterioration a "development".

# 4. THE NUMBERS OF OUR EYES

The need for numerals arises from the fact that our eyes are not able to register the quantity of a pile at a glance if it is larger than three. Our brain is able only to register 1, 2, and 3 without numerals. Look at the following picture:

Picture 38. Above three, the more lines are drawn side by side, the less it is possible to determine the number of the lines.

Let's look at the groups of lines one by one.

We can try this ourselves and be convinced that the amounts of 1-3 are immediately clear at a glance. **No counting, no numerals needed, our brain directly registers their amounts**. We do need a little more time with four or higher quantities. The direct "recognition of quantity" by our brain ends above more than three parts. Just watch yourself.

To avoid counting, one divides four in to '1+3', '2+2' or '3+1' partitions and the sum of the two numbers will tell us that one more than three lines are there.

Seeing 5 is more complicated. If we are still avoiding numerals, our eyes divide 5 into '2+3' partitions and we have to add the two parts.

At 6, we may take the product of '2x3' or add the partitions '2+2+2'. In the case of partitioning such as '1+2+2+1', we won't be able to tell the amount directly for there are more than three partitions. Our eyes can look at only one or – but more vaguely – two partitions at once.

Now look again at the picture and choose the group with 7 lines. If we are still not counting the lines one by one, if we try a glance we will immediately lose count when determining the quantity. Our attempts will get more and more impossible on trying larger amounts. We may try to build groups of 2 or 3, but 7 is more than 2x3. When our eyes move from one partition, we will lose the border of the previous partition.

So far, we have tried only groups with a few lines. Building partitions won't help anymore when counting larger groups of lines. The quantity of those can only be determined by counting the lines one by one:

Picture 39. The exact quantity of that many lines is not easy to estimate.

In these cases however, the lines must be counted one by one either out loud or silently and this is only possible by naming the numerals.

# 5. THE GROUPING

As we have seen previously, the lines must be grouped in some way for easy recognition of their quantity. We do the same today when we write 100000000 in this form: 100,000,000. There is more than one possibility for grouping. The ancient Egyptians, as did many other cultures, put the groups built of lines 1-9 below each other. See here the numerals 6, 7, 8 and 9 and later more detailed examples.

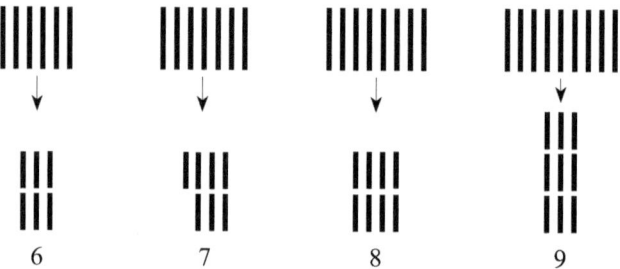

Picture 40. This way it was easy to read the numbers written with lines.

The Phoenicians used a different method of grouping numbers 1-9

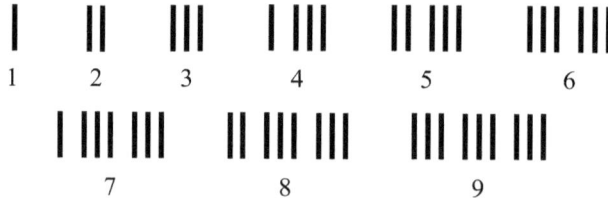

Picture 41, The arrangement of the Phoenicians looks less practical

One can build groups in other ways as well.

Most people used coal-heating at home during my childhood. The coal-man came with a horse-drawn carriage and hauled the coal to the basement in a basket on his back. He drew a stroke on the basement-wall at every turn with a piece of coal. He started grouping at every fifth stroke – as we have seen, no counting can happen without grouping – to have his numbers steadily presented.

At the fifth turn, he crossed the previous four lines with the fifth on the slant. In the following, he drew an oblique line after the 9th, 14th, 19th, and so forth at every 5th stroke, until he finished.

Look at number 18 written using the coal-man's method:

 = 18

Picture 42. The coal-man's number-writing on the basement-wall of my grandparents.

You just have to count the bundles, 3x5 = 15 and add the rest (3) to receive the result. Due to this grouping, the number of turns can be much more easily determined than by counting the lines one by one.

We can see the same grouping in a pub, where the skilled bartender records the orders of regular customers.

Picture 43. This way, the bartender does not have to count the strokes one by one.

This grouping happens in reality in the base 5 number system. The decimal system is a 2x base 5 system. We could build groups with 4 or 6 strokes also, depending on the system we are working in. It is important to build always the same size of groups for any particular calculation.

Picture 44. Numbers engraved into bone 20,000 years ago. The grouping by oblique strokes is clearly visible. It is part of a find from 1937 in the Czech Republic.

# 6. THE INDEPENDENT LINE

We can deduce everything that follows from the grouping method of the coal-man. Let's repeat the two possible ways of grouping by fives:

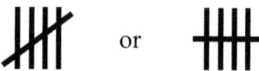

Picture 45. The main point is to cross over the bundle. The oblique stroke is used more often, because drawing a horizontal line needs far more care.

In reality, the line crossing over the four strokes represents by itself the number five perfectly. If there is a cross-over line, we just used it for crossing the four strokes. Therefore, it becomes unnecessary to draw the four perpendicular strokes before it.

We will see later that **the number 5 doesn't have a sign in this sign-system.** The five will only be expressed by these crossing-over lines. Therefore, we think of five when seeing a horizontal line and forget the four crossed over 'unnecessary' perpendiculars.

It is therefore enough to draw a horizontal line to write five. We only look for this.

Picture 46. The horizontal line tells already that this can only be a five.

We need only two numerals to write the numbers up to 9: the five is a horizontal — and the rest are made out of perpendicular I lines.

The numbers 6 or 8 can be written in the following ways:

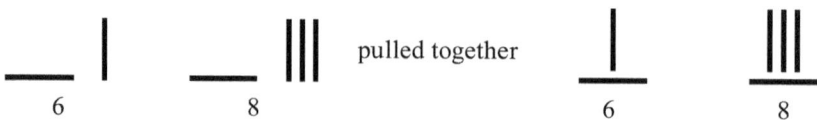

Picture 47. We have to write two perpendiculars for 7 and 4 for the number 9 over the horizontal line.

Let's see the numerals from 1-9 in the base 10 system:

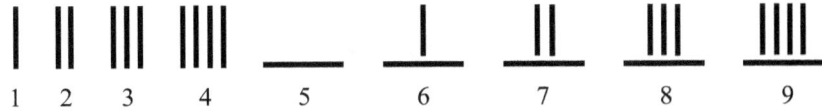

Picture 48. The numbers from 1-9

The same written with two Chinese modifications:

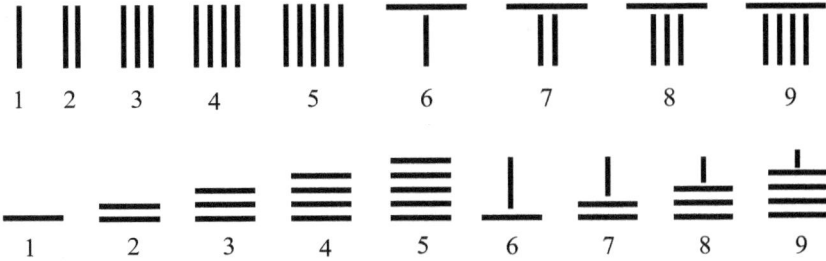

Picture 49. In the first row are the numerals of the previous pictures, but turned upside down. In the second row the same figures are turned through $90^0$. In this case, each five sign (the horizontals) has to stand perpendicular.

# 7. THE DOT

People have been drawing dots instead of the perpendicular lines from the very early past until today. See the numbers from the previous page written with dots instead of perpendicular lines:

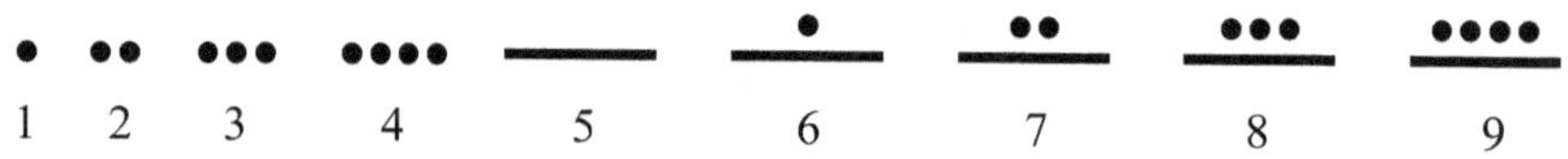

| • | •• | ••• | •••• | — | •̄ | ••̄ | •••̄ | ••••̄ |
|---|---|---|---|---|---|---|---|---|
| 1 | 2 | 3 | 4 | 5 | 6 | 7 | 8 | 9 |

The same again turned through $90^0$.

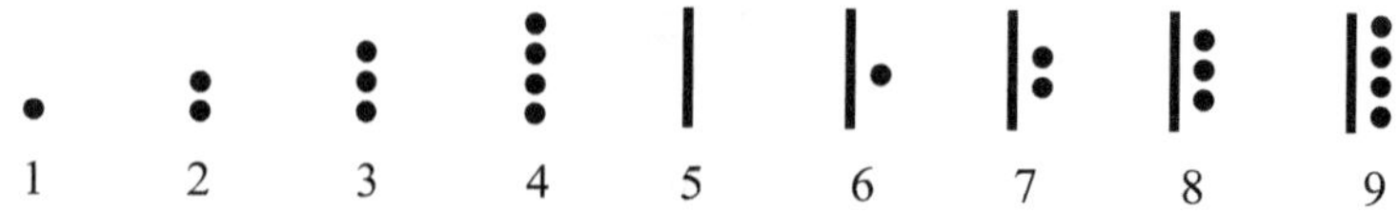

| • | : | ⋮ | ⋮ | | | | | | |
|---|---|---|---|---|---|---|---|---|
| 1 | 2 | 3 | 4 | 5 | 6 | 7 | 8 | 9 |

Picture 50. The dots may stand even at the left side of the lines.

The dot and the line were always exchangeable, if the given position system of numerals made this possible. For example: the Cretans exchanged dots and strokes randomly 3,500 years ago in their so-called "linear B" writing. It can be seen in the numbers below that decimal numerical values were sometimes written with dots, but at other times with strokes.

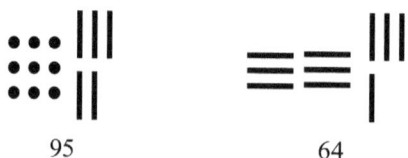

95                    64

Picture 51. The dot-line-change didn't create problems. If the perpendicular strokes mark the singular numbers, then the dots or lines before them must represent the next higher place-value in the decimal system, (even dots).

It seems that people preferred to use dots while writing by hand, but lines were easier to engrave or print. Chinese used rather strokes, because in this way the numbers could also be assembled from bamboo sticks.

It often happened that the numbers 1 – 9 were written only with dots. Here, the line - usually signifying five – got a further assignment: to mark a higher place-value.

See the numbers 1 – 9 written with dots only:

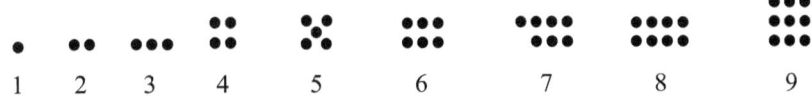

1    2    3    4    5    6    7    8    9

Picture 52. The arrangement of the dots varied greatly also for which we will show examples later. The lines are missing here; therefore, they certainly mark the number 10 or 100.

A large variety is possible in the system of dots and lines. There can't be a rigid rule for the arrangement in this system. Therefore every possible variety may have occurred during the long time of its use.

# 8. THE COMPLETE ANCIENT
# SIGN COLLECTION

We have seen three number-signs so far: dot, line and the line standing at right angles to the first line. There was a fourth line too: the 'long line' marking usually higher numbers, for example: a thousand. It was naturally a question of agreement as to which sign should represent which number. We see the basic (most often used) cases below:

| • | — | │ | —— |
|---|---|---|----|
| 1 | 5 | 1 | 100 |
| 10 | 10 | 10 | 1000 |
|   | 100 | 100 |   |

Picture 53. The four basic signs of the clear dot-line number writing and its usually assigned values.

These were the signs, and one could write all the necessary numbers with them. The values assigned to the signs varied, but it was only necessary to have an agreement within a certain system or culture.

# 9. THE ALTERNATING VALUES AND POSITIONS OF LINES AND DOTS.

Let's look again at the number-writing of the "coal-man". His horizontal line represents 5. Follow the form-changes of seven:

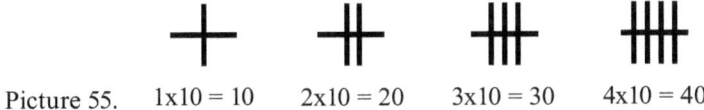

Picture 54. Variations on a theme.

The cross-over line of the "coal-man" has been used even as a "ten-folder" line. See below the Chinese "ten-folders":

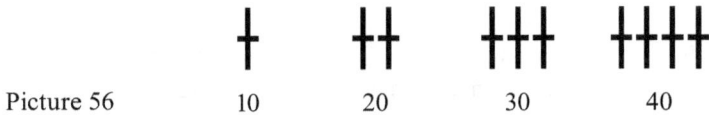

Picture 55.　1x10 = 10　　2x10 = 20　　3x10 = 30　　4x10 = 40

Chinese people sometimes ten-folded the singular numbers one by one while writing numbers in the column of tens:

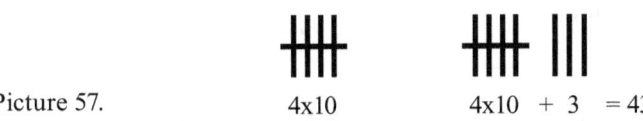

Picture 56　　10　　　20　　　30　　　40

In the ancient number-writing system preserved in Switzerland the horizontal lines are "ten-folding" in the same way as in the above Chinese numbers 40 and 43:

Picture 57.　　　4x10　　　　4x10 + 3 = 43

Ten-folding with line and using dots instead of the perpendicular strokes:

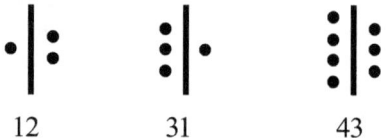

The line ten-folds the dots above it

12    31    43

*The same numbers written but with numerals turned through 90⁰:*

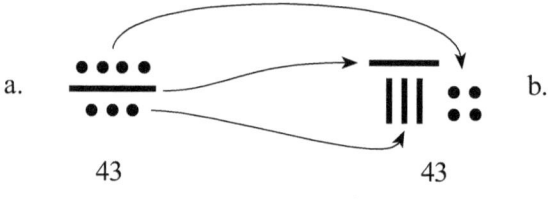

12       31       43

Picture 58 The reading is here 1x10-2, 3x10-1, 4x10-3. Ten means in Sumerian and in Hungarian as well "von" (-ven, -van). Therefore: 43 = négy – von – három, together: negyvenhárom. (We will explain in the part: Egyptian number-writing, why the "von" does not appear in the numbers from 10 – 30)

The Egyptians arranged the same numbers a little differently. Comparing both variants of 43, we can see that only the dots under the lines (the singulars) are changed for perpendicular strokes.

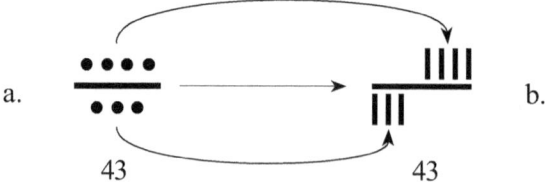

a.        43                              43        b.

"traditional" that means          the same number in
very ancient way of writing          Egyptian way.

Picture 59. It is a little-known fact that the Egyptian and Hungarian numerals are the mirror-images of each other. See page 298.

The number 43 (picture 'a' above) was also written with lines only:

a.        43        43        b.

Picture 60. It is clear that the dots have merely been changed to strokes.

It is unambiguous and follows from the previous pictures that the numbers in the pictures below are also only form-variants of each other.

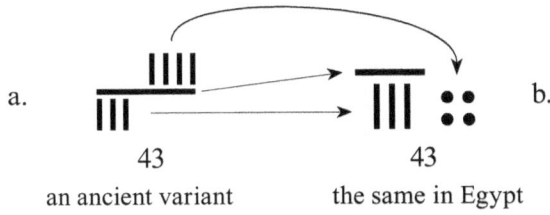

a.                                                    b.

43                             43

an ancient variant       the same in Egypt

Picture 61.

While omitting the ten-folding line and writing the numbers according to their local value one after the other, the number will stay unchanged.

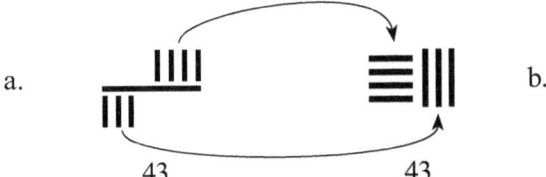

a.                                                 b.

43                             43

Picture 62. We see here again just a variation of the ancient numeral and saving one line, the horizontal one

The arrangement of the number in 'b' is identical with our modern writing. We can see it better by putting the number 433 afterwards:

a.                                              b.

4   3                       4   3 3

Picture 63. We see here again just a variation of the ancient numeral and saving one line, the horizontal one.

Let's turn the above number 'b' through 90⁰ then we arrive at the arrangement of the Mayans (it is **only the arrangement**, because below we present a number in decimal system, instead of the 20-based system of the Mayans).

            400

            30

             3

Picture 64. It is 433. An original Mayan number will follow in the next picture:

See number 639 in the base 20 system of the Mayans:

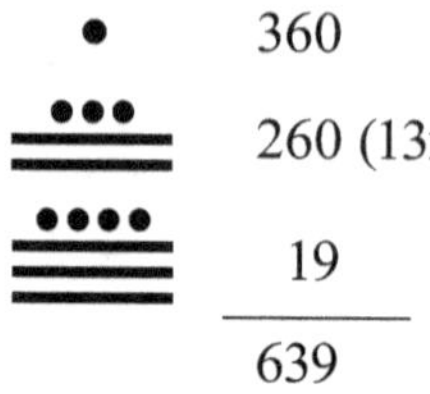

$$\bullet \quad 360$$

$$260\ (13\times20)$$

$$19$$

$$\overline{639}$$

Picture 65  Why is it 360 instead of 400? We will discover this in the chapter about Mayan writing (Page: 87)

Let's write the same number in the decimal system:

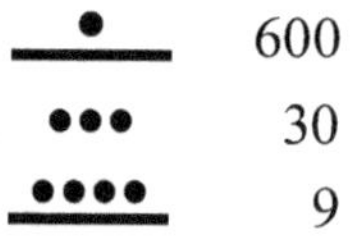

$$\bullet \quad 600$$

$$\bullet\bullet\bullet \quad 30$$

$$\bullet\bullet\bullet\bullet \quad 9$$

Picture 66

Now we turn again through $90^0$ and can see that we write our numbers in exactly the **same system** as previously shown

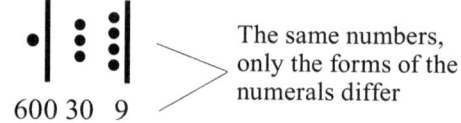

600 30 9

The same numbers, only the forms of the numerals differ

Picture 67  We pronounce the number exactly the way it is written: $600 - 30 - 9$, that is six hundred/thirty/nine

As you see, it stays the same. Turning the number to the left or to the right, using horizontal or perpendicular strokes, the number doesn't change.

Now, let's see the 'long-line'.

The long-line's role was mostly hundredth-folding. With time, this line became longer and longer in Egypt, but after the XXth dynasty it became shorter again. In many cases those were even curled.

$$\underline{\qquad}\ |$$

$$1 \times 100 = 100$$

$$\underline{\qquad}\ |||$$

$$3 \times 100 = 300$$

Picture 68 We say it in this sequence: one-hundred, three- hundred.

The Egyptians, using hieratic writing, made the line for 1.000 somewhat shorter, but put a dot under the right end of it.

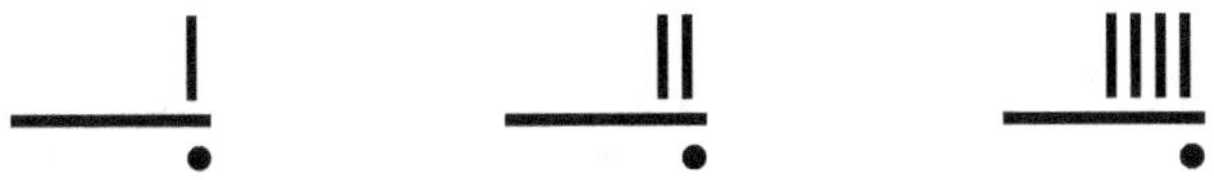

$$1 \times 1.000 = 1.000 \qquad 2 \times 1.000 = 2.000 \qquad 4 \times 1.000 = 4.000$$

Picture 69.

Follows the number 3,400 written this way (from the right to the left):

400       3.000

$= 3.400$

Picture 70.

Not only have the Egyptians written numbers this way. We can see below some examples of the same writing from different times and places. We know only the value of the Egyptian numbers for sure, because plenty of findings can be easily compared. In the case of the other two groups we can't tell the values exactly due to easily exchangeable lines for hundred and thousand. The system however, is undeniably the same:

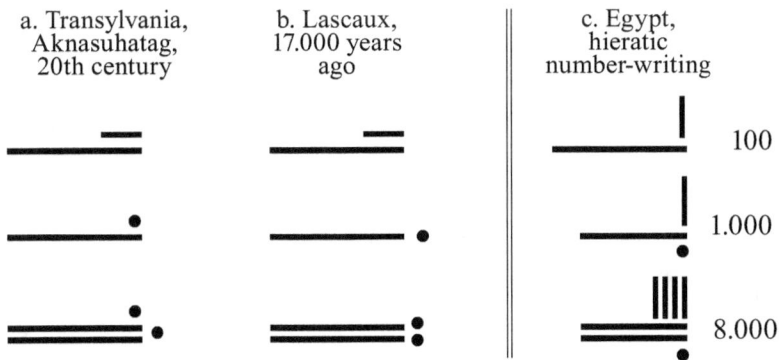

| a. Transylvania, Aknasuhatag, 20th century | b. Lascaux, 17.000 years ago | c. Egypt, hieratic number-writing |
|---|---|---|
| | | 100 |
| | | 1.000 |
| | | 8.000 |

Picture 71. We will explain later in the chapter about Egyptian number-writing why the last number on the right meant 8,000 (4 x 2,000) in Egypt.

These were merely selections, but very characteristic examples. We will see other solutions later as well.

Summary:

It is out of the question that number-writing with place-value could have been a medieval invention. What happened is that one of the previously shown number-writing arrangements became predominant. Since there is number-writing and people did not want to draw, for example, 5,864 dots or strokes in sequence, they had no choice but to write using the place-value system to be able to boil them down. This became possible by applying one of the possible and equivalent arrangements of dots and lines. In other words, there is no other way to compact the numbers than to write them applying the place-value system. [16] This is why the number-arrangement in Picture 67 is the most sympathetic for us, because we are used to it.

We therefore refute the scientific belief that writing numbers arranged by place-value was a big medieval intellectual invention. As we will see, the use of the presented number-writing methods and the application of dot/ line numerals can be placed in the early Ice Age, but there are no traces of uncertain beginnings, not even in the earliest finds.

---

16 Mathematical operations can easily be boiled down as we experience in many mathematical and physical deductions. Example: $10 \times 10 \times 10 \times 10 \times 10 \times 10 \times 10 = 10^7$

# 10. The '0' (zero)

It is a misconception that the concept of '0' (zero) had to be invented and its invention was a historical scientific event which happened not very long ago. This belief arose because nobody really examined closely the writings of the early ancestors, telling us instead that those signs are only primitive numerals, "rudimentary" number-writings.[17] As we could see earlier and will learn later, this is a total false view, but it is found in every scientific book and also in the schoolbooks. It is, however, easy to understand that the marking of '0' is not necessary in the dot/line number writing system.

Why not?

Because the form of the written sign determines the position value as seen in Picture 52. Please, look at it again. The point is that the position values can freely be adapted to the four signs and the arrangement of these signs with given position-value is again free within the frame of logical agreements. Therefore, if no value is assigned to a position, then no sign needs to mark it and nothing needs to be said by the reading of this position. Look at the example (and look again at Picture 62 as well).

Let's give the value 1 to the dot, 10 to the horizontal line and 100 to the perpendicular line. The numbers 400, 403, 430 and 433 will receive the following forms:

| 400 | 403 | 430 | 433 |

Picture 72. The forms of the signs tell us which position-value is missing from the number.

There are no tens and ones in 400, no tens in 403 and no ones in number 430. It is not necessary to mark this deficiency. It is visible that there is nothing on those positions of the numbers.

We repeat here again a very important observation. We don't read the number 400 as we write it now: **four hundred – zero ten – zero one;** we say nothing about the tens or ones, because those do not exist. We just say **four hundred** (see the four perpendicular lines above). We don't read 403 as **four hundred – zero ten – three**. Again, we just say what we write with

17 What could a rudimentary numeral look like? It is a numeral or not. Isn't it?

the dot-line method. Neither is 430 for us four hundred – thirty – zero, but just **four hundred – thirty and** we say nothing about the missing ones, since nothing is written on that position in the dot-line system.

In other words, we still pronounce the numbers as we wrote them in ancient "prehistoric" times.

Some disorders in number writing came up in relatively recent times. For example, in China the missing of the dot gave the most problems when writing numbers. They didn't have enough signs, and so they couldn't hand the position-values to the forms of the signs. They marked the "empty" positions between the numerals with inter-spaces. For example:

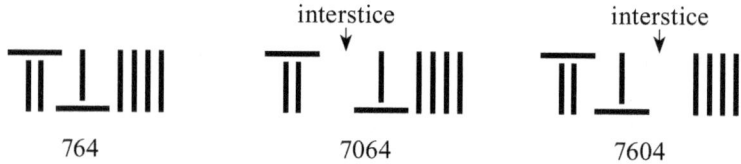

Picture 73. See more details in the chapter about Chinese number writing

If the case was not clear, then they wrote the position-value of the given number into the gap. For example: 萬 = 10.000:

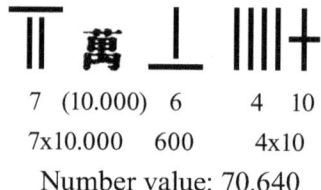

Number value: 70.640

*Written by the handed out position values:*

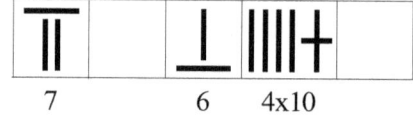

Picture 74. Text was mixed in between the numerals and the writing of the tens deteriorated, using two signs for it. However, following tradition nothing is written at the position of the ones, for there is no value there.

This way, however, made the Chinese number writing too complicated.

They started - probably because of misunderstandings, seemingly from the 9th century on – to mark the empty interstices with a circle: ○
See an especially large number:

|  |  |  |  |  |  |  |  |  |  |
|---|---|---|---|---|---|---|---|---|---|
| 1 | 9 | 5 | 5 | 1 | 1 | 9 | 6 | 8 | 0 |

*Let's see the same in a compressed form:*

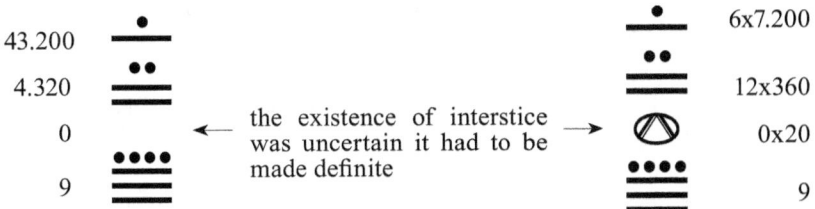

Picture 75. This number is almost 2 billion: 1,955,119,680. The signs here lost their value-marking role. One must be careful that the lines don't get mixed up and so they are written alternately in horizontal and perpendicular position. That's the reason for writing the five and the one above first horizontally than perpendicularly.

We could see that the zero made it clear only when not having any value on this position. Therefore, writing numbers by position-value never really depending on the use of zero. The number-writing was done marking the position-values before that.

The Mayans were also forced to mark the emptiness of the position-values. They didn't use the marking's role of the different forms of the signs either. The line marked for them only the fives and the dots the ones from 1-4. The numbers were written perpendicularly needing interspaces between them. It wasn't the same to write ☰(=15) or ☰ (=205). Thus, the interstices became occupied. One needed a new sign to mark the empty position-values and invented the following for this: ⬖

| | | | | |
|---|---|---|---|---|
| 43.200 | • | | • | 6x7.200 |
| 4.320 | •• | | •• | 12x360 |
| 0 | ← the existence of interstice was uncertain it had to be made definite → | ⬖ | 0x20 |
| 9 | •••• | | •••• | 9 |

Picture 76. The value of this number written in the base 20 system is 47,529

In India also, people used the circle ○ to mark the empty position-values, because there the dot-line signs above the number five became In India also, people used the circle to mark the empty position-values, because there the dot-line signs above the number five became increasingly ornamental due to very lax handwriting. The earlier clear rules became disturbed here also.

An interesting situation arose in Mesopotamia. The number signs of the Sumerians were the original archaic dot-line signs, but soon they were printed with sticks into clay and this method changed their forms. Furthermore, they counted in a base 60 system, and their writing became mixed. (See the chapter for more details about Sumerians number writing)

The number 1, normally a perpendicular line, changed to $\Upsilon$, and the number 10 to $\blacktriangleleft$. This last sign for 10 ($\blacktriangleleft$) if standing on the second position-value – meant 10x60 = 600, because of the base 60 numeric system. The number 610 therefore, was written as follows:

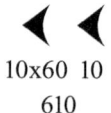

10x60  10

610

Picture 77. The Sumerian 610. Marking the zero is pointless in this case

It's clear that no zero is necessary here, because the sign already has a value of 10. The number 12 was written as $\blacktriangleleft \Upsilon \Upsilon$, which is the same as the earlier $-\|$ , $\underline{\|}$ , or $\underline{\bullet\bullet}$ . (The horizontal line in the Mayan system, mentioned before, had a value of 5 and not 10).

Also in Mesopotamia, knowledge about the ancient rules which regulated the position value according to the form of the signs seemed to fade. In many cases they also had to mark the empty positions of the numbers. Two kinds of solutions were developed. The zero inside of a number became two 450 slanted lines and for the zero at end of a number they used two 450 slanted long-tailed signs of number 10 (both doubled).

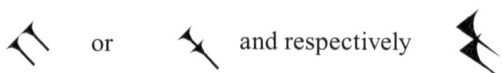

Picture 78. The zero in between and at the end of numbers. The zero at the end was not doubled in Babylon

Therefore, the requirement to mark the empty position in number writing has made things worse. The proof of this is the fact that no introduction of the substitute-sign zero was necessary as the form of the numerals determined definitely the position-value. This role of the numeral-forms is well demonstrated in the Egyptian hieroglyph number system.

The perpendicular line remained to mark the one in the hieroglyph number system, the traditional lines for the hundred were often curled and they used dots for the thousand, which were put on the top of flower-stalks. There are as many thousands as flower-stalks in the little heap. Look at the following around 5000-year-old examples for the numbers 409, 4006 and 2300:

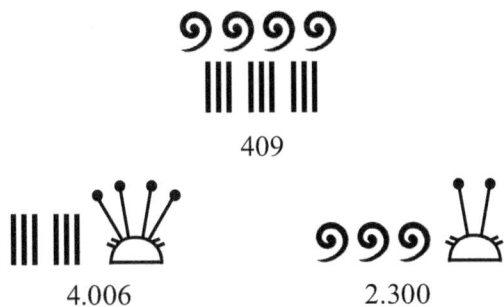

Picture 79 The Egyptian numbers above are to be read from the right to the left.

It is not easy to tell at a glance how many flowers are sticking up, therefore, in the case of seven, they in the case of seven already needed two heaps (4+3):

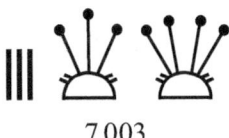

7.003

Picture 80 It is not just random that the figurative numerals seem to suggest a fairyland. However, their linkage to the dot-line number-writing is still recognizable.

We can clearly see here as well why it isn't necessary to mark a number-position without value. However, there is no doubt that this is a form of number writing using position-value.

# 11. SUMMARIES

In this second part of the book, we introduced the essentials of dot-line number writing and were able to prove three of our main statements:

1) Since there is number writing by whichever method, there is a numeral system as well.

2) Since number writing with the dot-line method exists, there is certainly number writing using position-values.

3) The necessary introduction of the '0' (zero) as a substitute sign appeared first after the worsening of the ancient clear numeral system as a result of the decline of the ancient organically built culture. Even the writing signs were deteriorating in some cultures, as you can read in my book "Signs Letters Alphabets".

Let us finally check the fourth statement, which says that number writing was practised for at least 30-40,000 years – not just for the last 5-6,000 years – and its use has been running in parallel with general literacy.

We can easily prove the continuous use of dot-line number writing for 30-40,000 years by looking at the archaeological findings.

# PART III

## THE MOST ANCIENT NUMERAL SIGNS

# 1. THE BASIC PRINCIPLE OF RESEARCH

I am convinced that many dot-line or line-line numbers are in the museums, on ancient objects, or on cave-walls, but nobody has yet recognized them as numbers just as happened in Lascaux (see page 76). They are probably using the signs – which build a closed system - for decoration, both scientists and visitors alike. Well, if people don't know what to look for, then their eyes won't see it and they can't be blamed for it.

The dot and the line are beyond doubt very humble signs and if one wants to decorate something then they like to put dots on it or draw lines onto it.

There however, is a dividing line at the first glance: are the signs put randomly or are they well-regulated. The difference is easily identifiable.

We see on the picture below randomly spaced dots.

Picture 81. This painting is around 17,000 years old (Pech Merle)

Seeing so many dots spread over this space, we wouldn't think of numbers. However, the "many-many dots" would mean something by itself.

Furthermore, the main essence of number writing is compaction and the multitude of dots in the previous picture contradicts this totally. Nevertheless, the artist certainly knew that 1 dot means 1 and two dots means two etc. Clearly visible random spots shouldn't raise any thoughts about numbers.

Picture 82. Random positioning is usually easy to differentiate from planned positioning.

Let's see some pictures with lines:

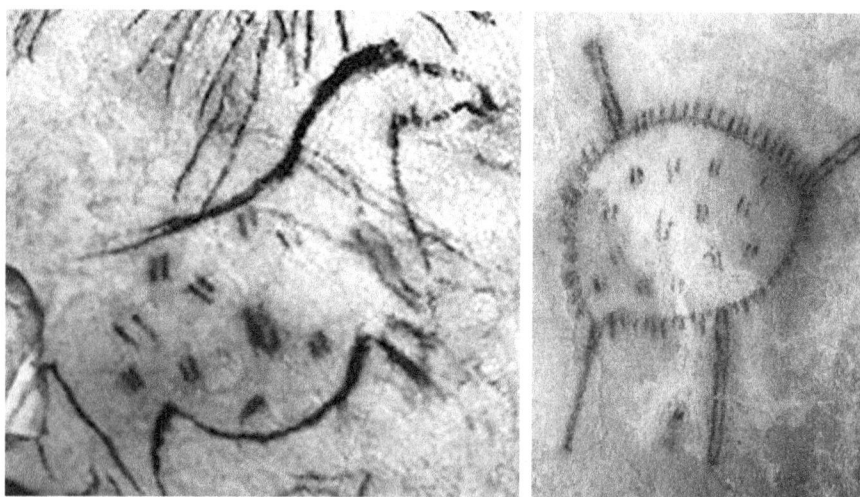

Picture 83. Short parallel strokes decorate the horse and the indefinable object. (both in Cueva de la Pileta)

We learned earlier that two parallel strokes can be read as the numerals 2, 10, 20 and 200. In these pictures however, nothing supports the probability of seeing numbers. The strokes could suggest the impression of fur. The right one could depict a drying stretched-out piece of felt. Seeing one or two dots below or above those double-strokes, (like these: ⚬ or ⚬⚬ ), could persuade us to see numerals.

Well, parallel strokes by themselves could present numbers based on the principle that 1 stroke = 1, 2 strokes = 2, etc. But there needs to be a certain regularity, parallelism and separation of the signs. On the find below the two conditions are not both fulfilled together, thus we can't say that on the stone below results or partial results of any counting appears despite the many apparently parallel engravings.

Picture 84. The age of this Australian find is estimated at 20-30,000 years

It's a quite different case if the regularity is clear.

We see on the picture below a wild pig with 8 strokes on it. We might find some connections in these lines,  but it is still not enough to make any generalisations.

Picture 85. Grotte de la Vache. A 13,000-year-old find from the Magdalenian culture.

Somebody engraved a wild pig and 8 parallel perpendicular strokes on its side. The lines belong together. Seeing those lines so accurately placed, we may certainly think that the artist knew how many lines he or she drew. especially since the lines on a piglet run horizontally, not crosswise.

Picture 86. Wild piglet.

Comparing the pictures we may well suspect that the strokes scratched on the piglet mean a numeral (number of the piglets?). But this is not certain.

We can be certain, as seen above, if we meet dots and strokes together in a group; even then, only if the coherence of dots and strokes or that of strokes standing at right angles to each other, is clear to see.[18] The importance of this attribute is clearly visible on the following pictures showing many dots and lines, but the most striking ones seem not be connected:

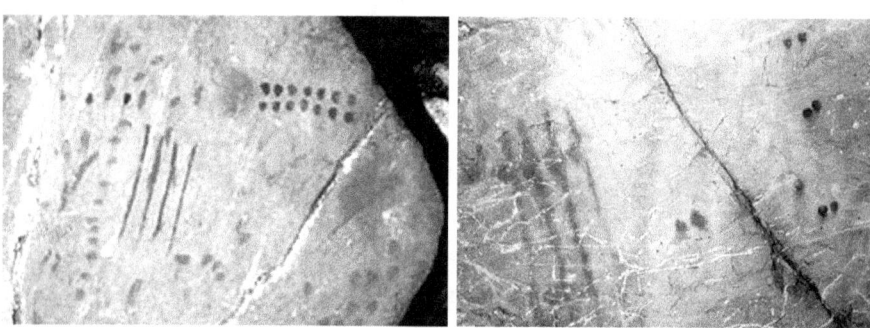

Picture 87. Left: 13-14.000 years old signs in the cave of Niaux. On the right: It looks similar to the left picture and may be of the same age, however its source is not known to me.

Note to the left picture: It is very possible that the two rows of seven dots in the upper right corner are intended to express a numeric value. Whoever drew the perpendicular lines certainly knew that they were drawing four of them. We can't say with certainty however, that they intended to draw a

---

18 In this case, however, it is certain that the number has been written using the position- value system.

numeral. Were the parallel rows of dots and the strokes close together, then we could be sure we were seeing a numeral written using the system of position-value. They don't belong together. The dots are painted, the lines are scratched and are not positioned nearby. It is essential, however, when writing numbers with dots and strokes that the components of a numeral be positioned in a clear group.

Picture on the right:

There are only dots and lines on this picture as well. All the dots but one are joined in pairs. No other relationships are visible. We see the four parallel lines as on the left picture, however no hints of them presenting a number. On the other hand, we can be fairly sure that the Egyptian signs below stand for numbers, since the dot-line systems on the find are unambiguous.

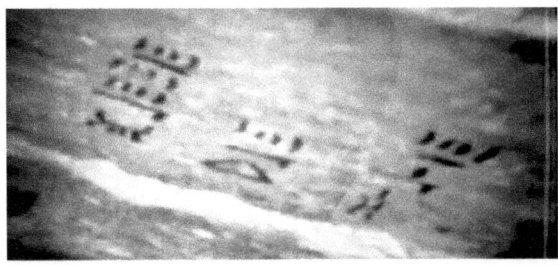

Picture 88. The ship found in a lined underground room next to the Cheops pyramid is today a frequently visited exhibit. A similar room also containing a ship has not been opened. Scientists drilled a hole through the stone-cover of the room and examined its contents through a telescopic instrument. We see in this photograph signs painted on the ship's coverage.

Someone painted those signs around 4,500 years ago onto the ship's board and this is not that far back in time looked from the point of view of our book. (See the chapter about Egyptian number-writing.)

Thus, the best way to receive good results in finding written numbers on ancient unknown finds is to look for dot-line sign-groups containing only a few signs. We won't consider as many possible finds this way, but in return our recognized numbers will be unambiguous. This precaution is necessary, because there are only a few or no comparable findings remaining from very old times. The number of finds dated from the more recent past has increased, thus our judgement has improved as well.

READING:

We should point out here that the dot-line and the clearness of the possible systems represented by them were rediscovered in 19th century. The Morse signs are built from dots and strokes.

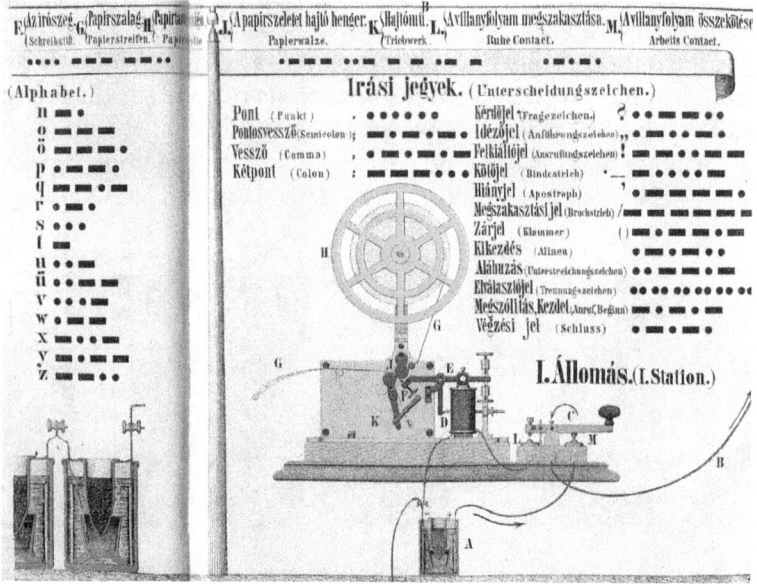

Picture 89. The signs of the Morse Code.(not including all signs)

Everybody, even Martians, can see at a glance that this sign-system was not designed for writing numbers. Since, going by the rules of number-writing, it would be nonsense to write 6 dots (●●●●●●) to mark one simple ● dot, such wordiness would be intolerable in number-writing.

The Morse Code doesn't represent the numerosity of a number-value. There is no connection between the number of signs and the value of the written number.

Picture 90. The Morse-numerals are not proportional to the value of the numbers they mark.

Therefore, there is some order in the numerals of the Morse Code, but it is an external one, one which is not related to the "nature" of the numbers. It is true, five is marked by five dots, but the match disappears by writing zero again with 5 strokes.

Picture 91. Samuel Finley Breese Morse (1791-1872) the inventor.
A drawing by Tibor Kaján.

# 2. INTRODUCTION TO THE ANCIENT NUMBERS

I made the following selections for this presentation in respect of having at least one group of dots with lines in selected find. As we said, the numerals may carry only dots or only lines. If two lines mean 10, then there is no need for dots. One added dot would mark 11. Writing down 8 dots would be enough, with no additional lines needed. On the whole, we just have to look out for the clear presence of the dot-line system. If it is there, then we are dealing with a number-writing culture. To put it another way, if they had an elaborate number-writing system, then we just have to look for the given value of the dots and lines, for the way in which they were put together and for the value of the number-positions (decimal, base 20 or some other system). Obviously – and this we definitely can't deny our ancestors – their goal was that if they were using this compound system for number-writing, then they built the system in such a way as to be able to write all the numbers they needed with it.

How could we even imagine that they were only able to write those numbers which are on the finds we are just now examining? Seeing a heart carved into an oak-tree and two letters (e.g. B. M.) engraved below it, we would never say that the person could only write these two letters. Since this is true, then the writer certainly learned these letters from persons who were able to write all letters.

# 3. . PONT D'ARC

30-32,000 years ago. Pont d'Arc cave, France.

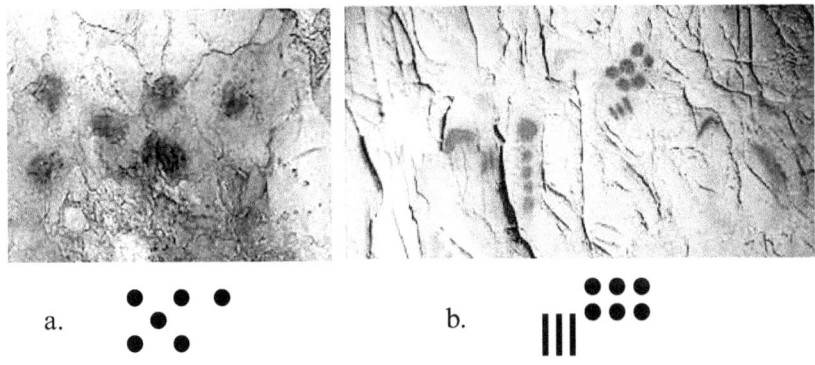

a. b.

Picture 92. Dot and line signs from the cave wall in Pont d'Arc

a.) The person who put 6 dots in this formation certainly knew how many dots he put. He counted. His chosen formation tells us that he had precise geometrical knowledge. We put the 5 dots in this formation onto our dice. We might even say that he must have had an important reason to put an additional sixth dot with the finely placed five.

b.) The formation of 3 perpendicular lines and 6 dots, seen in the picture, fulfils the requirement which we formulated above: we can first determine a group of dots and lines as a numeral if their relationship to each other in the group is clear.

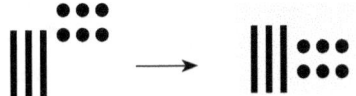

Picture 93. The two groups of signs are equal.

c.) The perpendicular row of dots in picture b gives no hint at its interpretation.

We can support all these with other quite obvious cultural data. In the first line, we have proof that there and then literacy was present too. Let's see the stone-slab found there, covered with writing signs:

Picture 94. The inscriptions on a stone slab in Pont d'Arc are around 30-32,000 years old.

There are several similar inscriptions on the external stone walls in the area of the cave of Pont d'Arc. This again points to a quite extended literacy at that time and in that area.[19]

The artistic value of pictures and drawings left behind points to the level of the culture producing them. Look at the picture below, found on the cave wall in Pont d'Arc.

Picture 95. Many more similarly high quality drawings and paintings survived on the walls of the cave.

Uneducated people cannot produce such fine, sensitive drawings. Only

19 See an extended presentation in my book "Signs, Letters, Alphabets" on pages 113-116 and 139-146.

somebody living in a high culture would be able to reach such high peaks of art. Imagine trying to reach this artistic level without special education. You can't do it without a good teacher and a great deal of practice.

Other similar finds from around the same time-period, but from another place some distance away tell us that it is not a "white raven", not a transient glance of intellect.

The cave Grotte Cosquer is far from Pont d'Arc, on the Spanish Riviera. The cave's entrance became closed because of the higher water-level at the end of Ice Age. Divers discovered the entrance and found many pictures on walls above the water-level. There are inscriptions on these two neighbouring pictures.

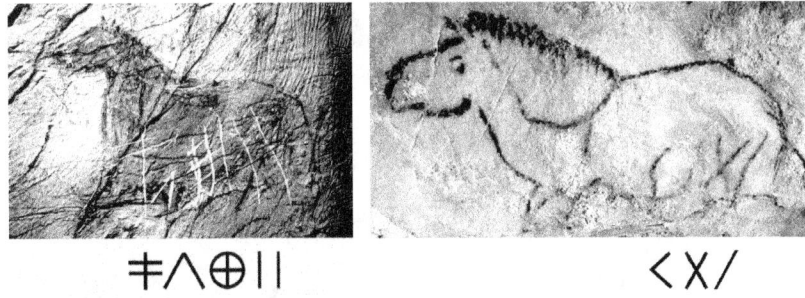

Picture 96. Two inscriptions in one place, both on horses. The pictures and the inscriptions on them were made around 25.000 years ago.

Dear reader, there are seven writing-signs in one place and that many signs already make up 1/3rd of an average sign-collection (Latin alphabet).[20] Furthermore, the signs remained well-known for the next 20-29,000 years.[21]

After all that, we would be surprised if we were told that people in those days could paint and write, but couldn't write numbers. However, written numbers luckily survived from that time as well as from earlier times, as we have already seen.

---

20 These are probably ideograms, but this is only a secondary question. We see that they were writing.

21 See "Signs, Letters, Alphabets"

# 4. LASCAUX

17,000-year-old finds in the cave of Lascaux, France

Let's look at the overall picture first. Even here there is a conspicuous compound sign-group visible in front of the face of the buffalo on the left.

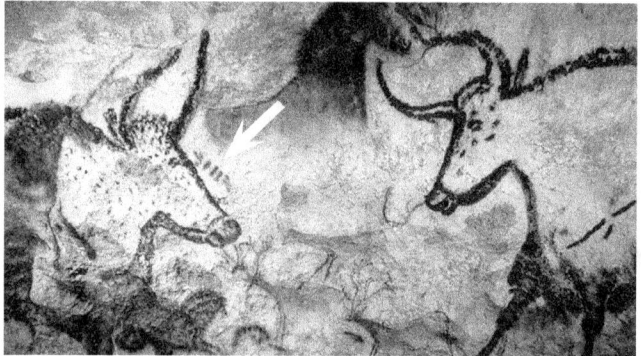

Picture 97. Detail from the cave.

Looking closer at the dots and lines, it is easy to recognize that those signs represent a numeral:

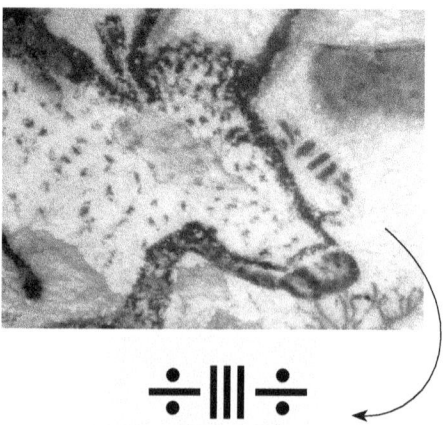

Picture 98. This painted numeral survived undamaged for 17,000 years.

This sign is indubitably a numeral. We could tell this even if it were the only one found in Lascaux. Furthermore, this one already forms an unquestionable proof of the general use of number-writing there at that time. Moreover, this method of number-writing survived into the 20th century in Europe. It is worth taking a peek at the pictures on page 125 and comparing the 17,000-year-old signs of Lascaux with the signs used in the 20th century.

In Lascaux however, this number wasn't the only one which survived. Look at the collection of signs visible on the cave wall:

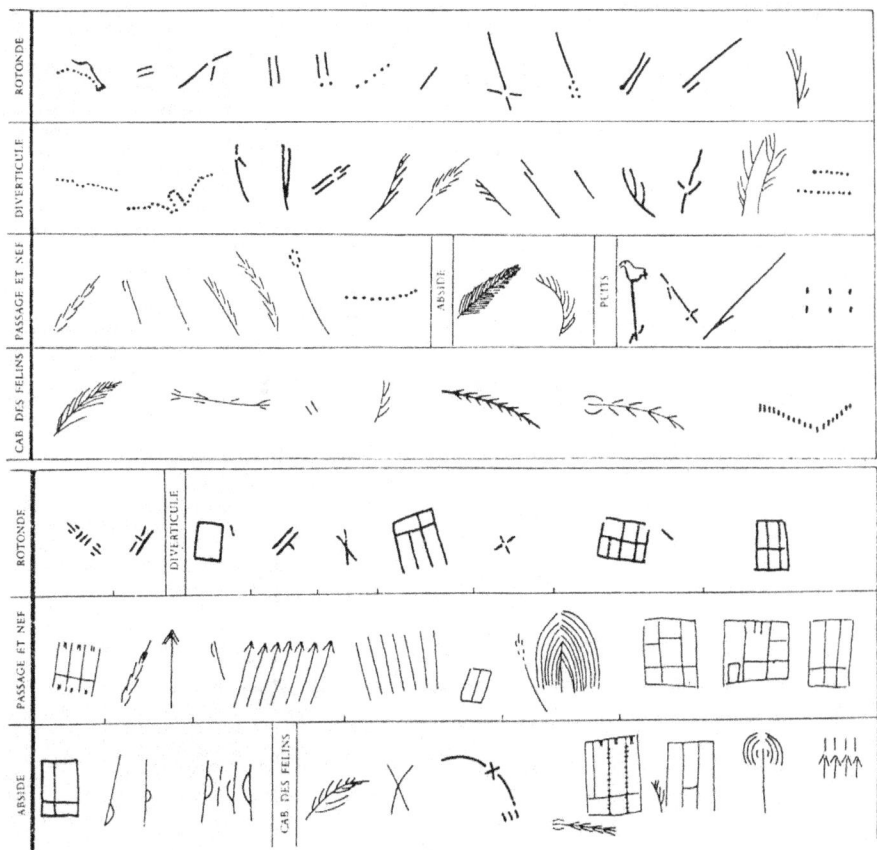

Picture 99. The collection of signs visible on the cave wall in Lascaux.
Georges Jean: Language de signes, Gallimard 1989, page 142.

It is quite interesting that dot-line signs are included beside several of the painted animals, just as we saw previously with the buffalo.

Picture 100. There are often dot- or stroke-signs beside the animals in Lascaux.

Below is a collection of the signs found in Lascaux, which clearly and visibly build a system:

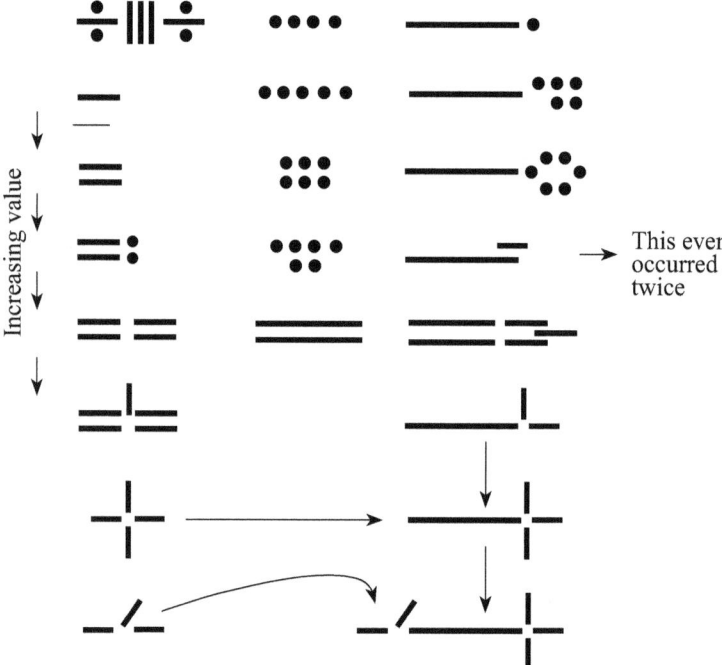

Picture 101. The steps of extensions and their components are marked with arrows.

There is a system in these signs. The extensions are logical creations as are their sub-parts which even appear together with other parts. Therefore they can only represent numeral values. They are numerals. Moreover, most of the signs reappear in many younger finds; some were still being used up until the 20th century.

It is worth analysing the following characteristic sign-group:

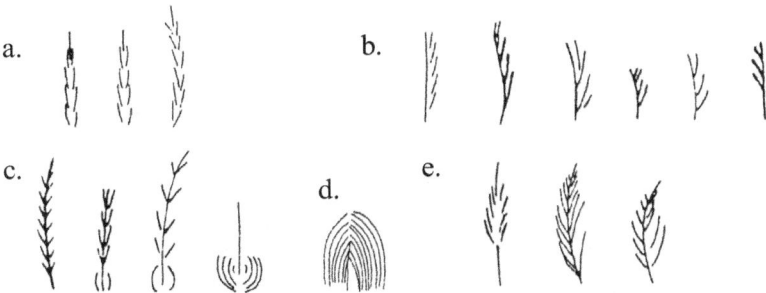

Picture 102. The signs in the groups "a,c,d", and maybe "b" are certainly not leaves

The first two signs in 'a' are identical, the third consists of several parts:

Picture 103. There is multiplication, unquestionably; fewer or more.

This kind of signs are engraved into a lamp in Lascaux, twice, in fact:

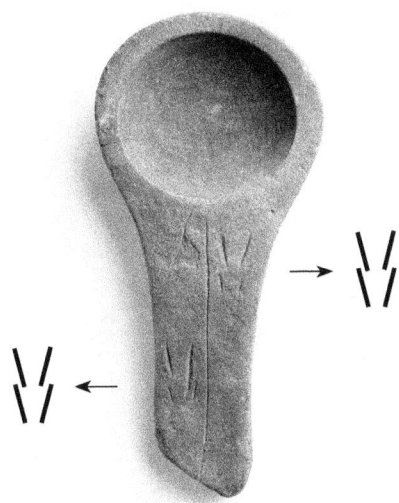

Picture 104. The handle of the lamp carved out of sandstone is divided by a line and on both sides is engraved the same "telescoped" sign.

My assumption is not unfounded, that if it was necessary to repeat and repeat again the horizontal pairs of strokes for the growing value of numbers, then line-pairs were pushed together after having been slanted a bit:

Picture 105. The row of signs on the left and the same row pushed together on the right.

The left- and the middle-sign of picture 100/a supports the above. The scribe started the left sign at the top with rather parallel strokes and then slanted them further down. Parallel and slanted line-pairs seem to mean the same for him.

Some of the signs have a middle-line, but others don't:

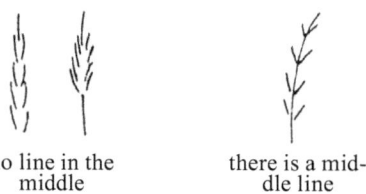

no line in the middle      there is a mid-dle line

Picture 106. It's obvious that the meaning of a sign-group is different if it has a line running through or if the line is only at the beginning or at the peak of it.

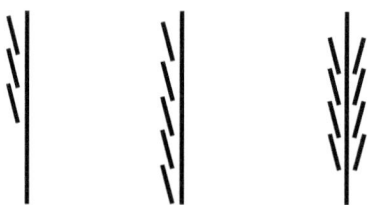

Picture 107. We can see double- and one-sided "palm leaves" in groups b) and c) of picture 102.

Note that those signs are either single- or double-sided and there are no signs among the findings which have e.g. 3 lines on the one and 5 on the other side. Moreover, the middle-line must play a major role in representing the value; it even can be extended by half-circles as on the following picture

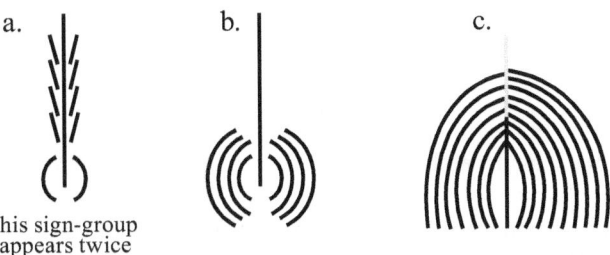

Picture 108. We see round brackets as signs drawn in front of each other.

The use of these brackets can't have happened just randomly, for the "palm-branch" with "parentheses" even appears twice on the wall. Those brackets surround only a perpendicular middle-line in picture b). Whatever we are dealing with, those brackets can only mean enhancement compared to the sign: if the sign represents a number, then the enhancement could mean thousand, ten-thousand, e.g. depending on the number of the brackets. We see the same on drawing c), only greatly exaggerated, and the person drew more brackets on the right side than on the other (mistake?)

The use of such brackets to express higher position-values was still a custom in Italy even during the early Roman empire. See a few examples from this time (more on page 356):

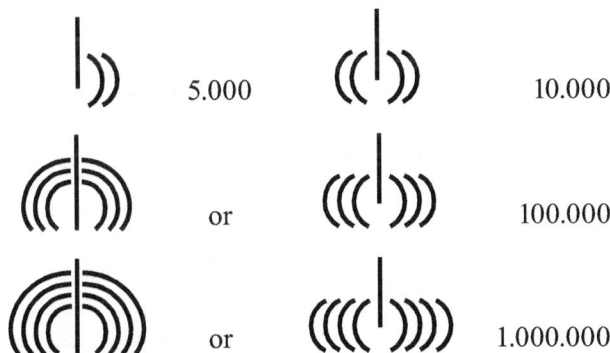

Picture 109. One bracket on each side of the line means 1,000, two brackets on each side: 10,000, three brackets 100,000 and so forth. If the brackets are only on one side, then half the amount was indicated: 500, 5000, 50,000 and so forth.

Having used the line with one bracket beside it for 1,000 (I) and with only one on one side I⊃ for 500, then the later "D" sign of the Romans for 500 is understandable. They just connected the bracket with the shortened perpendicular line:

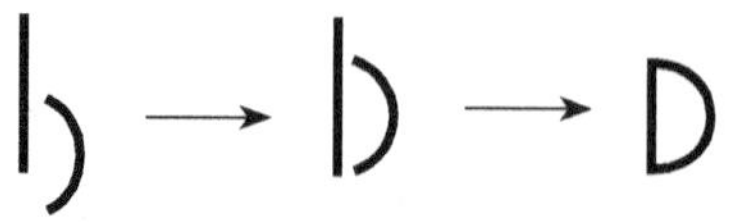

Picture 110. How the archaic sign of 500 became a D in Roman times.

This of course merely suggests the continuity between the signs in Lascaux and the Romans. The suggestion will be supported by finding another Lascaux-sign, numeral ÷|||÷ in Transylvania in the 20th century (see more on page 124). Here however, there are no brackets to find for the marking of position-value, which in reality are lines in half-circular form.

Further supports for continuity are the "palm leaves" used in old-Italian number-writing, but in this case the signs stand on the top. One can see in the following example that the half-circles (brackets) are identical with the fine lines of the "palm-leaves".

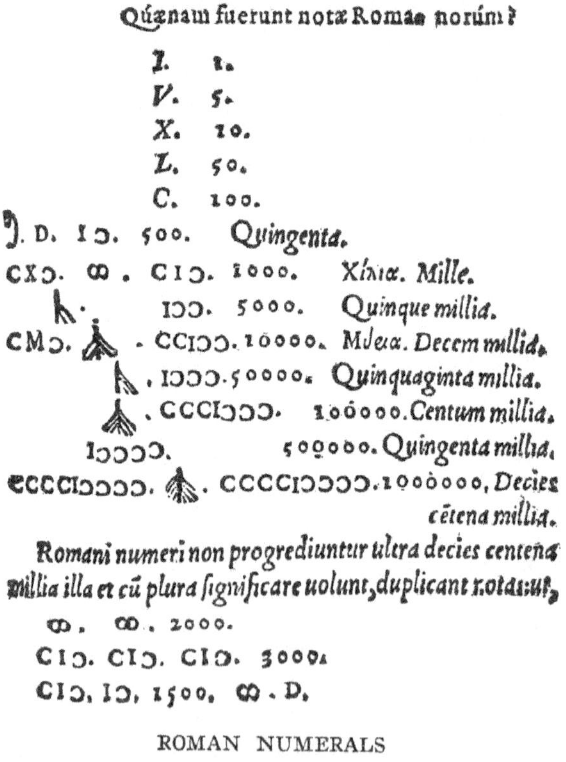

ROMAN NUMERALS

Picture 111. Roman numbers, noted by Fergius (Helvetian writer) in 1582.

We can exchange, therefore, the brackets with the "palm-leaf stalks", in other words, they mark the same values.

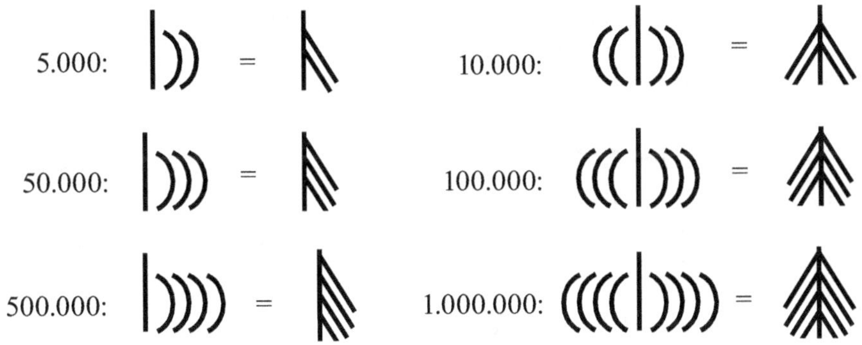

5.000: |)) = ⋀    10.000: ((|)) = ⋀

50.000: |))) = ⋀    100.000: (((|))) = ⋀

500.000: |)))) = ⋀    1.000.000: ((((|)))) = ⋀

Picture 112.

The Aztecs also used this "fish-bone" than "palm-leaf" form, but in a rather slipshod way. See the Aztec signs of 100, 200, 300 and 400:

100        200        300        400

Picture 113. In addition to these forms, they also marked the numbers 1-10 with dots as well.

This "fish-bone" number-writing was widespread in Asia, however if they used these signs, then they wrote every number this way. See more on page 374, but here a few Arabic samples.

1    2    3    4    5    6    ...    10    ...    100    ...    300

Picture 114. Arabic numbers, looking like Turkish military number-writing

Therefore, the brackets (half-circles) and the 'fish-bones' (palm-leaves) are two variants of the same number-writing method (bent and straight lines) used in Lascaux 17,000, and in Italy 2000 years ago. Let's now look again at the signs from Lascaux which were built from parallel lines. Pictures 101 and 105 ▬▬ ▭— and ▬ ▬ as well their loose "handwritten" form: ⊂⊂⊂⊂—, which is the same as ▬▬▬▬▬— .

The same signs were used in Asia Minor 2,500 years ago, but written from left to right.

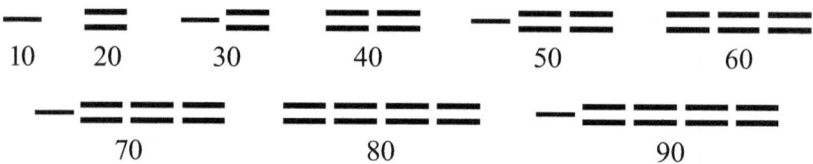

|  |  |  |  |  |  |
|---|---|---|---|---|---|
| 10 | 20 | 30 | 40 | 50 | 60 |

|  |  |  |
|---|---|---|
| 70 | 80 | 90 |

1Picture115. Each line has the value of ten according to this method.

A number written using the above method is found on a bilingually written stone plate in Hebron (Greek and Aramaic). It should be read from right to left:

$$||-=$$

32

Picture 116. Sign by sign from the right: 20+10+2 = 32

We can't figure out the number 100, but the method of number-writing is clear up to 99. It is easy to work out, because the only signs used are the horizontal and perpendicular lines. Using the same method, we also got the three numbers below right:

| $|\ |||-==$ | $||==$ | $|\ ||||\ |||-===$ |
|---|---|---|
| 54 | 42 | 77 |

Picture 117.

This is the line-line variant of dot-line number-writing, since if we change the perpendicular lines to dots, then the second number becomes identical with these ≡∶ or ══∶ numbers. However, there is an example of this kind of number ══∶ in Lascaux too.

Let's return finally to the most spectacular, previously shown, number from Lascaux.

We meet it at after 17,000 years, at the beginning of the 20th century in Transylvania. We don't need any further explanation to become convinced of the unbroken continuity. Just look at them.

| The number written in front of the bull's nose 17,000 years ago in Lascaux | A few numbers written in the salt-mines of Vizakna-Transylvania in the early 20th century |

Picture 118

Napoleon told his solders in Egypt, in front of the pyramids: "Forty centuries are looking down on you". Well, looking at the numbers of Vizakna, I can tell you, dear reader that "one hundred and seventy centuries are looking at you". This is four times more than the known duration of the Egyptian culture. However, this is not in Egypt, but in the Carpathian basin and it is our inheritance.

Summary:

Many signs found in Lascaux were recognized as numbers at the first glance because of the visible system of quantitative difference of the signs belonging to one group of forms. Obviously, these gradually extending signs only receive their meaning in number-writing. We can find these signs still used as numbers, even 14-17,000 years later.

This means that in the **time between** Lascaux and our century these signs have been continuously used for number-writing, despite the few finds which would prove our statement. If my shoes are in the same place in the morning where I put them in the evening, then generally I will think that they stayed in the same place until I saw them again in the morning. Especially if my socks with the dot-line clock on them lay over the shoes in the same way as I put them in the evening.

Let's think about it in a different way.

Luckily, not only one, but several different kinds of signs survived in Lascaux.

The fact that all the signs of the different kinds used in Lascaux popped up later as numbers proves unquestionably the continuity of use and tells us for sure that they were used in Lascaux as numbers as well. We can make the same statement not only in connection with Lascaux.

Having seen the numbers of Lascaux, it seems to us that the traces of several large cultures are visible there. The probability of this will be supported by the next chapter too. Some cultures left very specific trzces.

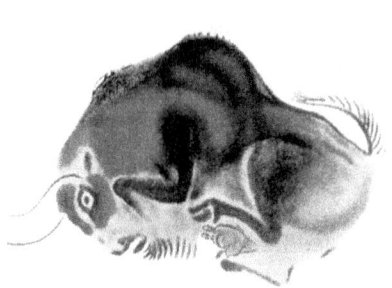

Zsolnay Wilmos, a known art historian wrote in *"The origin of Art"* (page 34, 2001): *"This is one of many art pieces supporting our statement that the Magdalene culture was the first of its kind in humanity's history, in its kind unrivalled, not repeatable, unreachable peak. Its art wasn't lower than that of Akhenaton's Egypt or Perikles' Athens".*

Zsolnay however, could not know about the latest finds, based on which his observation can be significantly extended: at least two, probably more great cultures existed over the last 40,000 years in Europe.

We can simply acknowledge their past presence based on some marvellous finds, but can't say much about the details yet.

# 5. A 22-16,000 YEARS OLD NUMBER

The La Pasiega cave, close to the village of Puente Viesgo in Spain, has four galleries, the longest of which measures 75 metres. Pictures of several animals and different signs decorate the cave's walls. According to scientists, these remains of human arts originate from at least two cultures: "Upper Solutrean" and "Lower Magdalenian" periods, but older objects were found there as well. In the end, the date of this culture was put at between 22,000 and 16,000 years ago.

One of the sign-groups remaining the in the cave looks like this:

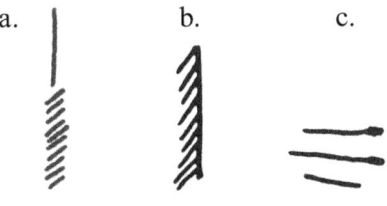

Picture 119.

It is probable that the three signs above are connected to number-writing, but we can't prove it. This is because there are no variants we could them compare with. Only the sign c) has a possible variant on the picture 121, on the next page.

Dots, however, are put in order in another sign-group in the cave:

Picture 120. a.) In the upper left corner, there is a loose, seemingly random, dot-group, but a carefully painted triple row of dots is directly below it. The number of dots in each line is: 13-15-15. b.) There are four carefully painted and placed dots above the horse's neck.

The dots-groups in the previous picture, however, are not built according to a dot-line or line-line system and according to our self-made rules we can't accept them as numerals. A written number is not the same thing as marking the amount of subjects with dots, although the latter is also the result of counting (e.g.: number of children, of passing days or the phase of the Moon, etc). This way, the above dots can be viewed as the results of some counting, but it is not number-writing based on a position-value system, for which we are seeking ancient proof.

Nevertheless, there is one sign in this cave, which meets all of our conditions. It consists of three strokes with four dots above them:

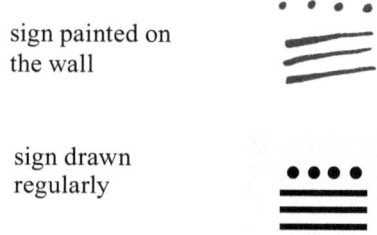

sign painted on
the wall

sign drawn
regularly

Picture 121. This is an acceptable regular numeral

The signs of all four sign-groups from Lascaux (differing by their forms) have been found in more recent times as well, even in the 20th century. The sign from La Pasiega has also been found nearer to our own time. Mayan people would read it as the number 19 probably just by looking at it. See below some more Mayan numbers and at the end a 16-22,000-year-old number:

| 14 | 15 | 17 | 18 | 19 |

Picture 122. This "layer-cake" form is unmistakable. Obviously, the base 20-base system belongs to it.

This is a very special variant of numbers built from dots and lines. The continuity between two such signs is unquestionable, not even with 16-20,000 years between them. (Sign c. in picture 119 – three horizontal lines - most probably represents number 15.)

Is this too long a time to think of continuity? I ask the dear reader to go back to the chapter "The continuity of the intellect", where we introduced several proofs for continuity in big steps. Here we can add some interesting historical data.

Historians divide the presence of the Mayans in Central America into three parts: the pre-classic (formative) period from 10,000BC–300AD, the classic period from 300 – 950, and finally the post-classic period from 950 – 1550. This means that the Mayans were in Central America for at least 12,000 years.

The La Pasiega cave was last visited by humans around 16,000 years ago. Four thousand years seem to be missing, however, the age of the two numbers above are only a rough estimate and certainly the point at which those numbers were put onto the wall was not the last time these numbers were used in Europe. In addition, this cave is only one place in Europe and it doesn't tell us much about what, where and when anything happened in Europe. Therefore, it is possible that the culture which used the same type of dot-line number-writing existed at the same time in both in Europe and in Central America. The number found in the La Pasiega cave proves even this possibility.

Of course, it is possible that number-writing with dots and lines in the base 20 system somehow reached the Mayans much later . Consequently, it is not certain that the Mayans used this number-writing system from the beginning of their 12,000 year presence in Central America. Furthermore, such a direct transfer was not necessary. In the absence of the data of events, any roundabout way may have served for the transfer of knowledge, but this does not weaken the proof that it happened. Again, this is not the only example of the long-lasting continuity of number-writing. Adding this type of numeral to the four different types of numerals found in Lascaux, we can already prove the long-time continuity of four different kinds of dot/line or line/line numerals over 17-20,000 years of unbroken human history.

This means that there existed several cultures 15-35,000 years ago in Europe which were not extinct; at least their number-writings still exist as well as their writing-signs. (See the Book: "Signs Letters Alphabets")

Because the number-writings are built uniformly on dot/line systems and nothing else can be found in the past, the cultures from 15-35,000 years ago may be viewed as the remnants or variants of a much much older culture. However those cultures are in reality still alive. By this I mean that we come again to the statement at beginning of this book: history is continuous; the division of it into epochs or eras is merely an illusion and exists only in the history books.

Finally, let's call for help from some more of the drawings on the cave-walls in La Pasiega

Picture 123. Drawings in La Pasiega: forms ending in small peaks with different decorations.

These forms are round and some of them (the 1st, 2nd, and 3rd certainly) look hollow; probably all of them are. The forms look contracted at the bottom compared to the widest part, therefore their material must be soft and it hangs down at one end. We could be forgiven for saying that they represent headgear, for then even as today objects did not exist with no imaginable purpose or form. All of them have a little peak at the top. We can say in conclusion that this kind of headgear still exists today:

Picture 124. Even the "antenna" is on its top has existed for the last 16-20,000 years

We call this a Basque cap, because it is part of their national dress and its use spread from them throughout Europe. La Pasiega, where the drawings were made 16-20,000 years ago, is the place where the Basque people have been living for at least 10,000 years. The search for numerals has come to an end also.

This has even provided an addition to the proof of the "continuity of intellect".

# 6. MAYAN AND OLMEC NUMBER WRITING

We have seen in the previous chapter that the form of a 16-20,000-year-old European numeral is identical with the form of the numbers written using the Mayan system in Central America. Now let's continue by looking at the world of Mayan numerals. Both the Mayan and Olmec people wrote numbers the same way (positional number-writing). Both used a quasi-vigesimal (base 20 number-system) but calculated using 360 as the second power of 20 instead of 400.

First, look at the numbers according to their position-value in our decimal system:

| 10 | 10x10 | 10x10x10 | 10x10x10x10 | 10x10x10x10x10 |
|----|-------|----------|-------------|----------------|
| **10** | **100** | **1.000** | **10.000** | **100.000** |

These would be the regular powers of 20 in a "vigesimal" system:

| 20 | 20x20 | 20x20x20 | 20x20x20x20 | 20x20x20x20x20 |
|----|-------|----------|-------------|----------------|
| **20** | **400** | **8.000** | **160.000** | **3.200.000** |

However, the Mayan and Olmec people didn't use 20x20 = 400, but used instead the number 360. They then carried this 360 over in the subsequent "powers" (as shown below):

| 20 | 20x20 | 360x20 | 360x20x20 | 360x20x20x20 |
|----|-------|--------|-----------|--------------|
| **20** | **360** | **7.200** | **144.000** | **2.880.000** |

It is worth saying a few things about the number **144,000** above. Interestingly, the traditional number of the "holy innocents" as mentioned in the Bible is just **144,000**.[22] In the case of such a large number, this can't be due to a misunderstanding or simply a random coincidence. You can't just randomly guess such a large number. Furthermore, this number receives

---

22 The children killed by King Herod are known as the "holy innocents". Their remembrance day is the 28th of December.

its sense only in the Mayan numeric system, which is in fact "irregular". If it was regular, then 144,000 would not point to a position-value. It can't be just pure chance. The number 144,000 can only have come from this irregular numeric system. We don't know how it originated nor how it got here, but the two threads certainly met each other somewhere in the distant past.

Even the mention of 144,000 is a further proof. This number can't be viewed as the exact number of the children being murdered; it is more an expression of exaggeration by using the "round" number of a next higher position-value. In the decimal system, we often say: "I've already called you a hundred times", "I haven't seen you for a thousand years", " I've told you a thousand times", "a million kisses" or "I've thought of you a million times. Therefore, to mention 144,000 is given a sense only because of the "irregular" base 20 numeric system.[23]

The Hungarian saying "one is nineteen and the other is one less than twenty " (in English we would say "six of one and half a dozen of the other" is probably a remnant of the base 20 number-system as well, since the number twenty only plays a special role in this system.

As we have seen previously, a c. 20,000-year-old numeral has been found painted on the wall of the La Pasiega cave in the Basque country and the Mayans, like the Olmec people, wrote their numbers with the same numerals even 20,000 years later. There are still many traces of the base 20 number-system in the numeral-names of different languages all over Europe.[24] The Celts counted in the vigesimal numeric system and their culture was once widespread. We speak about Celtic and Gallic territory when discussing Western Europe. Names of numbers containing a multiple of twenty are based on twenty in the Danish, Bretons, Welsh, Scottish, Gaelic, Spanish, and French languages, and even in some dialects of Albanian and Slovenian. The word for 20 is in French 'vingt', quatre-vingts=80 (4x20), six-vingts=60 (3x20), sept-vingts=140 (7x20) and huit-vingts = 160 (8x20).

Abraham Lincoln started his well-known Gettysburg Address in 1863 with the words "Four score and seven years ago..." meaning the year of the American Revolution in 1776, that is 4x20 = 80+7 years before. The word 'score' can be used in English (without a number prefix) to express 20, and we can find it in many places in the Authorized Version of the Bible.

An explanation for the "irregularity" of the Mayan system could come from one of their two Calendars. In one they had 18 months with 20 days = 360 days and the remaining 5 days were handled as a "5-day month". This

23 The number 144,000 appears in the Bible too, but in other connections not dealing with the "holy innocents"

24 According to some scientists, the language known as French today was largely influenced by the early Bas-Breton (lower Breton) language.

360 supposedly took the place of the regular 400 in their positional number-writing. I guess it was easier to calculate years with the number 360. (The years in their other calendar had merely 260 days and as a result the two calendars overlapped each other only every 52 years.)

Francisco de Montejo, the plunderer appointed by the king, led the first attack against the Mayans in 1526, but their independent state fell into European hands only in 1697. The Mayans had large libraries, but the over-zealous priests accompanying the invaders (robbers) had most of the scripts burned because they didn't understand them. Four codices survived and are now in the libraries of Dresden, Madrid and Paris.

Picture 125 The Mayan codex now protected in Dresden. There is dot-line number-writing visible on the left upper corner of the third page (counted from the right)[25].

We already mentioned previously that the Mayan-Olmec (in the following only Mayan) dot-line number-writing is in reality a dead end for number-writing. Namely, they used only two of the four possible signs (i.e., dot, line, a line perpendicular to it and the long-line) the dot and line. This had a major restricting influence. It is impossible to mark with the form or appearance of these two signs alone the position-value of the numbers. Therefore, they had two possible ways to do that: to put a gap between the numbers or invent a special sign for marking an "empty" position. Our forced (second best) solution today is to use "0". (The circle stands for "nothing", for emptiness).

---

25 There is an interesting question of what would happen if the Mayans decided to keep their historical finds in their own museums and one of them – several millions are still alive –simply just took one home. I guess he or she would land in prison for stealing an art treasure.

The Mayans already wrote the numbers perpendicularly one number under the other. This meant that they needed gaps between the numbers anyway and so were forced to use a sign for the empty space. This sign became a shell.

These shells marking the emptiness of a position-value were quite differently decorated. Often there would even be different shells on one page. They knew very well where to put the shells. (See Table II on the next page.)

The horizontal line had a value of 5 for the Mayans.

As we saw previously, the line with the value of 5 already marks a position. It is easier to recognize this if we handle the decimal system as a 2x5 system and the vigesimal as a 4x5 system. (In a hidden way, we still interpret our decimal system as a 2x5 system.) This is supported by the fact that in the vigesimal system the numbers are[26] written this way with the powers of 5, the powers of 2x5 and the powers of 3x5 up to 19, with one, two and three lines above each other:

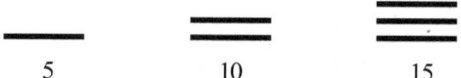

|  |  |  |
|:-:|:-:|:-:|
| 5 | 10 | 15 |

Picture 126. It is clearly visible that the only powers of number are the horizontal lines beside the dots.

Including zero, these are the Mayan numbers up to 20:

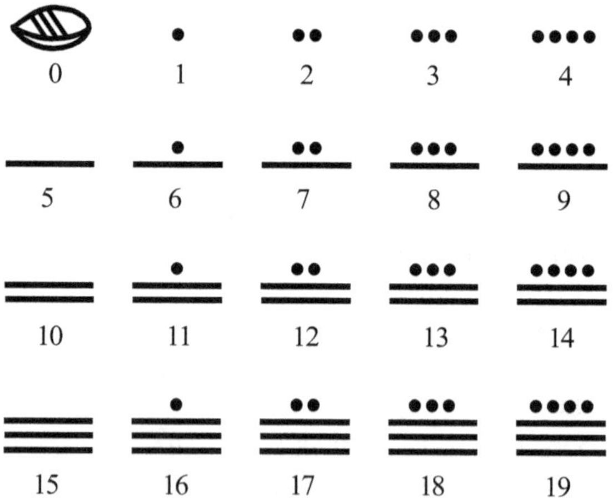

| 0 | 1 | 2 | 3 | 4 |
|:-:|:-:|:-:|:-:|:-:|
| 5 | 6 | 7 | 8 | 9 |
| 10 | 11 | 12 | 13 | 14 |
| 15 | 16 | 17 | 18 | 19 |

Picture 127. The Mayans needed this many separate number-names.

26 A little addendum to the base 5 numeric system: Research among the Indian tribes in America showed that out of 307 tribes, 146 used the decimal system, 106 the base 5 system and others used either the vigesimal or base 25 numeric systems.

# TABLE I

A little collection of the different shells used by the Mayans:

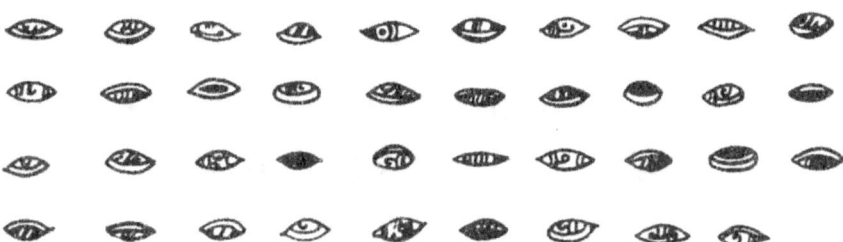

Below, we can see one page of the codex in Dresden. There are several variants of the shells (= sign for "0") even on this one page.

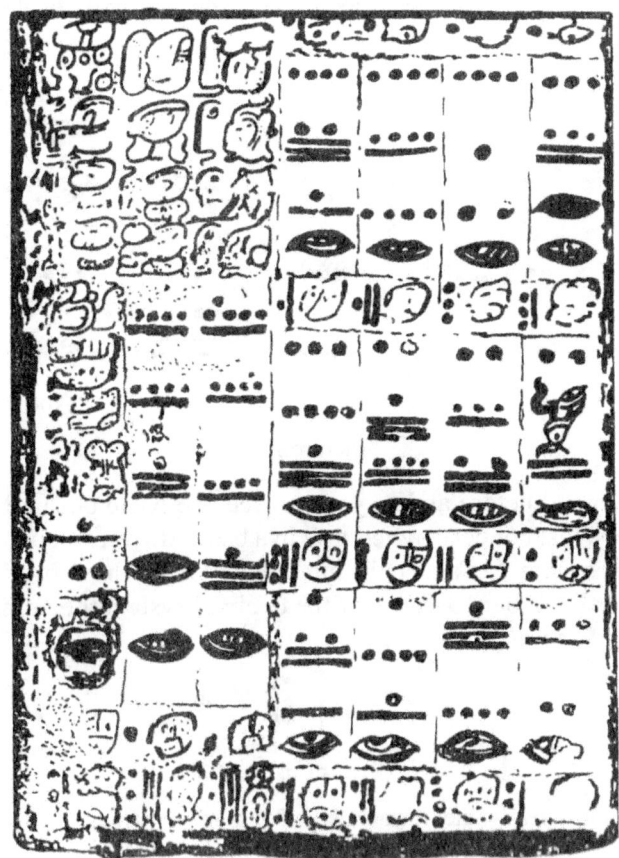

The Mayans wrote the numbers from above downwards according to their declining position-value. Here are some examples for reading of numbers: The dot has a value of '1' and the line '5'. The position values are standing on the left arranged by upwards increasing value.

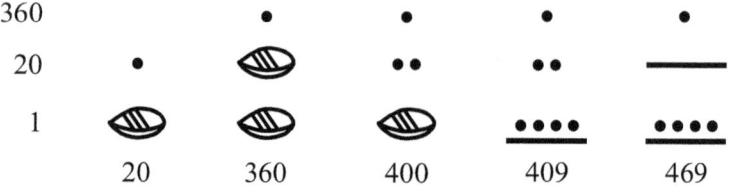

Picture 128. The Mayan number writing and reading is probably easily understandable from these samples.

Reading the following numbers, however, we start to become real experts; with time we could even become accustomed to their vigesimal system.

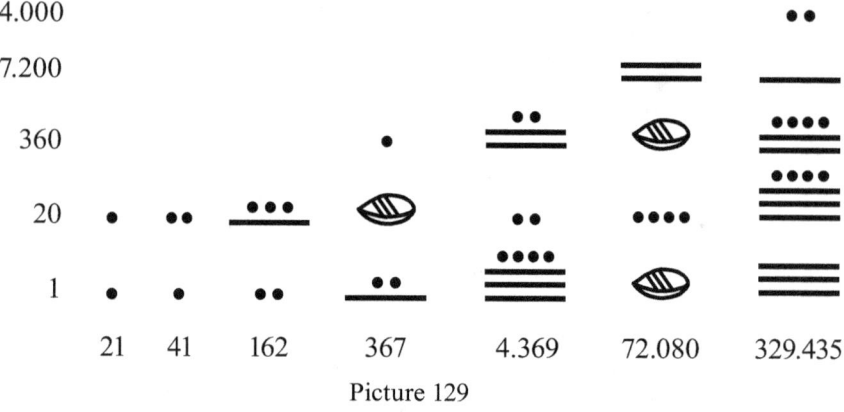

Picture 129

As mentioned before and as you can see, we write our numbers today the same way, just not perpendicularly in a base 20 system, but horizontally in a decimal system. Let's turn the number 469 from above by 900 and read it as a decimal number (because of the decimal system the number's value will naturally decline):

1 5 9

Picture 130. Clearly, the Mayan method of number writing was the same as our is today. There is no "evolution".

The Mayans did not insist on only writing the numbers horizontally. According to the space on the page, they occasionally put them perpendicularly as we can see in the upper part of the following picture.

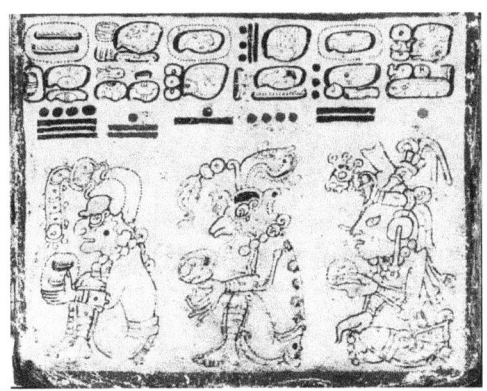

Picture 131. Details of one page in the Drezden codex.

Overall, the Mayans also created 19 different picturesque numerals (1-19) in their base 20 system, to be used in the powers of one position.

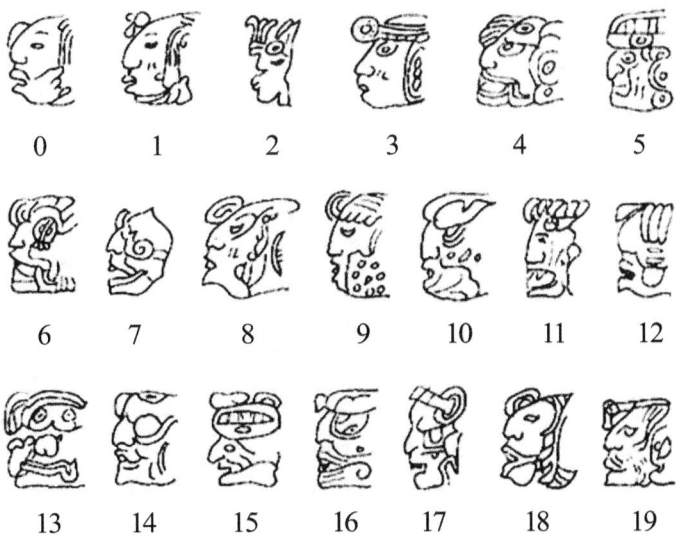

Picture 132. Every numeral shows a face. These are the faces of the Mayan Gods.

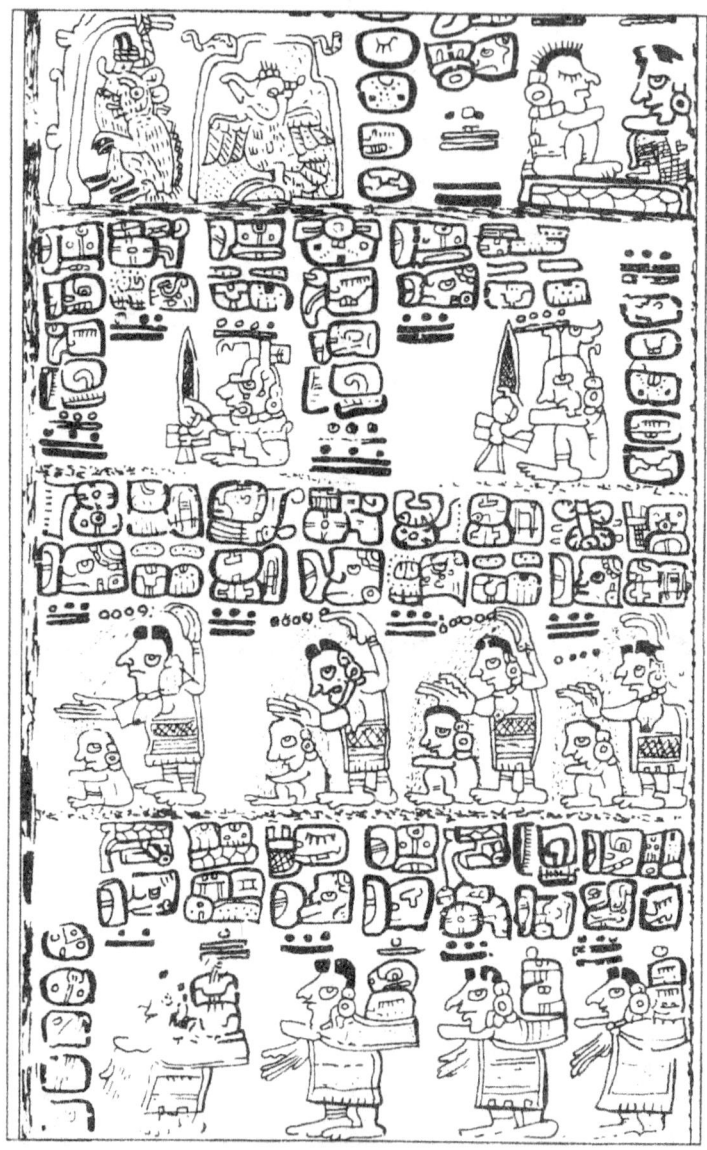

Picture 133. The 93rd page of the Mayan "Tro-Cortesianus" codex
(Madrid American Museum)

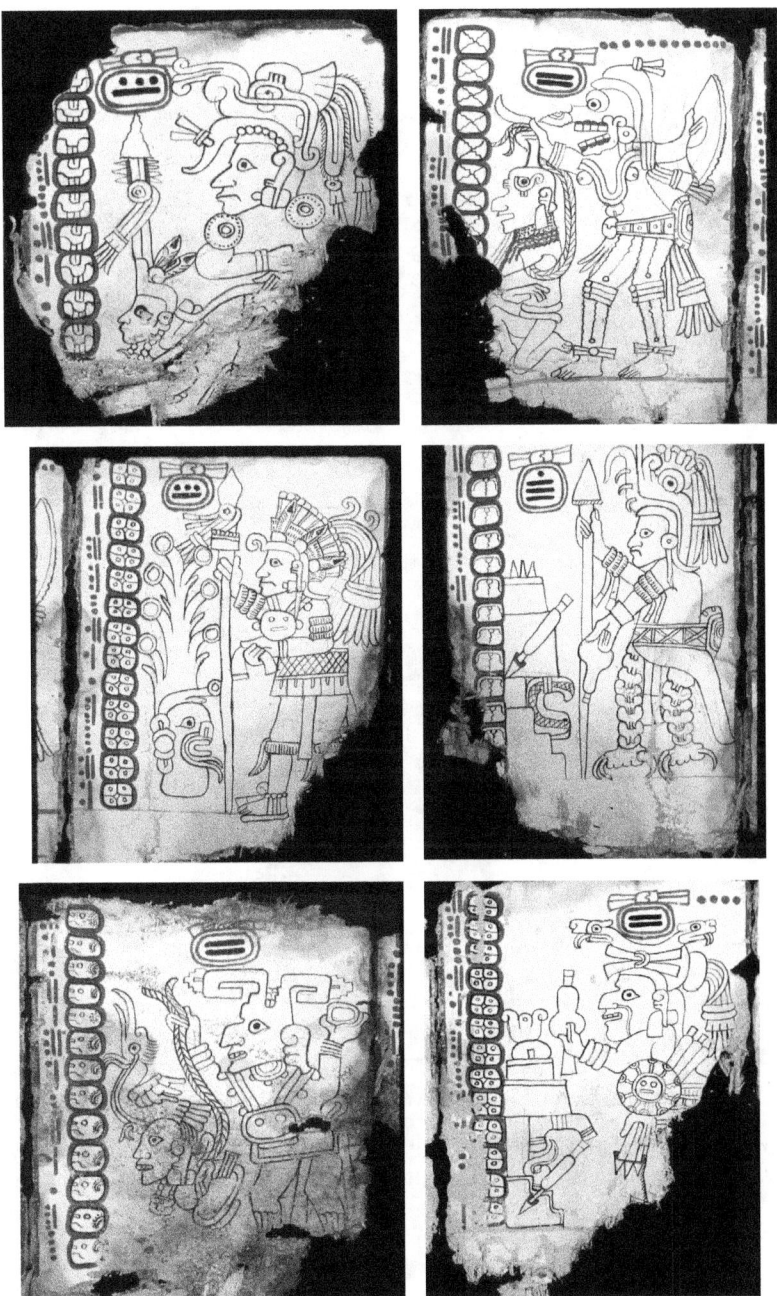

Picture 134. Details from the fragmentary Groiler codex. There are many numbers on the left side (and in one case on the right) of the pages.

## TABLE II.

Below is the structure of one Mayan calendar. This is special, because the year has 260 days in it divided into 13 months containing 20 days. They recorded the days in the following way:

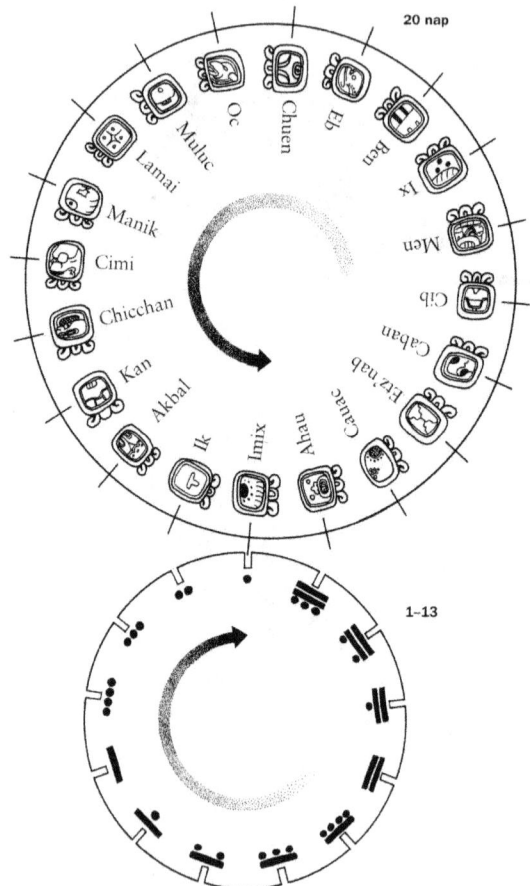

The picture is from the book *"The History of Writing"* by Andrew Robinson.

One could see the number of the months and the picture of the days on the drawing. See the day pointed out on the above calendar and some other days:

| 1 Imix | 8 Imix | 5 Chicchan | 1 Ix |

# TABLE III.

number carved in stone from Trez Zapotec:

108.000

2.160

320

16

_____

110.496

I leave it to the reader to decide whether the sum of the numbers is right or not.

## TABLE IIII

One of my readers, Tamás Lauringer, sent me the following picture, made by him at the Hágó of Verecke <Vɛrɛczkɛ> (Mountain-pass of Verecke). One sees on this column a number written the Mayan way. Probably somebody mad was having a joke. The engraving looks relatively young and is not a precise work. The Mayans often calculated with large numbers, but 23 billion (23,426,210,165) would be too large even for them. However, what could be the purpose of writing such a large number on a monument? It is interesting that people in our time know and use this kind of number writing. I include this picture to demonstrate that trials of falsifying old language-historical and other finds has become more common in our time. Unscrupulous falsification of history is even carried out in some countries by governmentally supported "scientific" circles (Slovak and Romanian among many others).

This script is not original Mayan, however, it is not a falsification. The person who created it did not intend us to believe that he had found an old original writing. He was just obviously playing a game.

# 7. A 10.000 YEARS OLD NUMBER FROM MESOPOTAMIA

There was a picture of a funny-looking perpendicular row of signs in an Italian newspaper presented on the Bolognese book market. The script under the picture said that this sign was written on a stone found in Mesopotamia and it is supposedly 10,000 years old. I drew the signs but I forgot to note the name of the newspaper. I hope that despite missing the exact source, the reader will look at it as an authentic find.

Picture 135. A 10,000-year-old perpendicular row of signs
found in Mesopotamia.

There may be not one, but two numbers standing above each other. The two signs above the six dots are not definitely numbers but the cross often turned up as a number in earlier times. The signs above the cross are definitely numbers.

# 8. TEPE YAHYA

Tepe Yahya is located within the territory of Iran. It has been inhabited for many thousands of years. There is uncertainty about the dating of the different archaeological layers; there is disagreement among the scientists. Thus, we have chosen 5-6,000 years as an average age for the oldest signs found there, because the place has been continuously inhabited for 7,500 years and its culture was also supposedly unbroken. We can't think furthermore, that everything was invented all at once at the time given by the archaeologist's dating for the oldest find.

Daniel Potts collected the signs on 400 ceramics in his study "The Potter's Marks of Tepe Yahya" in 1981.

Naturally, when we are speaking about literacy, then we are dealing with writing and number-writing as well.

Let's look at one comparative table out of Daniel Potts' study:

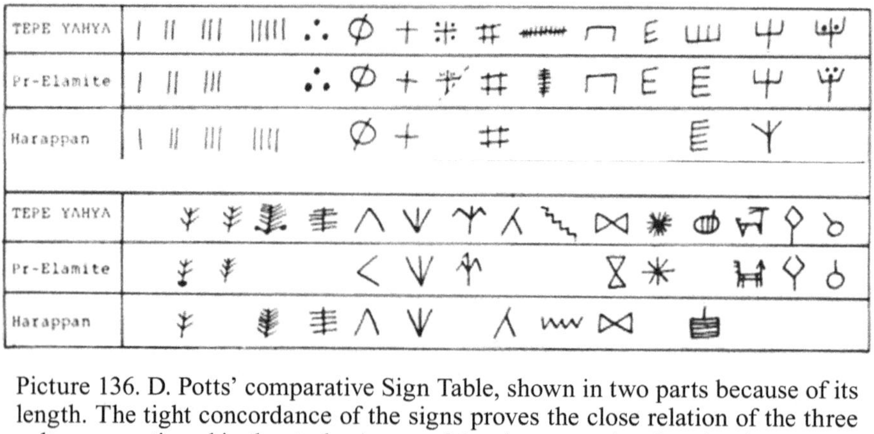

Picture 136. D. Potts' comparative Sign Table, shown in two parts because of its length. The tight concordance of the signs proves the close relation of the three cultures mentioned in the study. A part of the signs are related to the signs on the clay-tablets of Tordos-Vincsa, and therefore even to the Szekely (szekler) sign-collection (Rovás). Note that the "palm-leaves" are presented here as well.

On the next page, we can see the part of Daniel Potts' collection representing the dot-line signs found in Tepe Yahya.

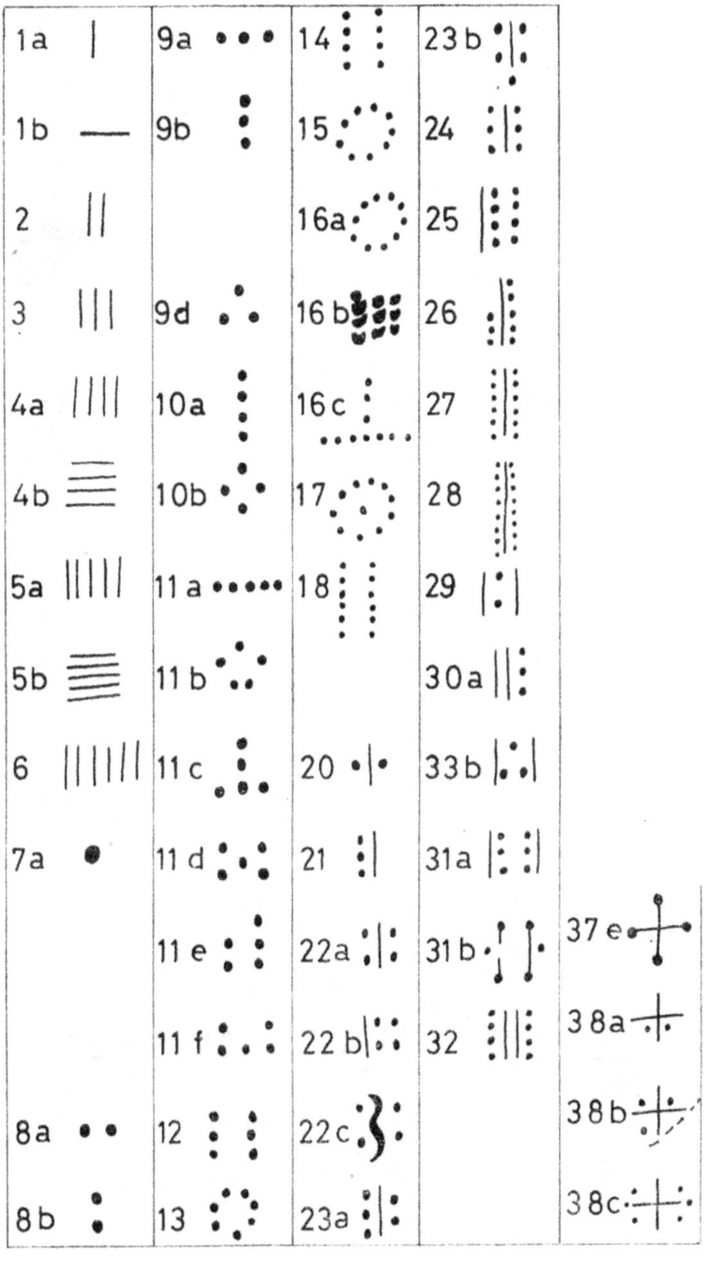

Picture 137 The numerals found in Tepe Yahya are around 5-6,000 years old.This picture is in the book *"Tartáriai leletek"* (Fings of Tatárlaka) by Makkay János (2990, Academic Publisher )

I don't know what criterion Mr Potts used to determine if a sign was written horizontally or perpendicularly? One can't easily tell looking at a ceramic pot, but it's the same anyway, whether I write 19 or 19 turned perpendicularly. In the case of same dot-line numbers, the value will change if we turn them head down and can't read them in relation to other signs.

There are numerals as well in Daniel Potts' second table. These numerals are separated by signs also known in the Székely sign-collection. Compared to these horizontally standing writing signs, the numerals are slanted, almost perpendicular:

Picture 138. Numeral – letter – numeral – letter from Tepe Yahya

From this, it is possible to suggest that these people often wrote their numerals perpendicularly, similarly to the Aramaic culture of the same Era (see page 117). But one certainly didn't draw a moustache perpendicularly as on the picture on the next page.

People probably freely used both directions of writing and often only dots to mark numbers below ten.

In the following there are the numbers written in Tepa Yahya, not only dots, and all turned into the usual horizontal form:

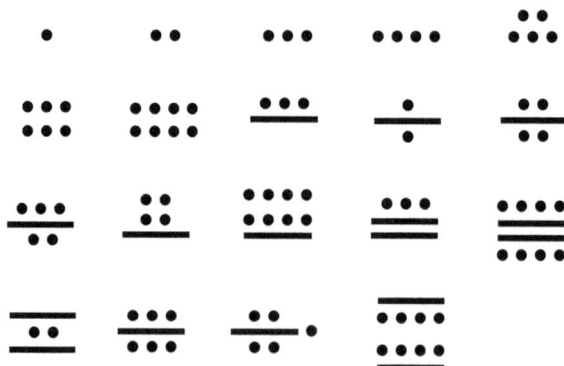

Picture 139. Dot-line numerals found in Tepe Yahya.

A special sign was found among the Tepe Yahyan Numbers. The scribe made a moustache from the line. Was it a custom to have such "shepherd" moustaches at that time? Did they call it a "moustache" for a joke?

22c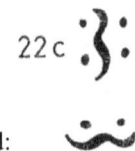

and turned:

Picture 140. This drawing will tell us that they saw a moustache looking at this line and as a joke, probably even called it that.

Dán Péter, a Hungarian shepherd, called the lines in his numbers written with the dot-stroke method a "moustache". Since he had one himself, he
saw in the line a moustache like that of the man in Tepe Yahya. (See page
158).
One found numerals written with dot-lines in Tepe Yahya, but even they
contained only lines. Daniel Potts presents those in his comparative tables.
The signs below must mean numbers, the coal-men of our times painted
them on our basement's wall.

*But there was a number 5 without the stroke across in Tepe Yahya*

Picture 141 Apparently, there was no strict rule regulating the writing of
numerals then.

Further we found the cross sign + with even legs. They used this as a
numeral as we can see from the variants of those in the previous chart.

Picture 142. Numerals like these were found in The Carpathian Basin
as well. See next page:

We show here in advance a few signs from page 124, which were still in use in the salt mine of Vízakna, Transylvania (now Romania) at the beginning of 20th century. Let's compare them with those of Tepe Yahya:

Tepe Yahya,
5.000 years ago

Transyvania
20th century

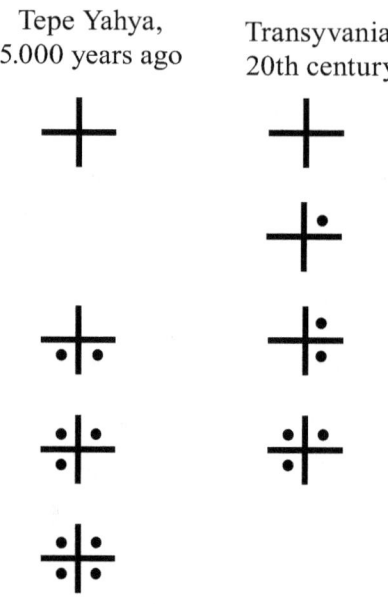

Picture 143. Almost 5,000 years is the difference.

There is no doubt; both are the same despite the 5,000 years difference. We can be sure that anything found in Tepe Yahya (the cross with one dot), was also in Transylvania (the cross with four dots). The growing number of the dots proves that those were numerals.

Picture 144. The collection of signs in table III of the book "The Potter's Marks of Tepe Yahya" by Daniel Potts proves that the ancient culture of that time was widely spread.

Picture 144.

# 9. A FIND IN COUNTY VAS <vash>

The find below, from Vas County, Hungary, originates from the early Copper Age, and is around 7,000 years old. The archaeologist drew what may be both sides of the find, but it could be one plate broken in the middle or two plates, however this uncertainty doesn't change the spectacle.

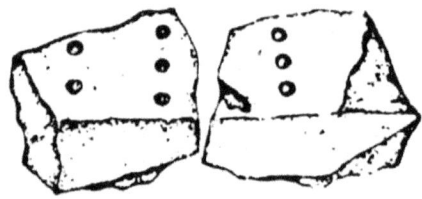

Picture 145. A find from early Copper Age. *"Relics of early Copper Age in Vas county, Hungary"* by Károlyi Mária. 1992, The Museums of Vas county, Szombathely.

The dots were placed regularly, random appearance can be ruled out. The person knew in advance how many dots he was going to mark. He had to plan the spots, where to put the dots. Otherwise, we can't imagine that a person living in such a high culture as he did wouldn't have been familiar with the dot-line number writing system, at that time the only one system broadly used.

The two signs regularly written:

Picture 146.

# 10. NUMERALS FROM COPPER AGE

We call the time 7,000-5,000 years ago the Copper Age. Makkay Sándor introduces spindle-buttons from the Copper Age in his book *"A tartáriai leletek"* (The finds of Tartaria, Table 39, 1990, Academic Publishers). He wrote under the picture: *"Engraved signs from a Copper Age spindle – buttons found in Hlinsko at Lipnik, Moravia"*. (Pavelcik 1983, 2nd and 3rd pictures"). Three of the spindle buttons carry numerals:

*Even two discs were enough to tell
that these signs are numerals.*

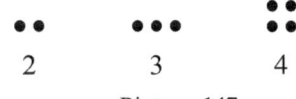

Picture 147

We can see further spindle buttons in the table of Makkay's book, but we will discuss them later.

It is worth looking at the following two spindle buttons. On these we meet again the 'palm leaves' seen before in Tepe Yahya. Let's compare the similar signs from Tepe Yahya with the signs on the buttons.

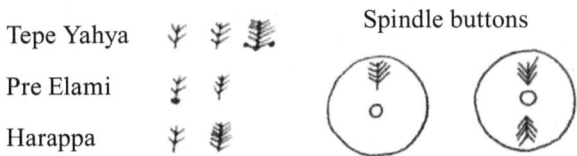

Picture 148

We have already discussed those little branches or 'fish-bones' on page 75.

# 11. NUMBERS FROM KÖRÖS CULTURE

The Körös (Vincsa-Tordos)-culture is mostly dated in our scientific literature to 6-8,000 years ago. Makkay Sándor presents in his previously mentioned book (on table 42) several signs being scratched into the surface of clay vases, found at Jela Sabac. We see in the following the numerals among those signs.

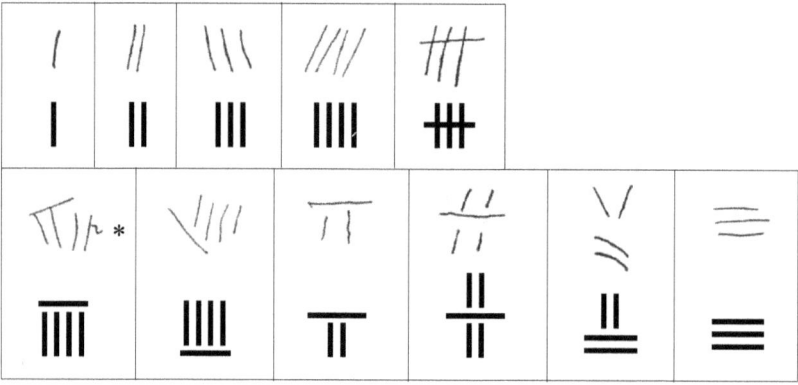

Picture 149. The c.6000-year-old numerals from Vincsa. Note: The first sign in the second row shows an additional small 'flag', which is probably a lapse of the drawer. It does not belong to the number. There are four perpendicular strokes below a horizontal one.

See the numbers on the following page too.
Even at this time the sign 'Λ' appears in the number-writing. There is more about it on page 201 in the chapter: Since when did Rovás numerals exist? See the two number-Rovás signs:

Picture 150. Number-Rovás signs from the Vincsa-Tordos culture.

# 12. NUMBERS FROM THE MIDDLE BRONZE AGE

Table 38 at the end of Makkay's book contains the following clay-pot fragments, on which there are signs written by the potter. (Makkay: "scratched-in signs on Middle Bronze-Age clay vases from Tirins., Döhl 1978, 1-2 figures"). Tirins is an important archaeological location from the Mycenaean culture of 3,400-3,200 years ago.

I think it should be clear to every reader that these signs are numbers and, based on what we have learned in this book (up to this point), we are able to read them:

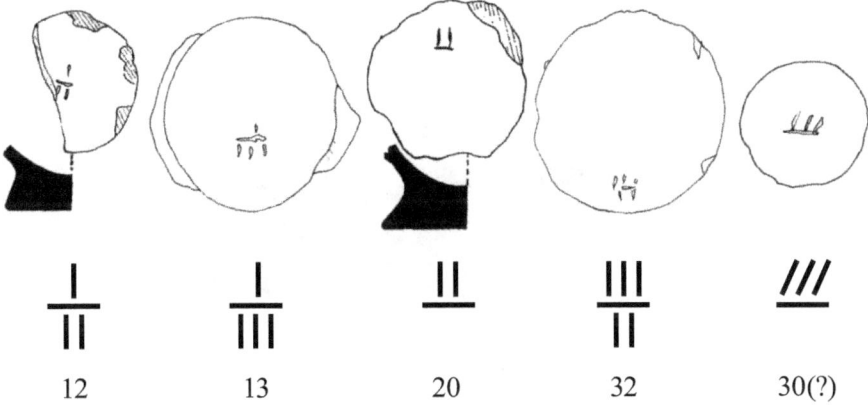

Picture 151. We discussed the interpretation of these signs in detail on page 49. Compare them with those on the previous page too: it is the same number-writing

Theoretically, the horizontal line could have a value of 100 also, but normally the line with a value of 100 is longer than that of a value of 10.

We put a question mark to the sign on the right end, because the potter drew the perpendicular lines clearly aslant and such lines always carry a certain message in number writing. The number-value remains 30 anyway, but we still have to figure out an additional message of the writing method.

# 13. SOME NUMERALS PRESUMED TO BE FROM BRONZE-AGE

The inscribed stone below was found on the Penobscot beach on Crow Island close to Deer Island. Crow Island is in the Atlantic Ocean, east of New-York.

According to the finder, the stone's inscription originates from the Bronze Age (5,500-3,500 and late Bronze Age even until 1,000BC).

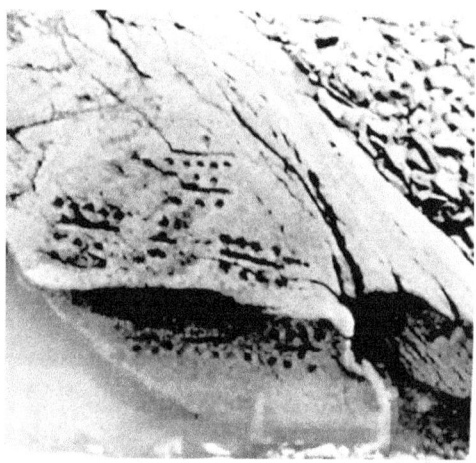

Picture 152. Unfortunately, the last row is in shadow and not readable.

The stone has several cracks which hamper the good readability. Two signs present 5 dots under and above the lines, which points to a duodecimal number system. It is not clear if the numbers standing below each other are independent, or are part of a larger number written perpendicularly. We now handle them separately, which would be like taking 8,724 as two numbers: 87 and 24. In case somebody proves their connection, then they could also be read together:

Picture 153. The numbers found on Crow Island.

# 14. NUMBER-SIGNS ON CORD (QUIPU)

Quipu is a characteristic feature in the history of number-writing. The numbers 3, 4 or 9 were marked by 3, 4 or 9 knots bound on a cord. In the decimal system, the local values follow each other on the cord. See the sign for number 4 (further pictures will follow on the next page):

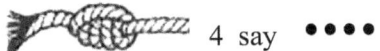 4 say ● ● ● ●

Picture 154. The knots bound on the cord represent here the dots

This number-writing tool and method vanished in the dim and distant past. It's not possible to turn up really old finds anymore, for the material decays very fast. However, two facts prove its continuous use since archaic times.

Firstly: The quipu was used from China to America, from the Hindus to the Aztecs, from Africa to Oceania and it is still being used in Guinea, Sunda Islands and Dutch Guyana. Its distribution over the whole world speaks for its use already in the Stone Age.

Secondly: the admonition of Lao-ce 1500BC,[27] "let people return to the cords with knots". This means, in China - far before Lao-ce's time – the line/line variant of the dot/line number-writing had been introduced (forcefully?). However people didn't like it, and couldn't live with it. Let's see when was it introduced. This will tell us the words of the minstrel Vei-can, recorded 4,650 years ago:[28] "once upon a time, the yellow emperor ordered to change, regulate or even invent things. There were Csü-sung and Cang-csi, who managed the structure of writing and with this they replaced the knotted cords". Therefore, the replacement of the knotted cords happened 4,650 years ago, but people did not forget them and Lao-ce asked 2,150 years later to reintroduce them.

Thus, we have an almost 5,000-year-old reference for the use of cord-writing and if it was widely used 5,000 years ago, then it must have been known much earlier as well.

---

27 One wise saying of Lao-ce: "The folk will stay hungry if a governor is very active and such people are not easy to govern."

28 Várkonyi Sándor: *The history of writing and the book,* page 75-76. Széphalom könyvműhely, 2001.

The quipu was "written" from 1 – 9, namely, one bound the knots on the cords according to the dots in the numerals:

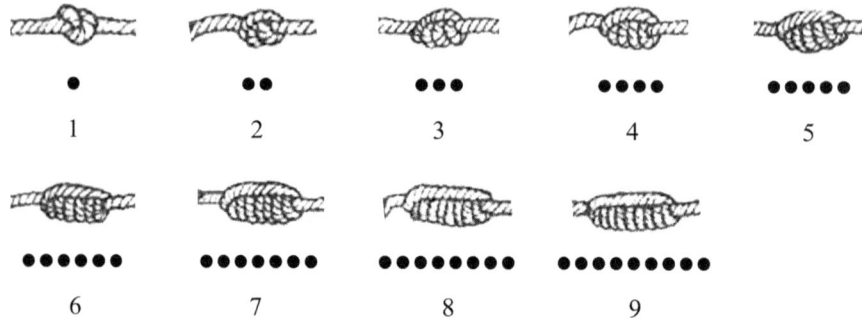

Picture 155. The row of numbers in knot-writing.

The numbers were knotted one after another according to their local value. We see the number 3,643 knotted on a cord:

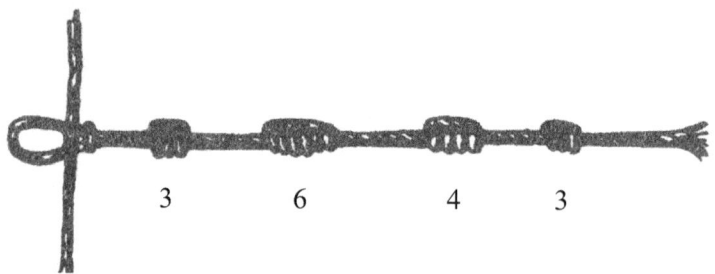

Picture 156. A number registered on a cord.[29] Therefore, only the registration of the number is special, but the number-writing itself was the same as it is today. Here we must accept again that there is no "evolution" visible.

They put every number on a separate cord and the numbers, belonging together, were strung on a common cord using the loops at the end of the cords. (You can see this loop in the picture above.) They made a bundle this way, as seen in the next picture. The sum of the numbers in the bundle was knotted on a separate cord added to the bundle.

If there was no number in a local position, then no knot was bound in this place. (The distance between the local positions always stayed the same).

29 Georges Ifrah: *Numbers,* Havril Press, London, 1998

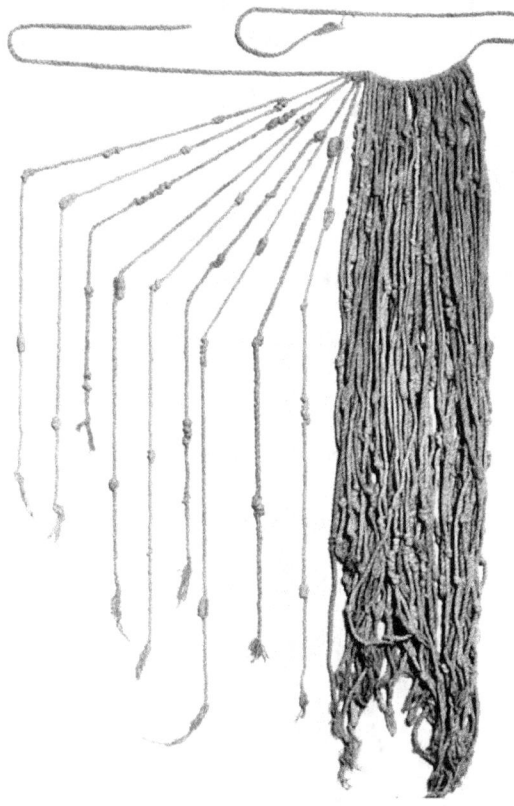

Picture 157. One bundle therefore represents a group of numbers belonging
to the same subject.

The colour of the cords was very important too. In South-America, people used yellow cords for data about gold, white for data about silver; the colour green was used for crops, purple for military and red was connected to numbers dealing with property. In India, at the national census in 1872, they used black cords for men, red for women, white for the boys and yellow for the girls.

Binding knots can be seen as one form of remembrance as is every other form of writing. Now we understand better the saying "let's make a knot in the handkerchief" if we want to remember something.

In some locations where they still use knot-writing, a workers leaving for vacation receive a cord with as many knots on it as the days their vacation is going to be. Once at home they just have to open one knot daily to know when it is the end of the vacation. Soldiers do the same by cutting off a centimetre per day from a measuring-tape before their vacation; or students write 'VACATION' on the blackboard and erase a letter daily from it.

Some people put punched fruit-kernels on a cord, others snakes or shells. The meaning stays the same.[30] It is interesting that the Mayans used shells for marking zero.

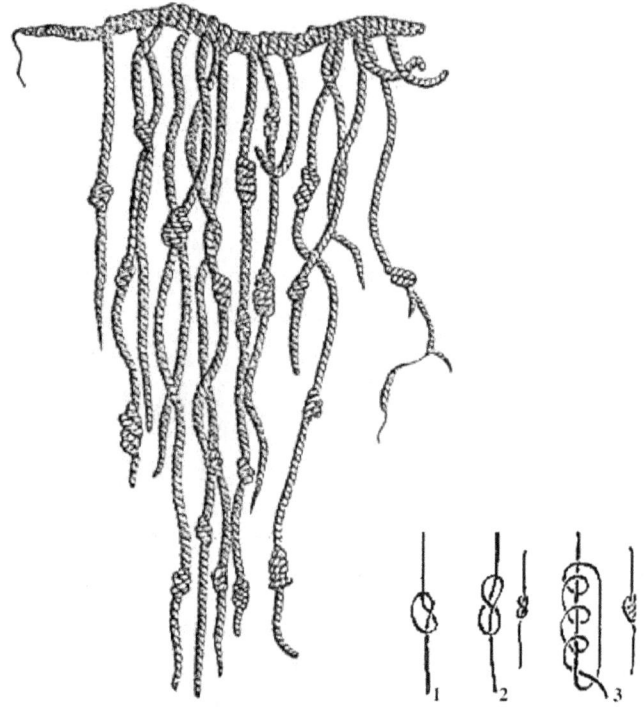

Picture 158. A typical quipu and the steps to bind knots on the right.

A characteristic feature of this kind of number-writing is that every cord marks only one number and numbers (cords) belonging to one object are strung together on one collecting cord (see next picture). We can see deep connections by just looking at this method.

---

30 The custom to use punched coins may be due to the possibility putting them on a cord.

Picture 159. An Inca imperial civil servant
with a quipu around 1613 [31]

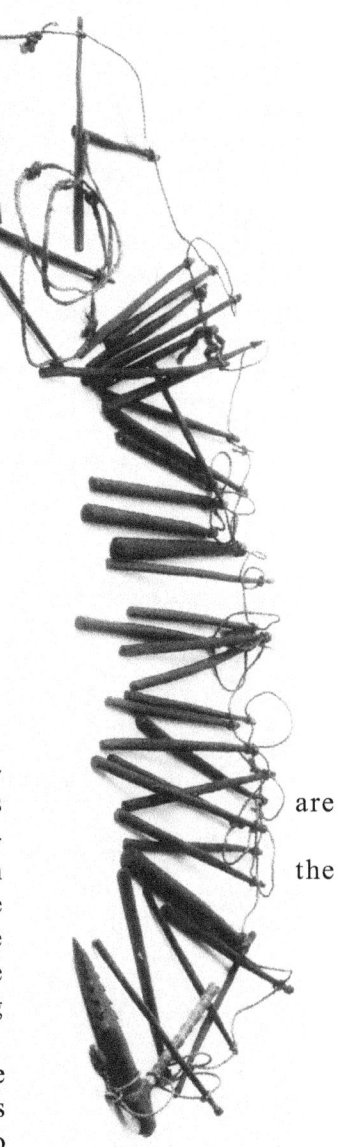

In the picture on the right, there are wooden plates hanging on a cord. The numbers are engraved on them. The system of keeping records is the same as that used in the quipu on the left. Both methods are descendants of dot-line number-writing and both write by marking the local values. It is evident that dots dominate knot-writing and lines are used for engraving onto wood.

The distribution of both methods is quite remarkable: Number-writing with knots was used on one half of our planet, engraving onto wood on the other. However, both together covered almost the whole of the Earth almost as if they had an agreement about it. The last picture was taken on the Islands of Torres Strait, but could have been taken anywhere in

Picture 160.
A collection of numbers engraved on wooden plates and strung on a cord from the islands of Torres Strait.

31 Andrew Robinson: *The Story of Writing* (Thames and Hudson, 1995)

Europe included Russia as well. This method of number-writing was used in England until 1812. A saying recalls to us its use in Hungary: "sok van a rovásodon" <shok van a rovaashodon> (you have much to answer for..). These all mean - because the method is the same - that it was one idea at the beginning, which separated into knots on cords and engravings on wood later on. There is however a large difference between the two methods. One can't apply many innovative changes while using quipus, making knots on ropes. This method was a dead end. On the other hand, the method of engraving the numbers gave its users many possibilities for innovation and experiment on the forms and order of the numbers. Our ancestors used these possibilities without restraint. (See more information about numbers engraved on sticks and lathes from page 169.)

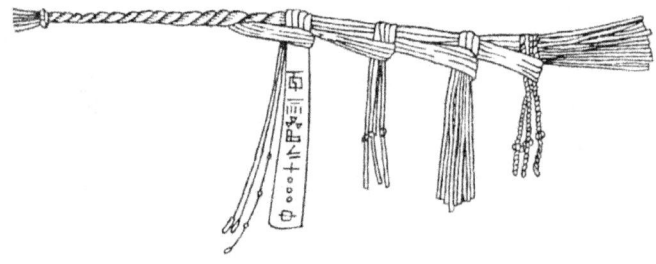

Picture 161. Donations to a temple are registered on this Japanese ketsujo (quipu). The name of the donor is written on the tape.

The signs made by knotting are the oldest signs used to record according to Chinese tradition. The Chinese bound knots out of straw or bulrush on cords made from bulrush or reed in order to record numeric data.

# 15. ARAMAIC NUMERALS

**A**ramaic people turned up on the stage of history around 4,000 years ago in Asia Minor. Aramaic became a widely used business language and later almost a world-language. It is still spoken in some communities. It was the mother tongue of Jesus, thus he certainly knew the following Aramaic number-writing:

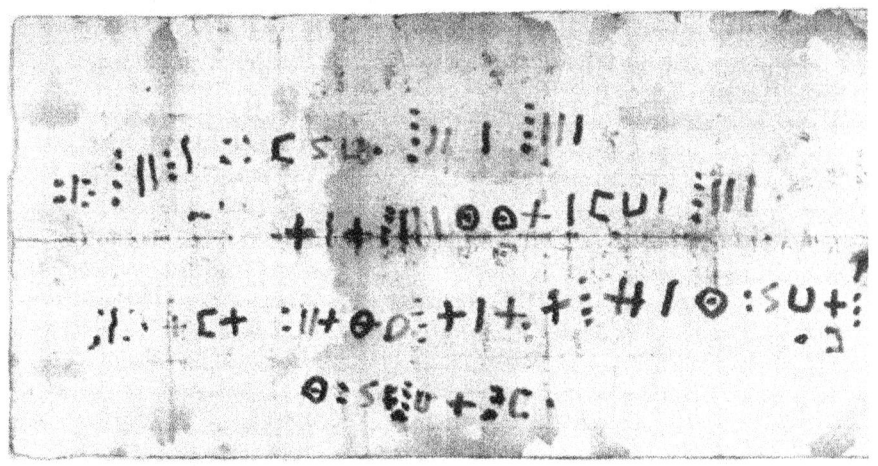

Picture 162. From the book The Story of Alphabets and Scripts, page 60 by Thames & Hudson, 2000.

The picture is dominated by dot-line signs. Some signs stand between "letters" of different, not Aramaic, alphabets. The Aramaic alphabet looked different:

Picture 163. The Aramaic alphabet is related to the Phoenician alphabet. Aramaic people made their writing signs quadratic on some occasions and these were taken over by the Hebrews. A variant of the Aramaic writing-signs became the Arabic alphabet.

The not dot-line signs of picture 162 in particular are supplementary signs for numeral-writing or are new numerals put in between the basic numeral signs. The latter is quite possible, because the cross with even legs and the circle with a dot in its middle turned up later as numerals in different places.

We can see it below that they placed the dot-line numerals in perpendicular position. See following the signs, which are certainly numerals in picture 162:

Picture 164. Aramaic numerals of the on picture 160, upper line.

# 16. SOUTH-AMERICA

Peru, Columbia, Ecuador, etc. offer many mysteries. Explanations are needed not only for the Nazca lines, but for heaps of objects, for the bones and skulls of unusual human beings. We see below an assortment of drinking vessels carved out of jade. There are dot-line inlaid numbers on the cups, which shine strongly in UV light. The large cup is covered with an exact astronomical chart. The inside of the cup is strongly magnetic, but this is only weakly measurable from the outside. Interestingly, the Orion constellation seemed the most important for the makers of these cups according to other objects made by them.

We can see the following dot-line numerals on the small cups:

Picture 165. We see the whole drinking assortment in the above picture. On the pictures below, there are 2 sets of 3 drinking cups with clearly readable numbers on them. The contents of the 8 cups fill the large one exactly, according to the description.

# 17. NUMERALS ON A ROCK IN MONGOLIA

The picture below is from the book Prehistoric Art by Paul G. Bahn[32] and made by Johann Tabbert in 1722. Tabbert copied the signs found on a rock while visiting Mongolia. The age of these signs is unknown.

Picture 166. Signs copied by Johann Tabbert from a rock while visiting Mongolia 1722.

There are three numerals among the signs on the rock:

Picture 167 The few only perpendicular lines and even the + sign as well could mark numbers. We won't get any more answers to these questions.

---

32 Cambridge University press, 1988.

# PART IIII
## THE CARPATHIAN BASIN
## OR ANCIENT TIMES IN THE 20th CENTURY

## 1. SEBESTYÉN GYULA

Gyula Sebestyén was born in Szentantalfa (Transylvania) in 1864 (then Hungary) and died in 1846 in Balatonszepezd. He was an ethnographer, a writer, a historian of literature and member of the Hungarian Scientific Academy. From 1898, he was the president of the Ethnographical Society and a member of the Kisfaludy society. He started a very successful voluntary ethnographic collecting movement.ment.

His major works are:

*The origin and name of the Széklers* (Budapest 1897)

*Minstrel songs* (Budapest 1902)

*Regősök (bards),* (Budapest 1902)

*Rovás and Rovás-writing (*1909)

*The authentic records of Hungarian Rovás-writing* (1915)

Picture 168. Sebestyén Gyula (1864-1946)

His book Rovás és rovásírás [33] (Rovás and rovás-writing) is extremely important with respect to number writing. He went all over the Carpathian Basin and collected most of the still traceable records written in the old ways. Many valuable proofs of our old culture would have been lost for ever if he hadn't done this "at the last moment".

---

33 *Sebestyén Gula: Rovás és Rovásírá*s, 1909, Magyar néprajzi társaság. Reprinted by Püski Publishers, 1999.

He covers number-writing in two places in his book. In the chapter Számrovás (number-rovás) he mainly describes the world of numbers engraved into wood. The most important of his records is the introduction of the method of writing numbers on sheep's heads practised by an old shepherd. This method is the 20th century survivor of the oldest number-writing we know. In the chapter Tulajdonjegy (property tag) he introduces the ancient dot-line number-writing, but is not able to explain the meaning of the signs. Because even the donors of these finds could no longer read the signs, he called them merely property-tags.

We will deal with these two very important topics from Sebestyén Gyula's book in the following sections.

# 2. DOT-LINE SIGNS IN SALT MINES

Sebestyén Gyula found dot-line signs in the salt-mines of Vizakna, Akna-Suhatag, Désakna and Aknaszlatina in Transylvania, part of Romania since 1918. These signs were made by the miners, and used as their identification numbers. Sebestyén Gyula writes: " these signs are recorded beside the workers' names in the alphabetical list of names in the mine-office". He notes furthermore that "the mine-manager Stépán could not tell him anything about the origin of the signs, since even the oldest workers had used these markers from the beginning." Thus, the management of the mine couldn't understand the meaning of the signs, but the oldest miners or their fathers certainly knew it, because those were inheritable signs. According to the management, the workers dictated the dot-line form of their signs themselves at registration s.

There was a "foreman" for every group of workers in the salt mine of Akna–Slatina, who received the numbers from the office and registered the workers. At Vízakna: "the overseer distributes and writes the signs into the company's register against the name of the miner".

The usage of the dot-line signs were advantageous, because the miners worked with a pickaxe and could easily engrave those into the salt-cubes.

Below are some salt-cubes with dot-line signs from Maros-Újvár:

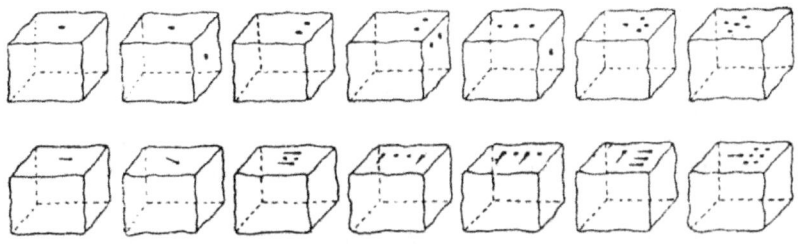

Picture 169. The drawing of Sebestyén Gyula. Salt-cubes with signs on them.

These dot-line signs were used continuously in the salt-mines discussed above until the early 20th century . So far as we know, they are no longer in use anywhere. The mine in Parajd was the latest to start paying the miners by the weight and not by the number of the salt-cubes.

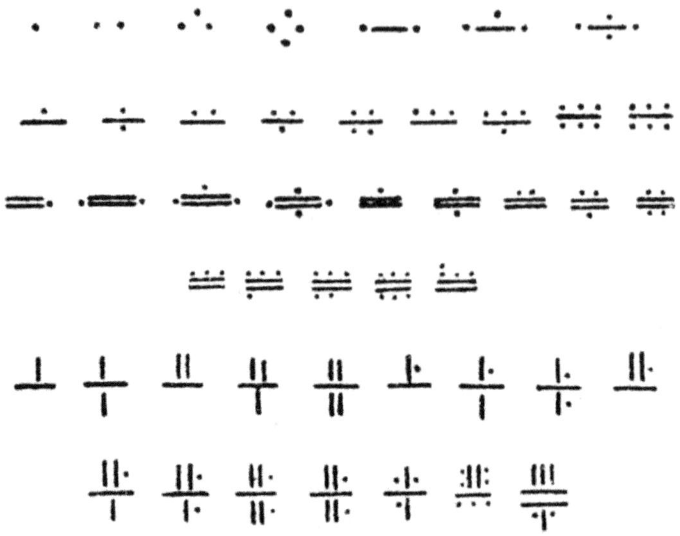

Picture 170. The signs of the salt-cutters in Vizakna. According to Sebestyén Gyula, 52 miners and 10-12 apprentices worked in this mine during wintertime, however there were many more in larger mines.

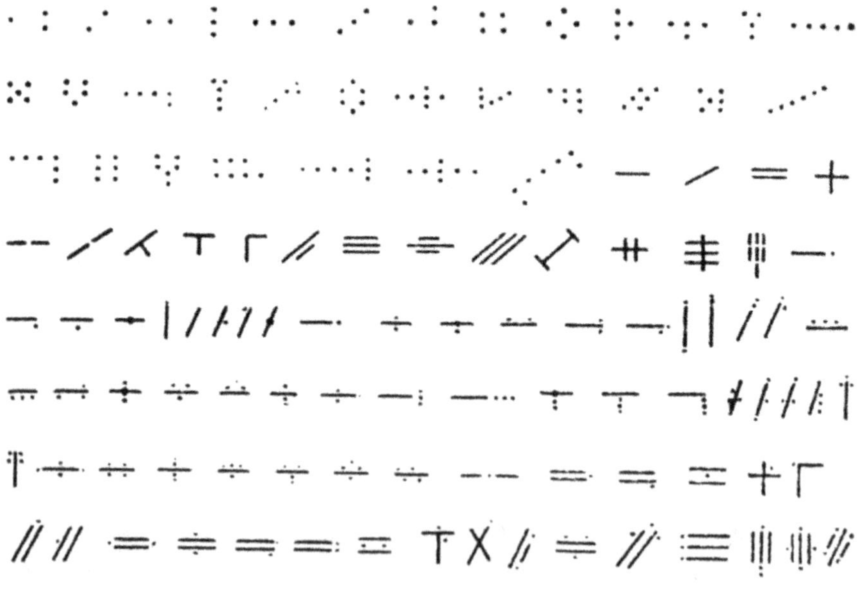

Picture 171. The salt-cutters' signs in Akna-Suhatag

# 3. THE SIGNS FROM VIZAKNA

We may well guess – having reached this far in the book – that the presented signs of the salt-cutters are regular numbers. These signs meet all the requirements, determined at the beginning of this book, for dot-line numerals. The workers in the salt-mines received serial numbers as an identifier according to some chronological (?) order. Already the fact that these signs can be put into a continuously extending order, qualifies them as numbers. [34] We see below only the dot-line signs from Vízakna. The expanding sequence is quite visible. We can even put in the signs missing from the collection of Sebestyén Gyula. We marked these assumed signs in the collection in grey:

**Detail A:**
only
dots and
lines

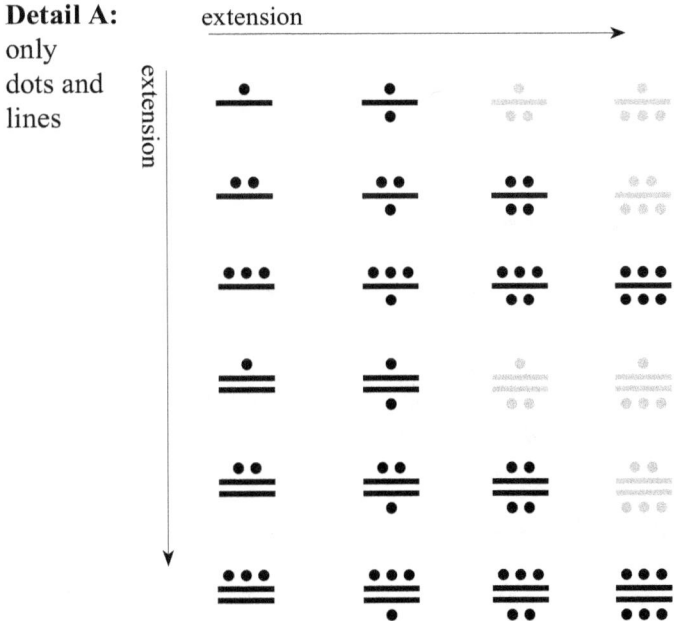

Picture 172. The signs of Vízakna in extending sequence

34 People looking for extraterrestrial individuals would be happy to find special sequences among incoming radio signals instead of only mechanical repetitions.

It is obvious that the signs written with dots / lines or with horizontal / perpendicular lines are numerals. Naturally, the row of numbers is incomplete in this case also:

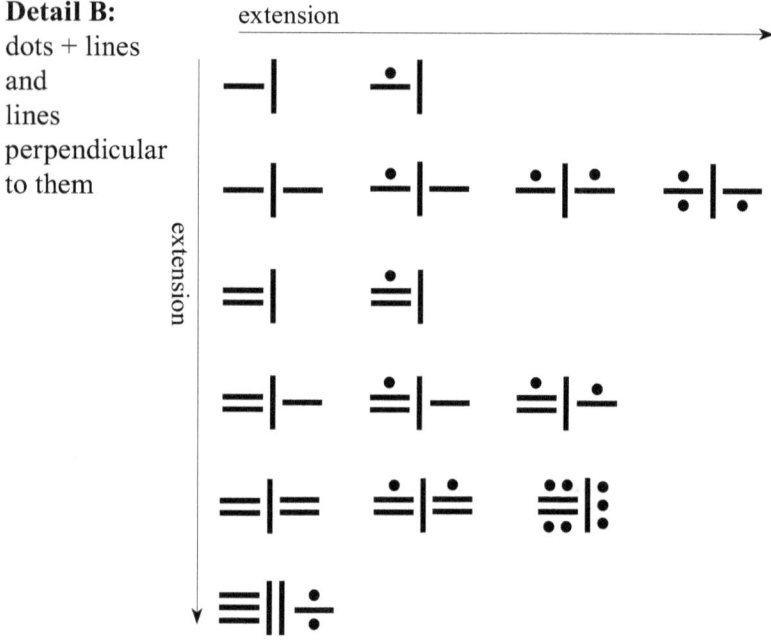

Picture 173. Other details of the signs in Vízakna

The details A and B contain all the kinds of signs which Sebestyén Gyula noted in Vízakna. We can see signs in the detail B which are built the same way as signs in Lascaux (see page 72).

One sign from Lascaux:

Picture 174. This 17,000 years old sign is built the same way as the 20th century signs in Vízakna.

There is no doubt that people in Vízakna used the 17,000-year-old number writing method from Lascaux. Therefore, people must have been writing numbers continuously using the same method for 17,000 years.

**Detail C:**
Dots at
the side

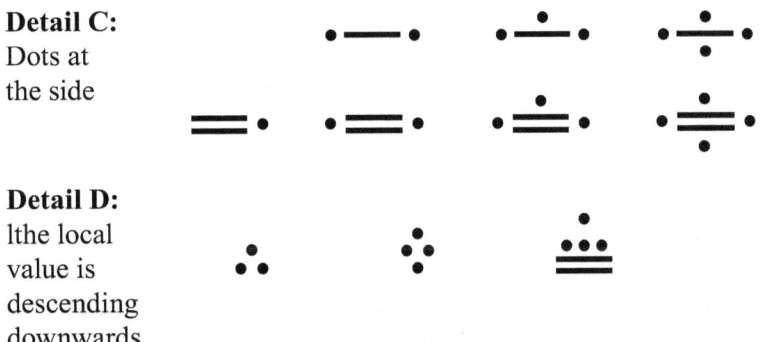

**Detail D:**
lthe local
value is
descending
downwards

Picture 175. Other typical details of the numbers in Vízakna

# 4. THE SIGNS FROM AKNA-SUHATAG

The signs in Akna-Suhatag can be grouped in a similar way to those in Vízakna. We have systematized here too only the signs containing dots and lines. The details A, B, C, and D contain the same kind of signs as in Vízakna. In Akna-Suhatag however, there is an additional fourth sign added to the group: a perpendicular-line, which is the long-line representing higher value.

Note: I made the slanted signs horizontal, because obliquity influences only the meaning of the sign (for example: minus). I left a gap in the middle of the "crossed over" signs: I drew ≠as=|= . There are several signs (made mostly from dots), where the drawers' individual inspiration took over and we may therefore be uncertain of the meaning from reading it. For this reason we left the guessing out from our discussion.

In other cases, I turned those signs through 1800, which made sense only in this position. I turned these Akna-Suhatag signs according to the very carefully written signs of Vízakna.

We assume copy-mistakes in some signs, so we don't deal with those either. There are still enough signs remaining to make a general picture of the signs' structure.

We won't differentiate between signs where the dots, above or below the line are put in the middle or on the side of the sign.

**Detail A:**
only dot
and
lines

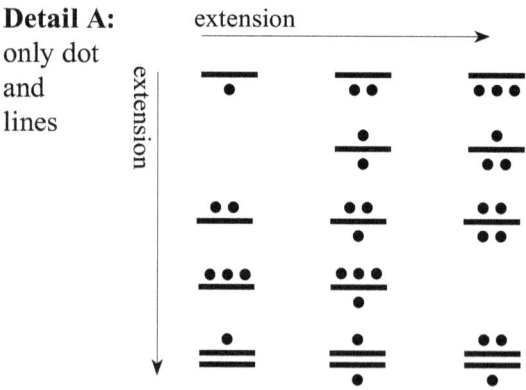

Picture 176. One part of Akna-Suhatg's typical signs.

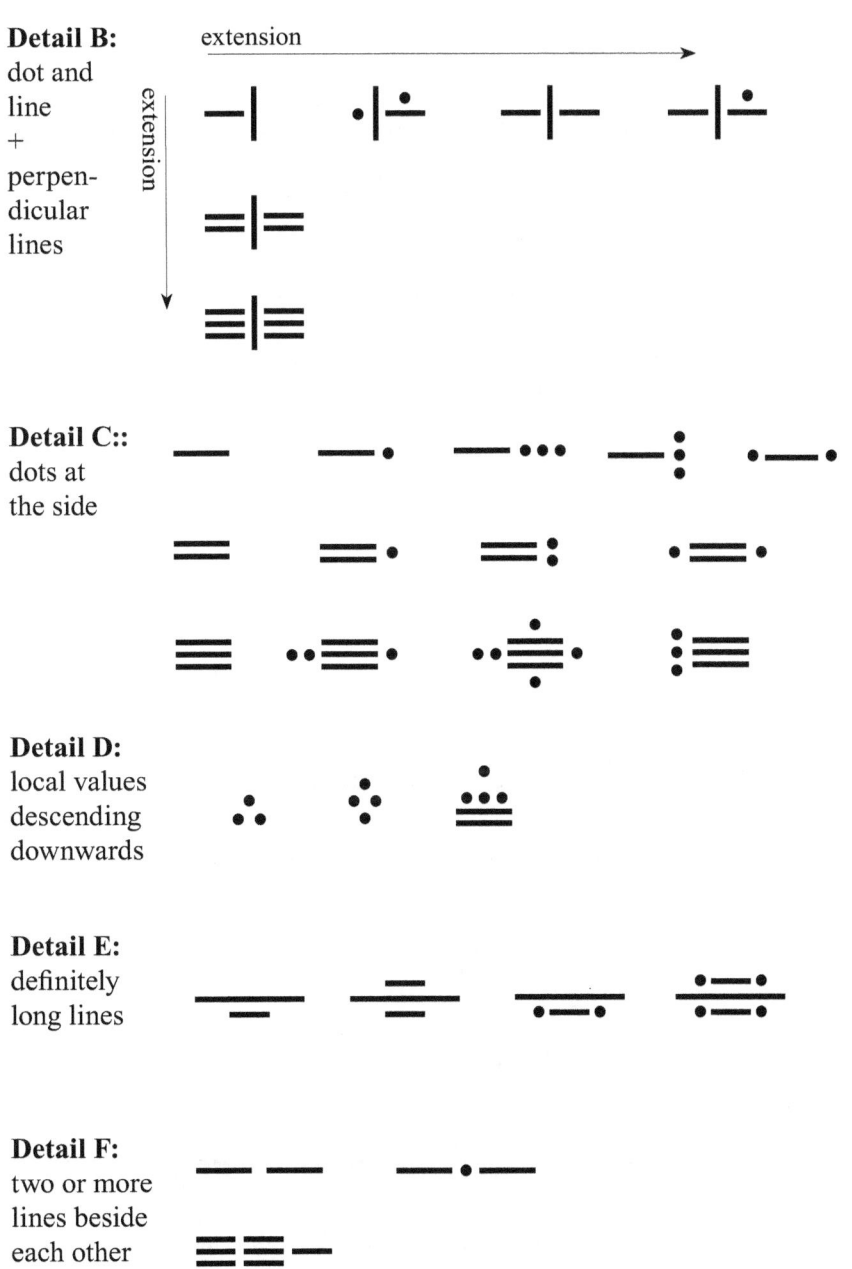

**Detail B:**
dot and
line
+
perpen-
dicular
lines

extension

extension

**Detail C::**
dots at
the side

**Detail D:**
local values
descending
downwards

**Detail E:**
definitely
long lines

**Detail F:**
two or more
lines beside
each other

Picture 177. Typical groups of signs in Akna-Suhatag

# 5. COMPARING THE SIGNS OF VÍZAKNA AND AKNA-SUHATAG

We compare here signs with an identical structure. Markings:

1.        without frame: signs from Vízakna
2.        in grey shading: only signs from Aka-Suhatag
3. ☐    put in a frame: signs occurring in both mines as recorded in the collection of Sebestyén Gyula.

**Detail A:**
only dots
and lines

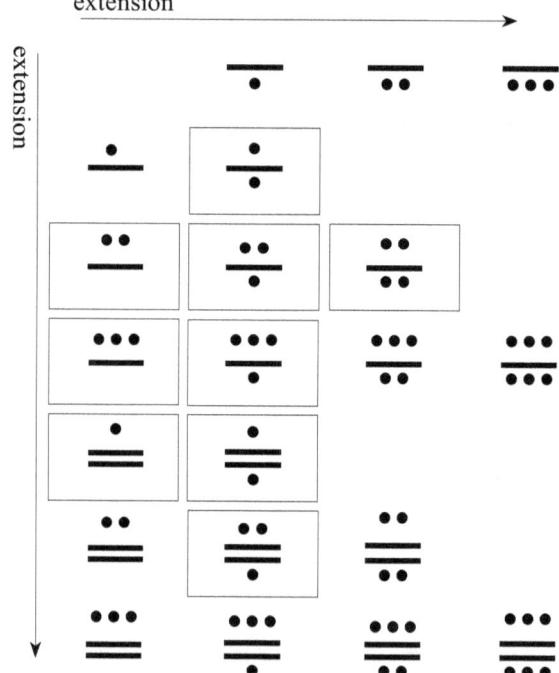

Picture 178. Comparing the signs of Vízakna and
Akna-Suhatag. Part I

The large overlap proves that both salt mines – Vízakna and Akna-Suhatag - used the same system of signs.

Markings:1.

1.   without frame: signs from Vízakna
2.   in grey shading: only signs from Aka-Suhatag
3.   put in a frame: signs occurring in both mines as recorded in the collection of Sebestyén Gyula.

**Detail B:**
dots and
lines
  +
lines
perpen-
dicular
to them

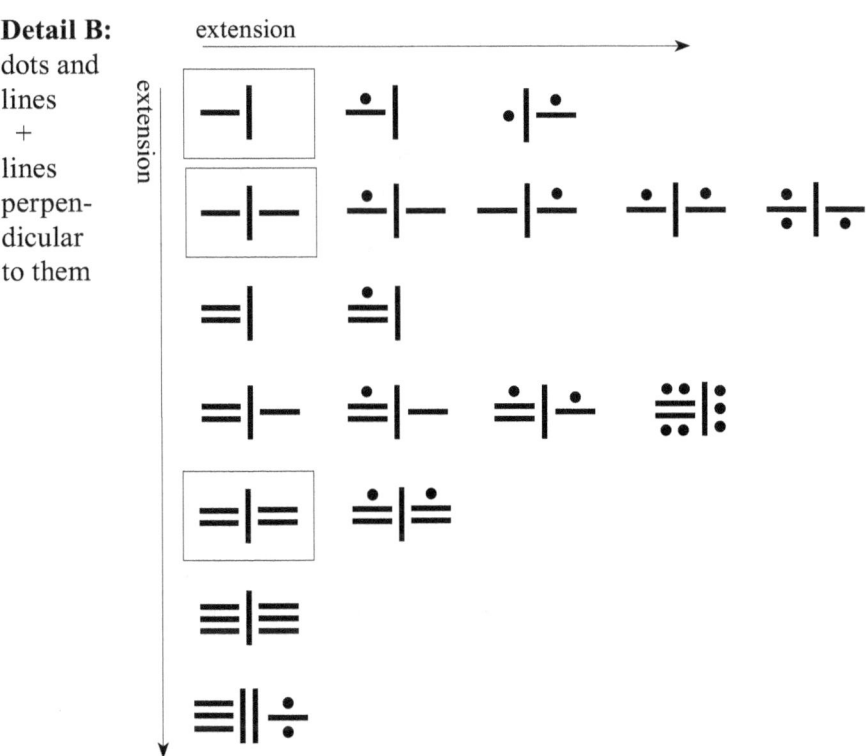

Picture 179.  Comparing the signs of Vízakna and
Akna-Suhatag. Part II.

There is no doubt that these kind of dot-line structures were widely used in Vízakna and Akna-Suhatag. The sign-rows are not complete in the different details, possibly due to scantiness of the notes made by Sebestyén gyula.

Markings:
1.             without frame: signs from Vízakna
2.             in grey shading: only signs from Aka-Suhatag
3.     ☐    in a frame: signs occurring in both mines as recorded in the collection of Sebestyén Gyula.

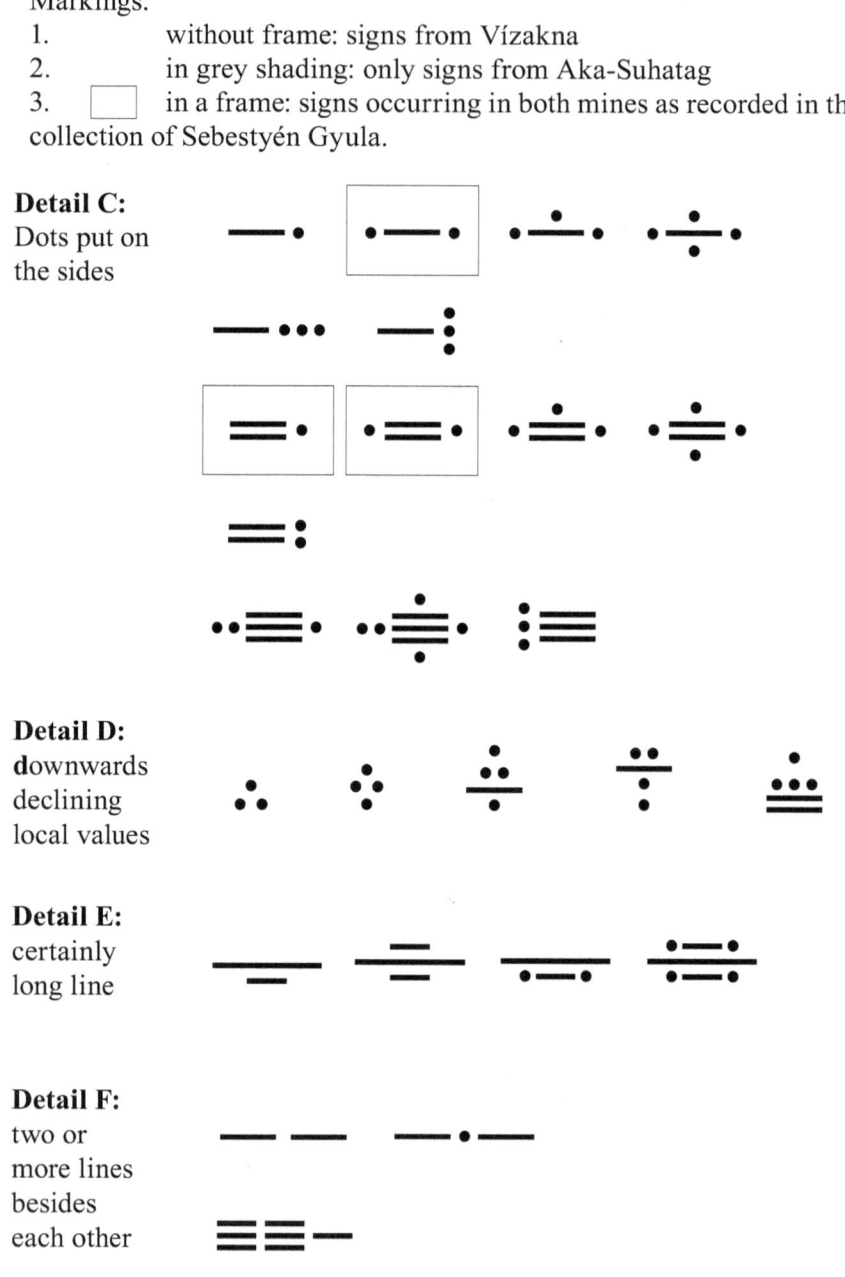

**Detail C:**
Dots put on the sides

**Detail D:**
downwards declining local values

**Detail E:**
certainly long line

**Detail F:**
two or more lines besides each other

Picture 180 . Comparing the signs of Vízakna and Akna-Suhatag. Part III.

# 6. PROBLEMS ARISING WHILE FOLLOWING CALCULATIONS

As yet, I am refraining from reading the signs of Vízakna and Akna-Suhatag. We could follow the rules presented in the chapter on page 49, but there are still some insoluble cases. This means that we still need to know something more, not visible on the written signs. It could be something they knew, but didn't write down.

As a result, I don't know the answers to several questions. First, I don't know in which numerical system those signs were written. Four dots could be written above, below or beside a line in the decimal system. Most are 3 dots and that points to a base 8 system. However, we can't say that there are examples for every possibility in both small collections.

The other problem is that while Sebetyén Gyula copied the numbers quite regularly, but he possibly did not care about the relation of the signs to each other. He didn't know about this problem. Certain constellations or positionings of the signs could even point to mathematical operations. But we can only rarely find such instructions. Our ancestors did often not tell us about the necessity of an operation. - They knew it and did it. Therefore, we can only be sure about the reading of old numeric signs if we know their complete system. Old numerals are often real puzzles for us.

To demonstrate the problems of reading and solving ancient number writing based merely on rare fragmental finds, I present a letter, written to me by Cseh Gyula. I post this original letter out of respect for him. Cseh Gyula promised to support me with his knowledge in writing this book, but unfortunately he died just as I started writing.

Cseh Gyula wrote the following letter as an introduction to our common work and in which he merely mentioned doing operations. We have to know that, according to this letter, people in some early cultures often drew lines perpendicularly to the horizontal base-numbers instead of dots. This way nothing changes if the numerals are turned through $90^0$.

Here follows the letter of Cseh Gyula. Note that he presents Chinese examples and because of this, lines were used instead of dots, (see page 44):

„Dear Csaba,

You have started an extremely complicated piece of work.

The undertaking is very difficult, because in many cases one didn't give operative instructions. The operations were sometimes determined by the location of the numbers. This, however, changed from school to school, indeed from "teacher" to "teacher".

Some examples (Chinese):

1. To multiply 152 by 16 was written in the following way:

152 |≡||

16 |⊥

2. In some cases even the colour of the written number decided the procedure.

3. He gives us the lesson in words, therefore the written numbers tell nothing mathematically:

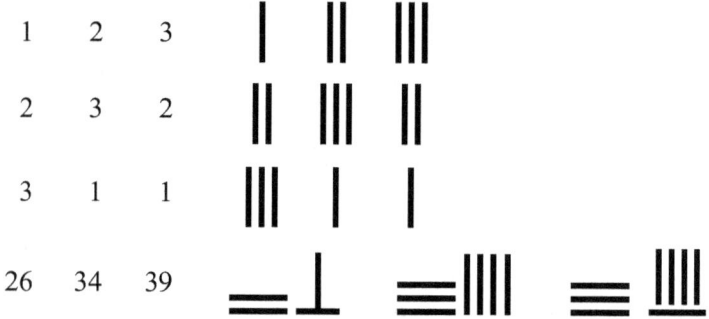

| 1 | 2 | 3 |
| 2 | 3 | 2 |
| 3 | 1 | 1 |
| 26 | 34 | 39 |

The text is as follows: "We received 39 tons of grain made up of 3 bundles with good, 2 bundles with medium and 1 bundle with poor harvest…" etc.

Looking just at the numbers, you can't find any operations. If there is no text, there is no solution.

4. In other cases, they marked the '+' numbers with the word "cheng" and the '−' numbers with the word "fu". In fact, there was a teacher who crossed the negative numbers with a slanted line: ⫲ = - 4.

Some scientists thought that "cheng" and "fu" were units of measurement, thus no result was received.

Note:

1. I was lucky to be able to solve the Sumerian-calculated $\sqrt{2}$ (it was a 3,500-year-old puzzle, even Simonyi could not solve it). I solved it using the base 60 system and discovered that people in ancient India then and there were using the base 12 system. Thus, their result was identical with that of the Sumerians, down to five decimal places if I calculated using the base 12 system.

2. It gets complicated because many times they wrote a self-constructed unit instead of the real value:

If one takes 9 "norms", the surplus is 11

If one takes 6 "norms", the shortage will be 16, etc. etc.

Finally, I came to the following deductions:

1. In most cases we should try to figure out what the writer wanted, to understand his sequence of ideas and then try to identify his method afterwards.

2. It is important to imagine oneself having their beliefs. Science for them is "sacred", "divine", etc. They have sacred numbers: 9 for the Chinese, 7 for the Egyptians and Sumerians and its integral multiples and 60. The holy numbers of the Hindus were changing (5, 6, 10, 12) and finally they stayed with 10.

Some Hungarian and Székely customs using numbers:

It is much more colourful than any other custom, because they always used the most suitable method for a specific task. People who used it agreed on the rules in advance. (We have to figure it out afterwards), It might be, for example, that the sign /// means -3 in Akna-Slatina.

While working with the numbers in Akna-Slatina, for example, I found a note, which said that: "by agreement" 25 people should have gone to a shift and cut 200 kg (or cubes..) of salt per capita. However in the end they just wrote down that 3 persons were missing and two had cut only 100 kg salt. You can calculate the amount of production and the money people get for their work this way as well, but a modern Sherlock Holmes does have a hard job.

3. These numerals do not always – indeed rather seldom – mark only one number, their position itself can already point to an operation. Here, we really need to know which numbers were written with ochre and which with coal (in the case of very old finds).

It is very important to know the sacred numbers, because we may get good results with the help of them. As an example: You can calculate and easily solve the measurement of the Cheops pyramid by working in the base 7 system.

I know, I couldn't help much and possibly increased the problems. However, …maybe!!

P.S.: The rules were handled as taboos in certain cultures or schools. Thus, we can relay on a detected rule only within certain limits. From this statement it follows that number-writing, customs of using numbers as science, building- and decorating-art always had and have an ethnic flavour. I have suggestions and some proofs that certain number theories (thesis of prime numbers, factorial, the sum of the first n integer numbers, the "Fibonacci sequence", etc) have been known for at least 5-10,000 years. Therefore, we should come down off our high horse and approach our ancestors with humility. And finally, there were certainly secret signs or instructions known only to selected people and which have deceived outsiders."

# 7. THE NUMBERS OF THE SHEPHERD SR. DÁN PÉTER, AROUND 1900

Sebestyén Gyula introduced the number sequence of Sr. Dán Péter by presenting the following examples (he drew the first number as a horizontal line. However, seeing the samples it will be evident that this line is turned perpendicular, if it stood above a hundred):

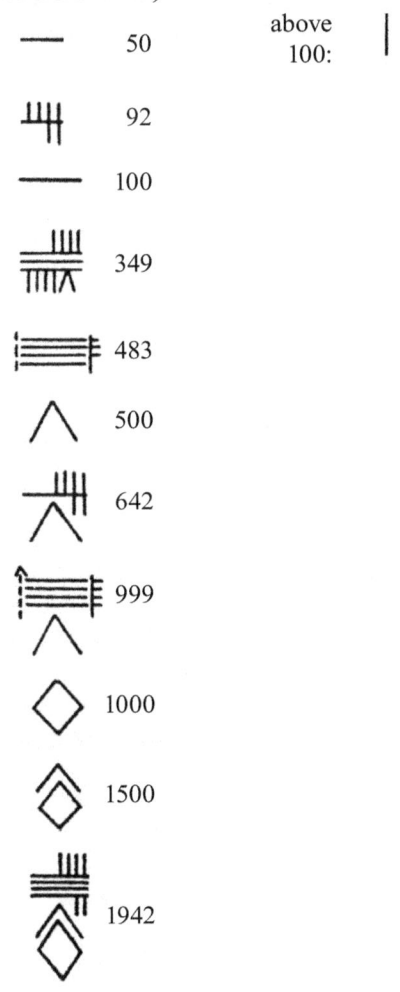

Picture 181.
The numerals
of the shepherd
Sr. Dán Péter

According to Sebestyén Gyula, the numbers start at 50, because Sr. Dán Péter wrote his numbers up to 50 using the Latin method. But apart from this, Sr. Dán Péter's method of writing numbers remains the very uniquely written record of number-writing history and it provides a special memorial to humanity's culture-history – for several reasons.

He learned this way of number writing as a shepherd boy in the 1840s on the pusztas around Hódmezővásárhely (Hungary). Later he moved to the puszta around Gyula and became one of the 35 shepherds of Count Károlyi György, where he kept using the number-writing he learned as a shepherd-boy. [35]

Sebestyén Gyula reported a quite interesting fact about the number-writing customs of shepherds: *"All the shepherds had to have a different kind of number-writing, because they accounted separately for the 1,500-1,600 sheep given to them for care."*

Therefore, a shepherd recognised his sheep by his own enumerating system. This means that the shepherds grazing close to other shepherds had to choose a number-writing system which was easy to differentiate from that of their neighbours. This demanded skilfulness and cleverness from the shepherds. Our shepherds therefore, must have been well- learned in writing and in the different forms of number-writing.

This doesn't mean that the 35 shepherds had as many different methods of number-writing. We did not find records about other shepherds' number-writing, but it is improbable that more then 4 shepherds were grazing in the close neighbourhood of each other, which reduced the numbers of different systems. One could achieve great diversity by adding a different value to the perpendicular or longer lines or write little strokes instead of dots, or by learning how to place signs with certain location value relative to the others. As we will see, the numeral-sequence of Sr. Dán Péter contains signs of 5, of 500 and 1,000, in addition to the ancient signs.

---

35  According to Sebestyén Gyula, Sr Dán Péter managed his shepherd-share very well. At the end he became the owner of 120 hold (170 acre) land.

# 8. THE MARKING OF SHEEPS

As Sebestyén Gyula tells it, Hungarian shepherds considered the blood-staining of a little lamb (cutting its ears several times) as unacceptable; they rather used soot or tar to mark and register the animals. Sebestyén Gyula noted: *"The shepherds on the earlier pusztas around Gyula – today the area around Csorvás and Kondoros – injured the animals' ears only when they had to grade them according to their fleece, or a mother sheep had to be marked for some reason after delivery."*

The procedure of "tarring" as Sr. Dán Péter told it to Sebestyén Gyula was as follows: *"When a lamb finally became a member of the group then its cheek became tarred together with its mother's. For this procedure, ashes and amber were put into a crock, and this tar was heated in a cup. The tar became liquid, and then the crock was put at the site of the corral's exit. Here, the shepherd - who was handling the wooden hook for tarring – while blocking the exit with his back would let mother and lamb through between his legs and mark them with a sign of the group."*

*This method, according to Sebestyén Gyula, has been known by the old shepherds in almost all sheep breeder areas.* (The custom is nowadays just a little plate pinched onto the ears.) The numerals were painted on their cheek as the only place on a lamb or sheep without long hairs.

Picture 182. Only the cheek of a sheep can be used for the placement of a durable sign.

# 9. THE NUMERALS OF SR. DÁN PETER

First let's recall the already analyzed ancient sign series:

1     10     100     1.000

Picture 183. The series of the archaic numerals.

The signs of Sr. Dán péter are identical with these ancient signs, but three additional signs were put between them (framed), which caused the values of the signs to change:

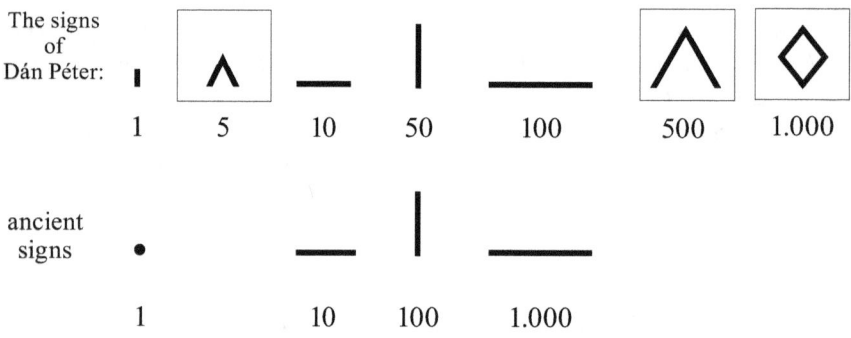

Picture 184 Comparison of the ancient signs with those of Sr. Dán Péter

It can be clearly seen that the numbers 5, 500 and 1,000 are simply inserted into the ancient numeral sequence. The short stroke marking 1 is not an insertion. It was easier to draw a short line than a dot with his tools.

Let's investigate the implemented signs.

## THE CONNECTION TO CRETE

The only other place where the sign ◊ can be found is Crete and there it has a value of 1,000 as well. (It was also used at a very special place 500 years ago as you will see later.) This makes the connection certain, since the basic signs were always quite variable. It can't just happen randomly that the form and meaning also correspond. Additionally, the short stroke means in Crete also the number 1 and the dot was used as 10.

It is important to know that the sign ◊ = 1,000 was only used in Crete until 3,900 years ago, when it changed to: ⟡. We never see the sign ◊ again later as numeral. However, we see it again 4,000 years later among the numerals of Sr. Dán Péter. Let's look at the relationship:

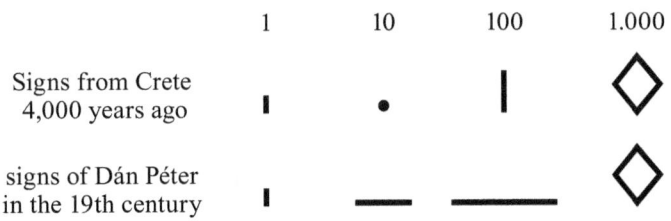

*However, the numerals from Crete changed 3,900 years ago to the following:*

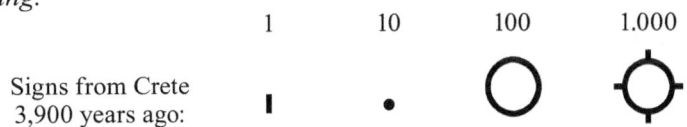

Picture 185. People from Crete liked draw the signs 1 and 100 a bit slanted, but this doesn't change the major point. See more on page 368.

There are several conclusions from the above. First: Even if people on the island of Crete stopped using the very characteristic sign ◊ = 3,900 years ago, the sign must have been used continuously somewhere as a numeral in order for it to become known to Sr. Dán Peter. If this sign had not existed somewhere unbroken, then the sign-series of Dán Péter couldn't contain it with the same value. There are two possible solutions to explain the connection with Crete:

## I. SOLUTION:

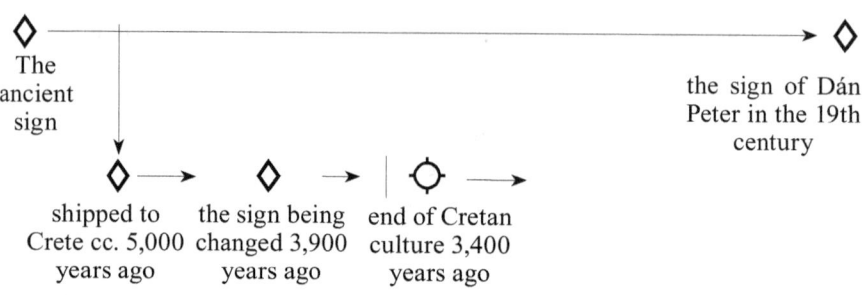

shipped to    the sign being    end of Cretan
Crete cc. 5,000   changed 3,900   culture 3,400
years ago       years ago      years ago

Picture 186. According to this, the sign moved to Crete, was changed and died out. The main branch however, died in the Carpathian Basin. only in the 20th century

The picture above shows that if the sign did not originate in Crete, then its story would have meandered to this faraway island and died out after a short time.

## II. SOLUTION:

The picture above shows that if the sign did not originate in Crete, then its story would have meandered to this faraway island and died out after a short time.

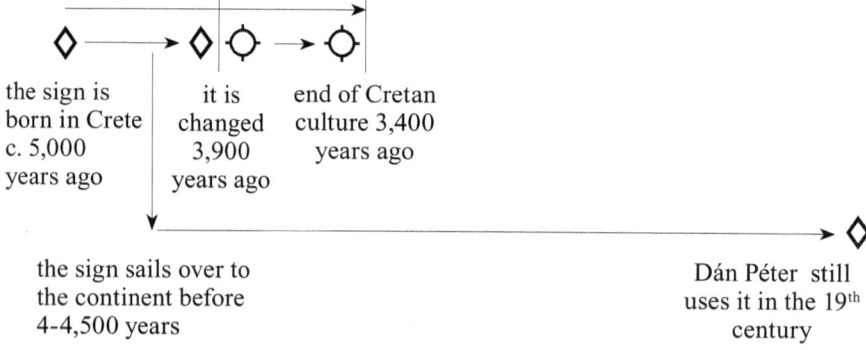

the sign is       it is     end of Cretan
born in Crete   changed   culture 3,400
c. 5,000       3,900      years ago
years ago    years ago

the sign sails over to               Dán Péter still
the continent before             uses it in the 19th
4-4,500 years                   century

Picture 187. If this sign was a Cretan invention, then it must have been taken somewhere else before the 4,000s in order to survive.

This picture demonstrates that if this sign was created in Crete, then it must have been taken somewhere else before the 4,000s, because after that a different sign was used for the value of 1,000 in Crete.

Consequently, even if this sign was an invention of the Cretans, it has been on the European continent for the last 4,000 years. The first solution sounds more probable to me, whereby the signs ◊ = 1,000 and the short stroke ▮ = 1 were invented somewhere else and spread to Crete. However it happened, Sr. Dán Péter was the last user of a 4-5,000-year-old numeral sign.

There is one more place where we can find the sign ◊ used as a numeral. The Aztecs marked the number 10 with it, but they did not use it to write the numbers 20, 30, 40, etc… This may mean that the sign ◊ is an old relic.

This is supported by the fact that their number writing is a mixture of several number systems. They even used "palm-twigs" as numerals, but every sign and every different variant has a clearly identifiable archetype. Why shouldn't this sign have an archetype too?

Historians put the brightest period of the Aztec culture into the 14-15th century, but this tells us nothing about the origin of their number-writing. They were quite aggressive and increased their territory conquering all the neighbours. In doing so, they incorporated all the spiritual goods of the conquered people. This may be the reason for their tangled number-writing.

## THE ETRUSCAN CONNECTION

The number-writing of Sr. Dán Peter is connected in many ways to that of the Etruscans. Firstly: the method of writing the numbers from 1-9.

Second connection: contrary to the ancient numeral-allocation (1-10-100-1000) Dán Peter used the same breakdown of his numbers as the Etruscans: 1-5-10-50-100-500-1,000. Further, the third similarity: the Etruscan 5 and 50 and Dán Péter's 5 and 500 are identical as we can see below:

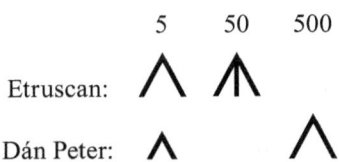

Picture 188 Dán Péter drew the small sign for 5 and a large for 500. Thus, one may not confuse them. The Etruscan signs have identical sizes, therefore the sign for 50 needed a perpendicular line in the middle.

I should point out that we can't discuss Latin numbers or their number-writing system here, because the Latin "culture" had no independent origin. Every element comes second or third hand.

Let's now look at the identity of the number-writing from 1-9.

The reading of the following number from Dán Péter's system is 109 (see on page 34 the lower part of the fourth number from above and the following examples):

$$\overline{\text{IIII}\Lambda}$$

The long line means 100, only 9 can stand below it and its form is a typical Etruscan numeral: IIIIΛ. It was read from right to left just as Dán Peter did. He should have written IX for 9, if he had been using the Latin numeral system. However, the difference between IX and IIIIΛ (both having the same value of 9) is a clinching argument.[36] See the part introducing the Etruscan numerals on page 352.

In summary:

The sign collection of Sr. Dán Péter is a mixture of three different numeral series. Its base is the ancient row of signs and in between are signs best known today as Etruscan and Cretan.

We even know the reason for this mixture. The shepherds had to make an effort to invent different enumerating systems in order to find the sheep belonging to their herds.

It is hard to believe that Dán Peter assembled his numerals on his own from ancient sign-collections. More probably, he learned them as a shepherd boy from one of the old herdsmen and even this herdsman probably learned them as a young boy from an old shepherd and so forth. If a herdsman had several shepherd-boys, then only one of them could inherit the old writing signs, while the others had to make changes to this enumerating system on becoming independent with their own herds, if they stayed close to the original one.

Very broad, basic changes did not happen even over thousands of years. As we see, there are merely two "guest" signs in Dán Péter's collection, in other words his base remained the ancient group of signs with its system of writing as well. Therefore, if a shepherd becoming independent was forced to find a different system from his neighbours', he just had to look around and borrow a sign in order to make his system

---

36  The Etruscan language is closely related to Hungarian. See *Etrusco: Una forma arcaica di ungherese* (Etruscan: is an ancient form of Hungarian) by Mario Alinei.

# 10. SR. DÁN PÉTER'S NUMBER-WRITING METHOD

1.) According to Sebestyén Gyula, Dán Péter wrote his numbers the Latin way. But he doesn't give any examples of this. Without examples the question arises of whether Sebestyén Gyula might have taken the Etruscan numbers for Latin? In other words, in all the given examples of larger numbers (109, 347) 1-9 is presented in the Etruscan way (see previous page). We can no longer find the answer to the question of whether Sebestyén Gyula erred or not.

2.) Dán Péter always wrote the numbers from 50-99 onto the right side of the sheep's cheek. The base of these numbers is 50, the short horizontal line:

50

3) the singles are always under the line, the tens up to 99 (60, 70, 80, 90) were positioned above the line of 50.

66                    82                    99

4.) The sign for 100, a long horizontal line, crosses both sides of the sheep's cheek; naturally the 200, 300, and 400 does the same. (500 is different):

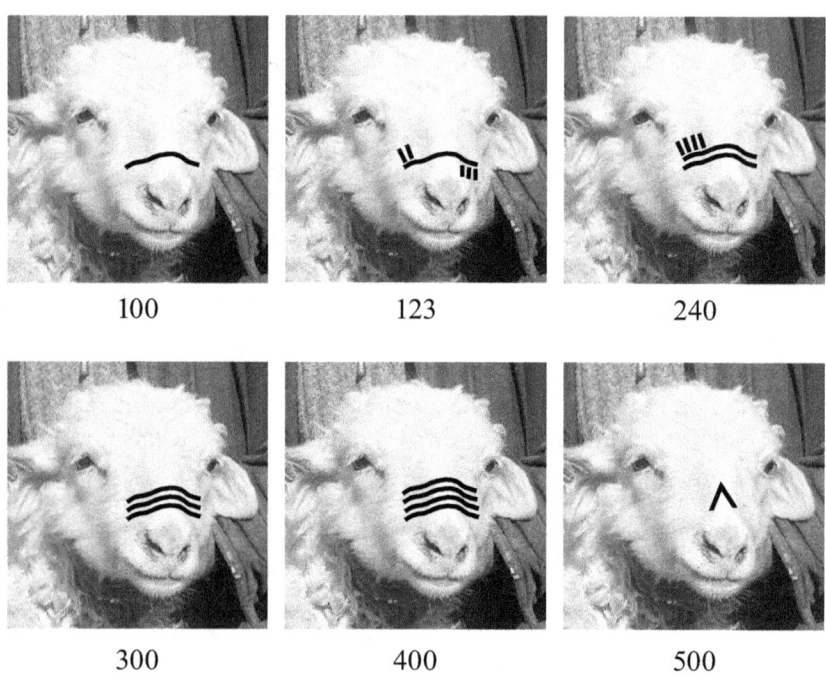

|     100     |     123     |     240     |
| 300 | 400 | 500 |

5.) After 149 comes 150, namely 100 + 50, but 50 has a different sign, which is drawn to the side of the 100 line, and is turned by 90% instead above the 100:

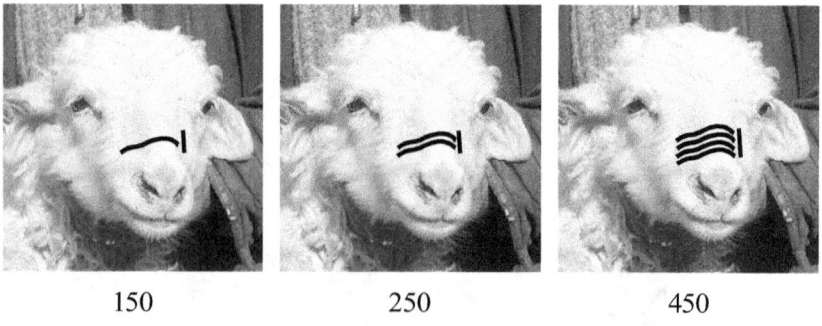

|     150     |     250     |     450     |

6.) The short lines of the tens, turned back by 90 degrees again, are placed after the 50 on the left side

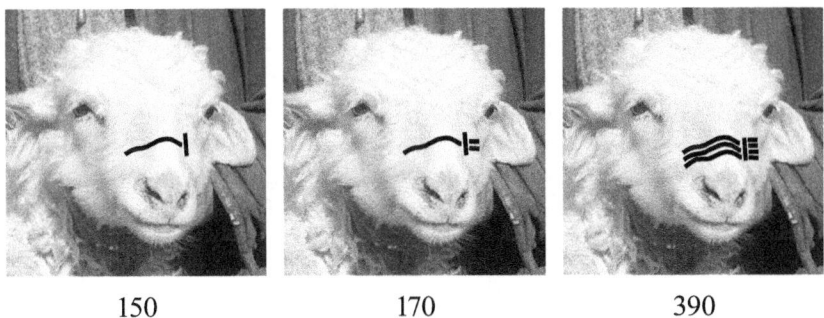

150                    170                    390

7.) The singles are placed on the right side if the 50 was drawn on the left. The singles will be written the Etruscan way as elsewhere but perpendicularly:

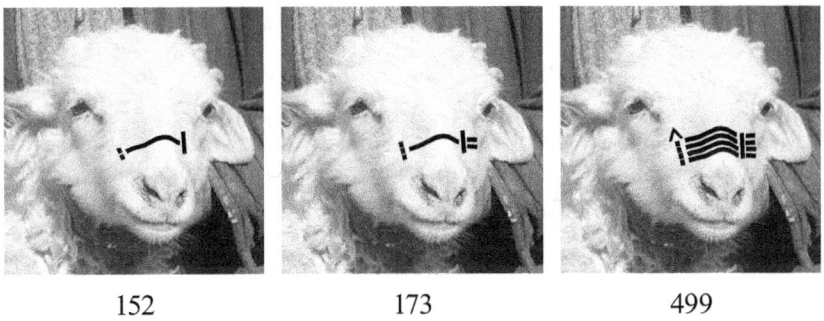

152                    173                    499

Let's stop here for a moment to gain a better understanding of the change. The tens are above the long line, see 240, but if the next ten is added, the shorter line of 50 is placed to the left from the hundredths (see 250). The next ten will be put to the left of 50 and the singles to the right of the hundredths (see 263):

40
||||  ———— +10 ————→   50   +10      3      50
====                  ====|        ====|— 10
200                   200           200
240                   250           263

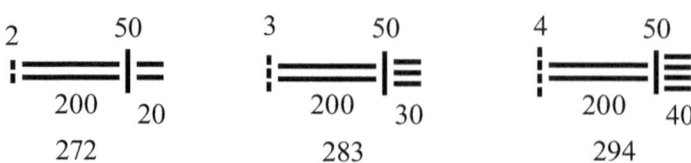

272                    283                    294

After stepping over the numbers 199, 299 and 399, we have to add one more line for the next hundred and with this the old order returns: the next four tens will be placed above and the singles below the lines of the hundreds. However, the 50 pushes the shorter lines of the tens again to the one and the singles to the other side:

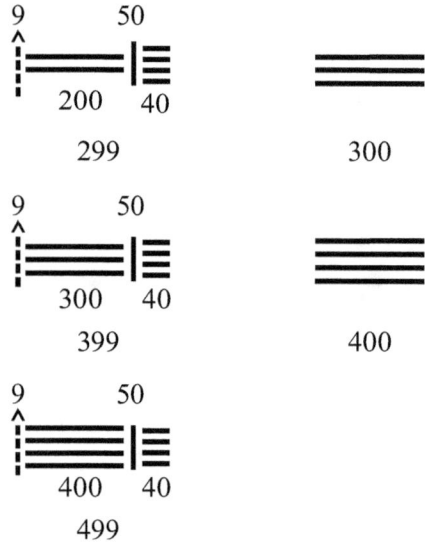

8) We go to 500 after this and everything starts from the beginning. See the numbers 582, 840 and so forth until 999. See the next page:

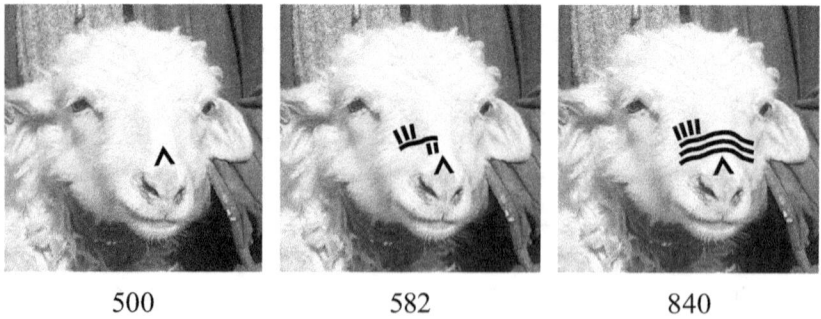

500                    582                    840

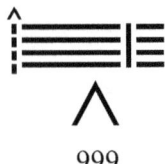

999

9.) The next higher base number is ◊ (1,000) and the numbers will be written above this sign from 1-1,499 in the same way as for 1-499

For example:

50

3 ┃ — 10

◊ 1.000

1.063

2

50 — 10

◊ 1.000

1.262

This way, we arrive at 1,499:

◊

1.499

The next number is 1,500, written with 500 and 1,000 together

For example

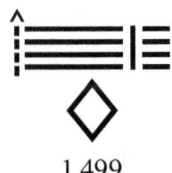

1.500

1.600

1.784

11.) Following the above system we arrive at 1,999

1.999

To write 2,000, if it is necessary, we just double the sign for 1,000.

2.000

This big a number, or an even greater number was seldom necessary for a shepherd, since there were perhaps never more than 1,600 sheep in a herd.

It is worth looking again at the Cretan number writing.

People wrote the number 1,403 this way earlier than 4,500 years ago in Knossos and on the right is the same number used by Sr. Dán Péter in the 19th century.

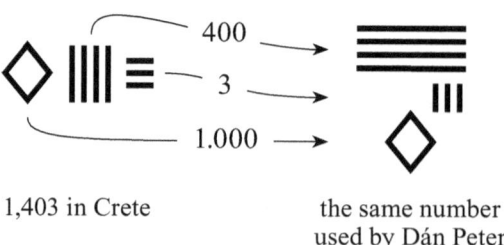

1,403 in Crete

the same number
used by Dán Peter

We can see that here they only rearranged the order of interrelated sign-groups contained in one number, and nothing else. Dán Péter tried arranging his numbers in a more perpendicular way, for he had to match the form of a sheep's cheek. The signs in Crete stood horizontally and they could become longer without a problem. Despite this, they often wrote the tens under the other numbers to make it more compact.

The previous example shows the spiritual connection between the number-writing in Crete and that of the Hungarian shepherd. Some particulars are totally identical, but not all. The values assigned to the signs and their placing in the numbers vary more often (see Cretan number-writing on page 370.

The numbers of Dán Péter are more complicated compared to the ancient or Cretan writing methods, because he also used 5, 50 and 500 as base-numbers, adding new signs for 5 and 500. The discussed sign of 1,000 is inserted here and there. The reason for this variety is understood. There had to be variant numbering systems in the neighbouring herds and this must have been a kind of spiritual competition between the shepherds. Dán Péter's numbering system - at least - used all the possibilities of the ancient dot-line number system.

It is unfortunate that no records have been found of the number-writing methods of the other shepherds. Maybe it still will be, since Sebestyén Gyula wrote, that the procedure of tarring has been well known to all old shepherds all over the country. However, tarring was the method of writing numbers onto the sheep's cheek.

The records of Sebestyén Gula preserved a Hungarian spiritual treasure, unparalleled in the world and we hope that this book will help to make it better known along with the numerals of Vizakna and Akna-Suhatag.

I am happy to have been able to fully unfold this spiritual treasure (containing 11 numerals) and thankful to Sebestyén Gyula for the introduction of Sr. Dán Péter's number writing method (around 1909). The total unfolding of this method was only possible because Sr. Dán Péter knew very well which numbers he had to present for his system to be understood.

# TABLE Λ.

Sr. Dán Péter's base numbers over 50 and some examples for them:

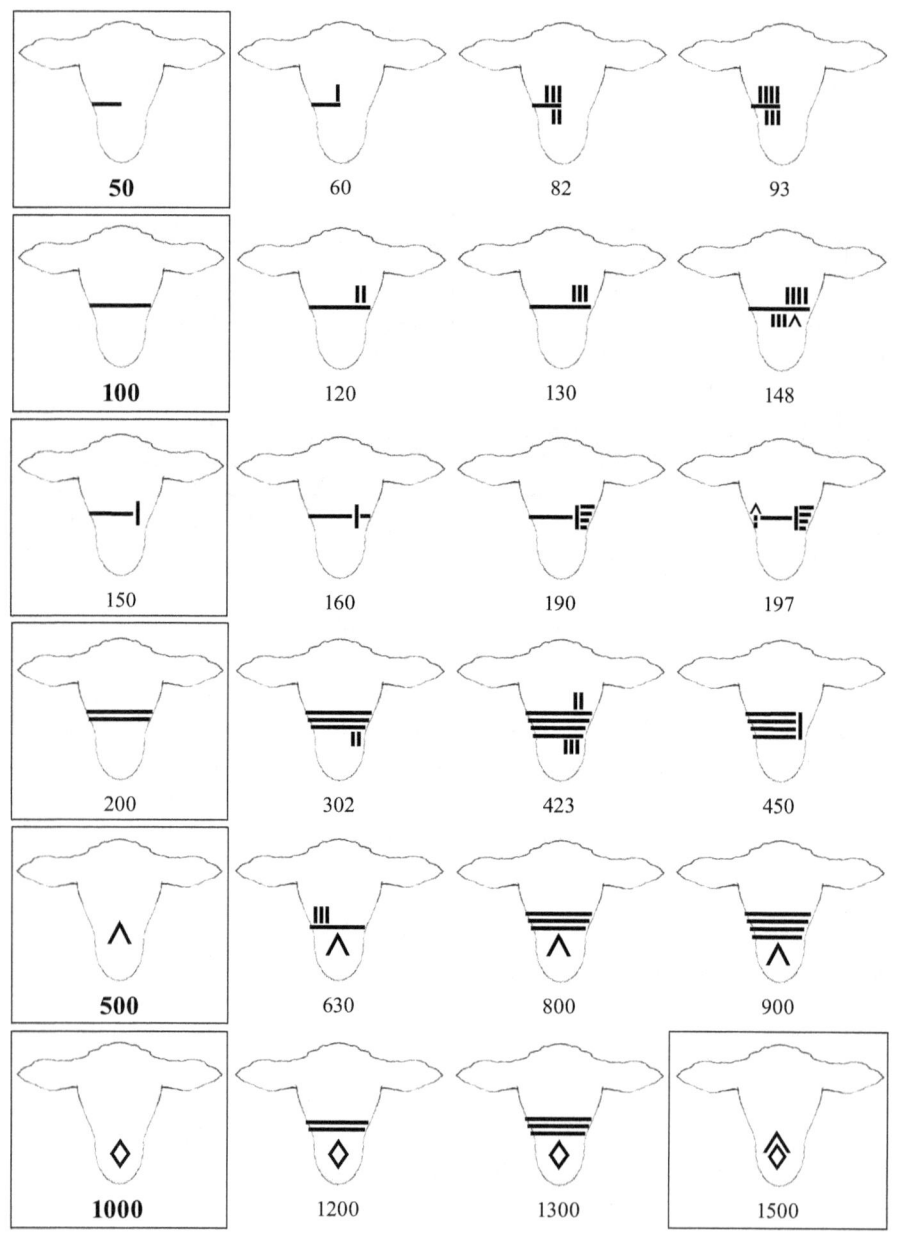

# IΛ. TABLE 4/1. (Etruscan 6 = IΛ)

Teaching table for the number-writing of Sr. Dán Péter

Sebestyén Gyula didn't present examples for the numbers 1-49. He noted thatDán Péter used the Latin method for writing those numbers.

We see below the numbers 50 – 99 which were drawn only onto the sheep's right cheek. In his system, the sign for the number 50 was a horizontal line from 50-99 and only became a perpendicular line on the left side of the hundreds after 150, as we will see later.

| ― | I | II | III | IIII | Λ | IΛ | IIΛ | IIIΛ | IIIIΛ |
|---|---|----|-----|------|---|----|-----|------|-------|
| 50 | 51 | 52 | 53 | 54 | 55 | 56 | 57 | 58 | 59 |

| I | I / I | I / II | I / III | I / IIII | I / Λ | I / IΛ | I / IIΛ | I / IIIΛ | I / IIIIΛ |
|---|-------|--------|---------|----------|-------|--------|---------|----------|-----------|
| 60 | 61 | 62 | 63 | 64 | 65 | 66 | 67 | 68 | 69 |

| II | II / I | II / II | II / III | II / IIII | II / Λ | II / IΛ | II / IIΛ | II / IIIΛ | II / IIIIΛ |
|----|--------|---------|----------|-----------|--------|---------|----------|-----------|------------|
| 70 | 71 | 72 | 73 | 74 | 75 | 76 | 77 | 78 | 79 |

| III | III / I | III / II | III / III | III / IIII | III / Λ | III / IΛ | III / IIΛ | III / IIIΛ | III / IIIIΛ |
|-----|---------|----------|-----------|------------|---------|----------|-----------|------------|-------------|
| 80 | 81 | 82 | 83 | 84 | 85 | 86 | 87 | 88 | 89 |

| IIII | IIII / I | IIII / II | IIII / III | IIII / IIII | IIII / Λ | IIII / IΛ | IIII / IIΛ | IIII / IIIΛ | IIII / IIIIΛ |
|------|----------|-----------|------------|-------------|----------|-----------|------------|-------------|--------------|
| 90 | 91 | 92 | 93 | 94 | 95 | 96 | 97 | 98 | 99 |

IΛ. TABLE 4/2

In the next numbers, 100 is written with a long horizontal line:

| | | | | |
|---|---|---|---|---|
| ⎯⎯ | ⎯⎯ I | ⎯⎯ II | ⎯⎯ III | ⎯⎯ IIII |
| 100 | 101 | 102 | 103 | 104 |
| ⎯⎯ Λ | ⎯⎯ IΛ | ⎯⎯ IIΛ | ⎯⎯ IIIΛ | ⎯⎯ IIIIΛ |
| 105 | 106 | 107 | 108 | 109 |
| 110 | 111 | 112 | 113 | 114 |
| 115 | 116 | 117 | 118 | 119 |
| 120 | 130 | 140 | | |

The singles in the lower line are the same as above

Now the 50 appears perpendicularly on the left end of 100 while the singles move to the right side and are also written perpendicularly.

| | | | | |
|---|---|---|---|---|
| 150 | 151 | 152 | 153 | 154 |
| 155 | 156 | 157 | 158 | 159 |

After 159, a ten is drawn horizontally to the left of the 50: 50 + 10 = 60 and so so forth 70, 80, and 90.

## IΛ. TABLE 4/3

| 160 | 161 | 162 | stb. |

| 170 | 171 | 172 | stb. |

| 180 | 181 | 182 | stb. |

| 190 | 191 | 192 | stb. |

The singles continue as before between 150-159

The numbers at 200, 300 and 400 are written as before between 101 and 199 but now there are two, three or four hundreds (long horizontal lines) in the numbers.

200      for example      280      282

300      for example      320      349

400      for example      440      450

in this way we continue to 499:

499

IΛ. Table 4/4

From here, 500 is represented by the large sign Λ . From 501 to 999, the numbers are written as for 101-499, but with the 500 standing below it:

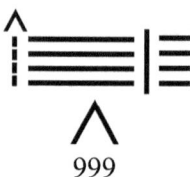

999

999 Once again we repeat for the succeeding numbers: everything re-starts as for 101-149, but at 1,000 and 1,500, the previously introduced sign of 1,000 will be put under the numbers.

1.000          1.500

Let's look at one more example, the date of the Hungarian revolution 1848 against the Habsburg oppression.

1848

Picture 189. Shepherds on the Lowland.
Sr. Dán Péter was probably wearing the same dress.

# 11. THE MOUSTACHE OF TEPE TAHYA

There is one number among Daniel Potts' signs registered in Tepe Tahya and the arranged drawings which look like a moustache (see page 102). A middle line intercepted by two pairs of dots. The line is drawn as a moustache. Let us look at it again.

22 c

Turned to
horizontal

It is a real
numeral

Picture 190. The line is the moustache

Dán Péter called the line drawn across the sheep's cheek a "**moustache**". The picture below demonstrates why this line was called a moustache: it looks like one. This might well explain why somebody painted the line as a moustache in Tepe Tahya 5-6.000 years ago:

Picture 191. It is understandable seeing this picture why the name moustache came about. The serial numbers of the sheep were painted on the sheep's cheek on the Hungarian pusztas and probably in Tepe Tahya as well.

Only the long line of 100 and the one-sided 50 of Dán Péter's signs can be called a "moustache". Therefore, this sign must have marked a local value of 100, probably also in Tepe Tahya, because it must have been painted across the cheek. A value of ten would have made the numbering in such a system too puzzling for a larger herd  because of the very complicated forms. The value 1,000 was too big, because herds with over 1,600 animals are not manageable.

Thus, the "moustache"-number of Tepe Tahya would actually have looked as shown here:

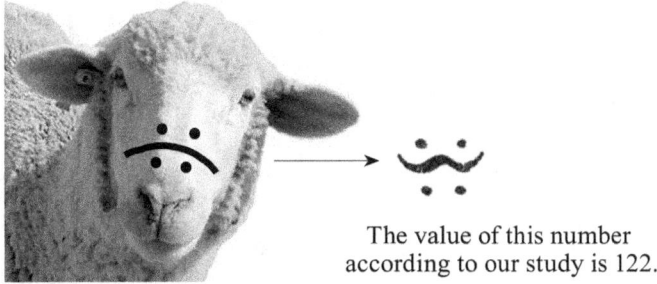

The value of this number according to our study is 122.

Picture 192. A sheep with its serial number in Tepe Tahya.

This gives us important possible culture-historical data. We have made something visible, although no archaeological finds could give us proof of it. The next two pictures demonstrate what we might have seen in Tepe Tahya walking through a flock of sheep.

Picture 193. Two sheeps from Tepe Tahya (numbers 121 and 241)

Picture 194. How a flock of sheep might have looked 6,000 years ago around Tepe Tahya. The numbers look like  number-plates.

Unfortunately, there are not enough numbers remaining from Tepe Tahya to reconstruct every form in the whole number series. Dán Péter presented fewer numbers; however those contained and explained the system very well. In any case, there were enough to explain the meaning of the "moustache".

The custom was to be expected since there were sheep and shepherds. Sheep with thick fleece turned up supposedly 8,000 years ago in Iran, but they were certainly bred for their meat before that. The use of numbers was also certainly not invented for this purpose and happened long before it was implemented for the registration of sheep. Thus, we have gone back again at least 10,000 years in human cultural history.

# 12. NUMBERS ON EARS

Shepherds, therefore, were the grand masters of number-writing and developed their own methods while attempting to outdo each other. The ancient number-writing principles make it possible to create very different looking number-signs by just small changes in the signs or their allocated value. Using lines instead of dots is one way to make a different picture.

Sebestyén Gyula shows us examples of a very special numbering system from Kiliti (Somogy County). Determining and registering the years is very important for the cross-breeding of sheep. This kind of marking has to be permanent, so the ears are cut. There were two signs: a "v" shaped notch on the edge of the ears – called a "goat-nail", fairy stone – and a hole in the middle of the ear. The numbering-method was as follows:

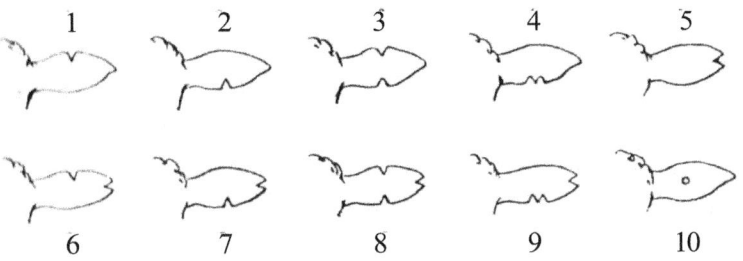

Picture 195. The dating of the year on the sheep's ear in Kiliti.

We receive the following numbers, taking the notch as a line, the hole as a dot and marking the middle of the ear with a thin line:

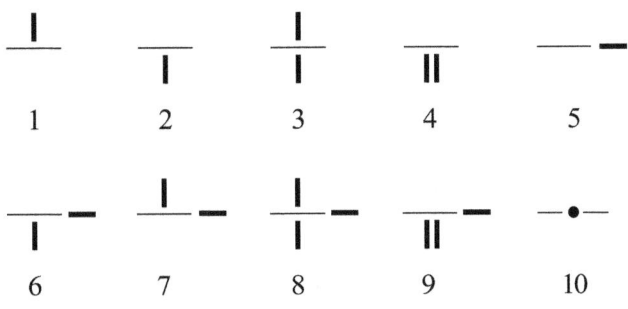

Picture 196.

As Sebestxén Gyula says, based on Tömörkény István's report, the shepherds generally tried to create their own number-writing system: "due to this custom, most of the shepherds around Szeged even didn't know the numerals of their colleagues".

We see below a number-system used in the surroundings of Szeged. We can see the same signs as before ("goat-nails" and the hole in the middle of the ears) but applied quite differently. The numbers 1-99 are on the left ear and the same signs on the right ear mean the numbers 100-199. Again on the left ear, but with one added sign, come the numbers 200-299 and on the right 300-399. The number-system can be easily increased by this method. Here it makes a difference whether the "goat-nail" is to the right, in the middle or to the left.

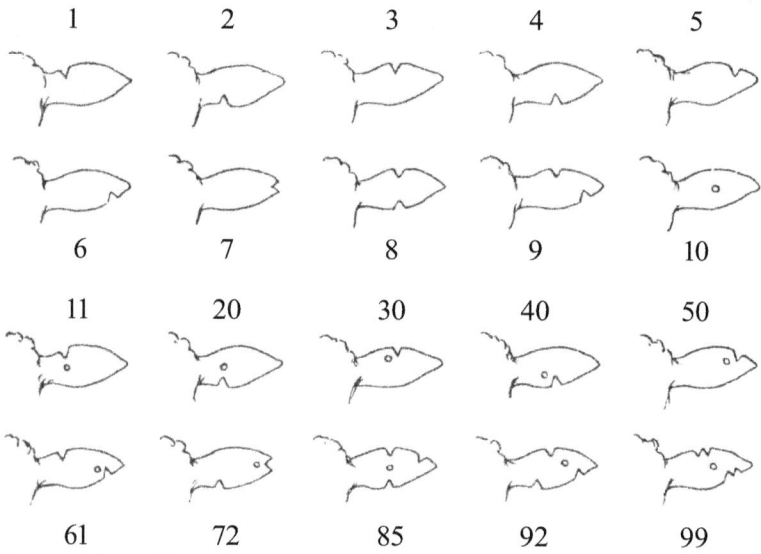

*From 100 – 199 is the same as above, but on the other ear:*

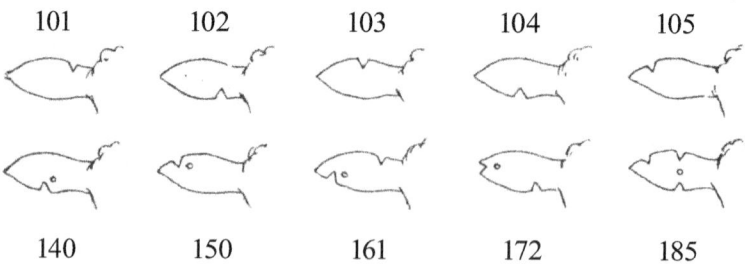

Picture 197. The numbering-system on sheep' ears used by a shepherd in the surroundings of Szeged.

# 13. THE HUNGARIAN SHEPHERD AND LITERACY

First of all, we need to clear up an over-700-year-old misunderstanding: Kézai Simon wrote in his Chronicle around 1282 that *"the Szeklers mingled with the Vlachs of Transylvania, as they say, and used their writing-signs"*. (I: IV.6. §.)

Let's compare this statement with that of Makkai János *"There is no toponymic [37] proof of any Romanian presence before 1242 in Transylvania, except in Fogarasföld and Hátszeg. The first place-name (toponym) appeared in the southern part of Hunyad County[38] around 1360 as the still-existing place name of Nucsoara.)*[39]

Therefore, the Romanians appeared first after Kézai's death in Transylvania and his statement about Vlachs could not meant Romanians.

The misunderstanding happened much later. The words "vlach", "blach, "whale", "wallach" and oláh are the dialectical variants of the same word.

But what does it mean?

The Slovak vocabulary, as the storehouse of ancient Hungarian words, tells us that "valach" means shepherd. Thus, Romanians were given this name in the 13th-14th century, because they were moving around mostly as shepherds scattered over the Carpathian Mountains and into Transylvania.

We understand from the report of Kézay, made from hearsay, that the Szeklers used the writing of the shepherds. This kind of writing is better known as the writing of shepherds.

Were those shepherds speaking Hungarian or "Romanian" language?

This question can be answered by asking and answering two other questions.

First: well, show me an old piece of writing done with Szekler Rovás signs but in the Romanian language!

---

37 Toponymia: investigation of place-names.

38 However, there are the beautiful finds of Szekler Rovás in Transylvania which are 100 years older : Writing on the tower in Alsószentmihály; the writing of Vargyas and writing on a ring from Kenyérér.

39 Interview with Makkai Laszló by Potó János: The settlement of Transylvania
(internet)

Second: Please show me an old piece of writing done with Szekler Rovás, but originating from outside the Carpathian Basin.[40] We can follow the origins of the Rovás quite easily. Let's quote the relevant words of Sebestyén Gyula again:

*"Comparing the Rovás-writings of the 'nationalities' in The Carpathian Basin with those of the Hungarians we can see that they borrowed not only the name [41] Rovás but the Rovás signs as well from us. The number-series of the nationalities point to the Hungarian origin of their number writing: The series of the Ruthenian numbers contains the signs of the old-Hungarian 100 and the new oblique sign of 5. One can even prove that the old sign of the 50 was used by them as well. The number-sequence of the Wlachs (Called Romanian today) looks identical with that of the Ruthenians, only their value is different: here the 5 means 50, the 50 means 500 and the 100 means 1000. This value-difference remains characteristic of the entire Vlach number writing. However it still can't be seen as an original character, because the tenfolding of the signs wasn't caused by the technique of the Vlach 'rovás'. It was caused by the two lined Székely Rovás, in which the straight line of the tenth 'one' has been engraved onto the next page and also caused the sign X (ten) to be tenfolded. We are supposed to believe that the Vlachs originally drew the Hungarian number-signs as did their Ruthenian neighbours, because the sign for 1,000 in the Vlach numeric row is identical to the 100 sign of the old Hungarian Number-series". ...".*

Occasionally, Slavs will be named as the rovás teacher of the Hungarians.

In the Slav literature, according to Sebestyén Gyula, *"Delič recorded the sequence of the simple number-row from the Rovás of a trucker moving bricks. The following signs were on this sample:*

**XXXXXXXXXXXXXXXXIIIIIII'•••**

*and meant 1678 bricks as told by the trucker"* (Delič therefore was told by the trucker about the number, he couldn't read it by himself). Here the dot (•) meant: 1, the short perpendicular line ('): 5, the long line (I): 10 and

40 There is one location, but 'Vlachs' never came around there. Varga Csaba:

41 The samples of Sebestyén Gyula: The Rovás is called by the Serbs: 'rābos', in Bulgarian: 'rábos', along the beach in Croatian: 'rabos', in Montenegro: 'rabus', in little-Russian: 'ravas', in Kaj-Horváth: 'rovás' as well as in Old-Czech and in the Tót of Hungary.

the X: 100. Therefore, the signs at the end: ¹••• = 8 (5+3).

The value of the number marked by Delič is 1678. But it is a deterio-rated number-writing. The trucker no longer knew about any position value larger than 100. Thus he had to write 16 X (16 times the sign for 100) in order to reach 1,600. If he had had an order for 3,249 bricks (a real number needed for a house), then he could only have written this number in the fol-lowing way:

XXXXXXXXXXXXXXXXXXXXXXXXXXXXXXXXXIIII¹••••

The point here is that the trucker would have to repeat his 100 sign until infinity. It did not enter his mind to introduce a one higher position-value, or if he did, he couldn't manage to do it. He just got accustomed to this very tiresome way of writing and reading of numbers.

On the other hand, we have seen how splendidly Hungarian shepherds managed number-writing. They often did not even know the system of their colleagues, as in the case of ear-marking, but invented different methods of number-writing using merely two signs: the 'v'-form cut into the ears ("goat-nails") and the round hole in the middle of the ears (puncture). They possessed a much deeper knowledge, that of the ancient number-writing, using position-value. This knowledge was certainly not known by the Slav-ic trucker described by Delič. Apart from this example, we can't speak of much Slavic influence in the Carpathian Basin. However, we will note one more thing.

There are two kinds of signs in the number of the trucker. The X (in his case with a value of 100) is identical with the Roman/Etruscan sign for 10. However, the end of the number is 78

IIIIIII¹•••    and readable, if divided    I III III ¹•••
                 in sections, as earlier:

That means, 78 is written with the ancient signs; it could have been easily read in ancient times. We have seen these signs (¹••• = 8, 5+3) in the book in Tepe Tahya, in Mexico and in the salt-mines of Transylvania.... As on all the ancient finds: :

If we also read the number 10 of the trucker into the picture, then we can be sure, and the rearrangement of the signs shows us, that the trucker's number (18 = ¹•••) looks finally as if it had been written using the ancient method:

Picture 198. We have seen plenty of samples earlier for such rearrangements.

It is clear that the trucker's number-writing up to ten differs from the archaic method only in the arrangement of the signs. We can only receive a different reading if we give different values to the signs - like the trucker using the Roman X with a value of 100. Values up to ten usually don't cause any problem.

We can definitely state after the above that the number-writing of the Slav trucker is the inheritance from Pont d'Arc, Lascaux, Tepe Tahya, Salt-mines in Transylvania and of Sr. Dán Péter. However, it has deteriorated in his case. It wasn't his fault, but that of his cultural surroundings having lost the connection with the ancient culture, while in Hungarian the connection is still kept right up to today, due to its language based on organic thinking.

The deterioration of number-writing is detectable around the Carpathian Basin radiating in all directions. We will see examples of this later on.

Sebestyén Gyula wrote about the presence of Rovás on the peripheries of the Carpathian Basin: " ..the Ruthenian lumbermen use the Székler's rovás-table (45. sz 8.), the Tóts of Turócz county write the sign for 50 in the old Hungarian number-row and the Germans living in Tolna county use the sign of 100 in the same series of signs. The tenfolding of the number-signs turned up in Bosnian, telling us that the Rovás of the Szeklers has spread as well as that of the Hungarians". "We can even verify that the original number-system of the Vlach Rovás (originally Hungarian) has been disturbed [...] by the Csángó type of Rovás."…. "A detailed study of the Rovás used by the different nationalities of Hungary would certainly point to various cultural borrowings."

What do we see?

There is a central region with a common culture, the Carpathian Basin, and its culture has a definite influence on people at the periphery who have arrived from outside to the region. We see further that the number-writing in the Carpathian Basin (Akna Suhatag, Vízakna, Sr. Dán Péter) is a clear, unbroken continuation of the ancient culture. That means it is not due to an 'influence', it is in itself original, ancient. Thus, the language, the written finds and other characteristics tell us that, at least as far as number-writing is concerned, no major happenings disturbed the developing culture of the Carpathian Basin during the past thousands of years, right up to the beginning of the 20th century. There were plenty of influences however; the bringers of those influences were never so numerous, that they endangered the ancient culture. It remained alive. To put it plainly: what is now called Hungarian culture is the ancient culture of the Carpathian Basin and it belongs, according to the history of writing, to the Pont d'Arc-Lascaux range. I suppose it is the immediate precursor of the Magdalenian [42] culture, which means the latter isn't finished yet.

The Carpathian Basin had been suffering strong Roman attacks against its original culture, but fortunately only at its western end (Pannonia). Due to this partial occupation, the original writing and number-writing, along with the language, were able to survive, which speaks for their strength. The Romans fought the Pannon people for 45 years, before they were able to reach the limes of the River Danube and the 5 Roman divisions were unable to cross it east/northwards for 400 years. As the Romans left, Pannonians continued using their different Hun(gar) dialects as they still do today. The people living east of the Danube called their home Pannonia "Dunántúl" (Transdanubia) and we still call it that.

A counter-example: the ancient culture of the Southern -Slavic people has fallen apart. They suffered many occupations and cultural influences from south-east and south-west.

This old southern Slav number-row best demonstrates the many influences they suffered:

Picture 199. A row of old-Slav numerals. These numerals could only come into being because somebody (probably a pope of their church) intended to invent a new system of number-writing. But he wasn't a very knowledgeable man. He collected and invented several signs but took no notice of rationality and lucidity.

42  It is counted from 19,000- 11,000 years before us.

This row of numerals offers the greatest possible mess. The numbers 20 and 50 are written with the Roman numerals X (10) and L (50) but for the 10 the double VV, the Roman 5 (2x5=10), is used. The single numbers are the usual lines, but written in a strange way. The 40 is a double cross. The 50 is a circle and the 100 has a dot in it. (It could have come from Old-Greek, but at that time it had a value of 10).[43]

Sebestyén Gyula noted as a matter of curiosity that the numeral 100 with the little circle in the middle meant 100 garas <garash> (small change, like the 'cent' of today) and a double circle with a dot in the middle meant 100 gold.

The row of numbers above is certainly artificial and its inventor did not care about its usefulness or easy recognition. Its invention could have had the aim to cover up, or to hide the old system. (Method and Cyril pursued a similar goal by introducing their artificial Macedonian church language and writing, creating with this a 'Slav' cultural community of quite different population groups.)

Well, this is how a distorted number set looks. By contrast, we can see how significant the steadiness of the number-writing in the Carpathian Basin was until the 20th century, preserving the ancient method, but not just here, as we will see later.

## Note

In the following chapters, the word "rovás" <rovaash> will often be used. What does it mean?

The word "rov" is a dialectical variant of the Ancient Greek "graph", which meant originally "scratch" or better known as "engrave". We might also say "rovology" instead of "graphology" for the science of writing.

"Rovás writing" is the general, even internationally used term for the ancient, but still alive writing system using the ancient Scythian-Hungarian-Szekler set of signs. Those are similar, but nowhere near identical to, and much older than, the German runic signs. Many of these signs are part of the Ancient Egyptian demotic alphabet.[44]

In this book, we deal with "number-rovás" and mean engraving numbers into wood, tallies, or just using these signs for registration of numeric values. The signs of "number-rovás" are designed to be easily engravable even into bisected tallies, mainly into wood.

---

43  See more: I. Schullenburg: *Botenstöcke bei Südslaven* (Zeitschrift für Ethnologie,)

44  See more about this in the book: Signs Letters-Alphabets by Caba Varga (2001)

# 14. THE HUNGARIAN-ETRUSCAN NUMBER-WRITING

The Hungarian and Etruscan Number-writing systems are identical apart from the late Etruscan changes. Let's look at the comparison.

Georges Ifrah writes in his book "Numbers" on page 197: *"A.P. Ninni (1899) noted that the farmers and shepherds in Toscana were still using the numerals below in the 19th century":*

Etruscan:  | Λ  X  Λ  Ж  Ж  ⋇

1　5　10　50　100　500　1.000

Picture 200. The old Etruscan numerals, still used in Toscana by the farmers and shepherds at the end of the 19th century.

Investigating the old Hungarian number-writing, Sebestyén Gyula came to the following result: *Investigating the old Hungarian number-writing, Sebestyén Gyula came to the following result: "We were looking for the original Hungarian number-writing in the middle of the Alföld (low-land) and believed we had found it. I let 'row'(carve) the numbers by several old shepherds in Hódmezővásárhelyen (Fullai István (79), Czuczi István (83) and Marozsán Mihály(85), who all knew the numerals) 'row' the numbers in picture 22."* We can see below picture 22 from the book by Sebestyén Gyula: We can see below picture 22 from the book by Sebestyén Gyula:

Picture 201. (picture 22 by Sebestyén Gyula) *"The old number-row of the Hungarian Rovás*

*/"/.../The number-sequence above - which can be found sporadically all around where Hungarians have lived – must be connected with the general history of Rovás based on the ancient history of the development of engraved number-signs. Its ancient age is, surprisingly, supported by the fact that the Etruscan 50 is among its signs. Because of the great distance of time, we can't go into solving the problems of the historical connection.."*

The three shepherds from Hodmezővásárhely don't speak in this text about the sign for 1,000. However Sebestyén Gyula found it in the neighbouring county Békés, around the settlements of Gyula, Csorvás and Kondoros. See the drawing of a tally stick from this area:

Picture 202. Sebesthén Gyula's picture 23. "The new number-row of the Hungarian Rovás"

The number 1,000 appears here in its full Etruscan form. There are interesting changes: the 50 appears as a somewhat wider and the 100 as a wide cut. The 5 looks like /, its one line (∧) is missing. We take here the older form, the ∧. Let's compare now the Etruscan numerals and those being used on the Hungarian Alföld (Low-land) during the 19th century.

| Etruscan | I | ∧ | X | ∧ | X | ⋈ | ✳ |
|---|---|---|---|---|---|---|---|
| | 1 | 5 | 10 | 50 | 100 | 500 | 1.000 |
| Hungarian | I | ∧ | X | ∧ | X | | ✳ |
| | 1 | 5 | 10 | 50 | 100 | | 1.000 |

Picture 203. Etruscan and 19th century Hungarian numerals

The sign for 500 is missing from the Hungarian row of signs. However, we shouldn't forget that this is only a sign-list from the 19th century and we can be sure that there was a sign earlier for 500 as well. This is a "bisecting" sign-row, which means that it contains not only the numerals 1, 10, 100, and 1,000, the signs of the full local values, but even the values in between: 1, 5, 10, 50, 100, 500, 1.000. Sr. Dán Péter used such a "bisecting" number-row. Furthermore, since the Hungarian sign-row contains signs for 5 and 50, it is unthinkable that the 500 was missing. It is possible that there was no need to use it in some locations.

A further proof: the Vlach number-row, which originated from the Hungarian, does contain a sign for 500. In this however, the value of the signs has been tenfolded and therefore the singles are marked with a dot (szúrás), a hole. It is inconsistent also, because they are using the same sign for 5 and

50.

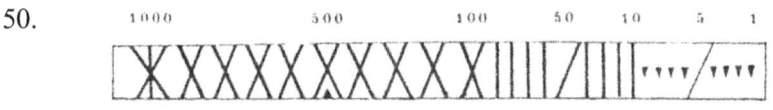

Picture 204. "The Vlach numeral-row in Transylvania." Picture 47 in Sebestyén Gyula's book

However, the Vlachs became a larger population, originally around 1360 by moving into Transylvania, and learned the Hungarian number-writing as used at that time. The Vlach number-row still had a sign for 500 in the 19th century, differing by one little notch from the 100 sign. The Hungarian numeral-row of the 14th-15th century certainly had a 500-sign, but we don't know what it looked like. The Vlach and the Hungarian number-row also deteriorated with time, but we can assume that the Hungarian 500 sign was built similarly from an X.

The Ruthenian numerals, also originating from the Hungarian, support this thesis as well:

Picture 205. Numerals of the Ruthenian Rovás. Picture 46 in Sebestyén's book.

The Ruthenians changed the signs for 500 and 1,000 somewhat by putting them into frames, but in this case the point is that they had a sign for 500.

Unlike the three shepherds, Sr. Dán Péter also learned number-writing at around Hodmezővásárhely and acquired an older sample of numbers containing the large Etruscan sign for 5 (Λ) as 500. Its form is related to the Etruscan 5, 50 and 500. Thus, Dán Péter's number-row preserved the oldest known numerals in the Carpathian basin.

The writing-direction varied in the 19th century. The writing with Székely Rovás is done from the right to the left, thus number-writing did happen certainly in the same direction as in the Etruscan. The finds of the 19th century show a loosening of customs. Numbers were occasionally written from left to right.

Here are some examples for number-Rovás in Sebestyén Gyula's book, picture 30. We see in the following picture numbers graved by a shepherd , from Szentes at the top and at the lower row. The number in the middle is from a shepherd at Hodmezővásárhely.

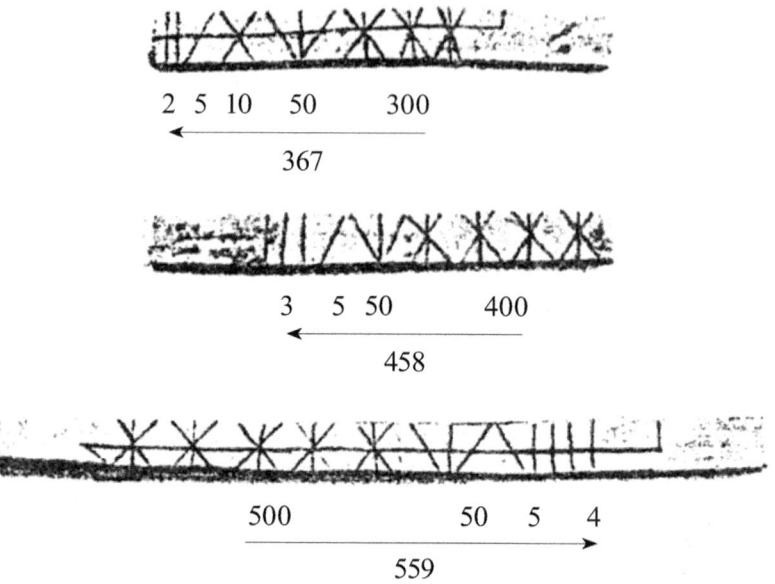

2  5  10  50     300

367

3   5 50        400

458

500          50  5  4

559

Picture 206. The two numbers above are written from right to left, as usual,
the third number as seen by the arrow, is written the reverse (new) way.

The writer of the last number (559) didn't know the sign for 500 anymore. The point of using the "bisecting" numbers was that one had to write
a number only 4 times before the position-value "changed". Here he wrote
5 x 100 instead of the forgotten 500 sign.

Changes were happening not only in Transylvanian-Ruthenian-Vlach
number writing. More modifications happened in Dalmatian:

I    Λ    X    И    .
1    5    10   50   100

Picture 207. Dalmatian numerals.

The sign for 50 is the modification of Λ (5), or that the sign '/' (which
is again a modification of Λ) received supporters on both sides, becoming
like a mirrored N. [45] The dot came back into the picture as the sign for 100.

---

45 The supporting line on both sides is seemingly the faith of the sign '/'. The sign of
the sound 'r' appears in these two forms ('/' and 'И') in the Székely and the demotic
alphabet as well.

See three Dalmatian tallies (to be read from the right to the left):

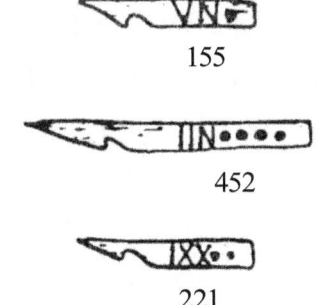

155

452

221

Picture 208. Dalmatian tallies. (Georges Ifrah: Numbers, page195.)

# 15. THE ORIGIN OF THE NUMERALS CARVED INTO STICKS.

The Romans couldn't have brought us the numerals and the way of number-writing discussed above. True, at the beginning of the Roman times, Etruscan and Roman number-writing were identical, since the Romans learned it from the Etruscans. However, by the time the Romans finally occupied Pannonia, they had radically changed their numerals and number-writing method (see the chapter about Roman numerals). The destructive Roman influence acted mainly in Pannonia and for short periods in Transylvania (shortly before the empire's collapse) after going around the - for them dangerous - Low-land (Alföld) and defeating the Dahas (a Hungarian-speaking group of Scythians).

Our most significant finds of ancient number-writing originate from the area avoided by the Romans.

The Pannonians continued using their Somogyi, Vasi and Kórógyi (Counties in Pannonia) Hun dialects after the collapse of the Roman Empire and the country started to be called Hungary and its inhabitants Hungarians around the 5th century. A partial takeover of Latin numerals happened in official and international scripts as well as by using Latin in the Church. The population, especially in the villages or the shepherds, never got accustomed to the Latin numerals.

"Arabic" numerals turned up here in the beginning of the 15th century [46]and replaced our old numerals largely first at the end of the 19th century.

However, the world (culture) of the Etruscans had already declined 2,000 years before (absorbed by Rome). It is possible that several of them migrated to the Carpathian Basin, but certainly not in high enough numbers to be able to teach Hungarians their number-writing system. There are no historical notes about it.

On the other hand, we know that earlier they left the Carpathian basin:
*".... the ancestors of the Etruscans, the so called "Villanovan" people[47],*

46   Numbers were written onto the church-wall in Kalotaszeg / Magyarvalkó, Transylvania, discovered by the mathematician-historian Tóth Sándor. (Filep László.-Bereznai Gyula: History of number-writing, 1999, Filium)

47  The Villanovian settlements appear in the 10-11th century BC along the Tyrrhenian coast and later on the inner territory of the peninsula between the Rivers Tiberis and Arno in the counties Toscana and Emilia-Romagna of today

*(the members known as the carriers of the predecessor culture of the Etruscans) originate from the Carpathian Basin. This fact has been discovered by the US archaeologist Hugh Hencken and later research strengthened his statement.*[48]

Therefore, it is most likely that over the last 3,000 years the Etruscans weren't able to bring the number-writing here. On the contrary: they took it from the Carpathian Basin to the Italian peninsula.

Nevertheless, there is something which suddenly spoils this peaceful picture. The Csuvas <choovash> number-row looks identical to the Hungarian number-row and to the Etruscan as well. Moreover, they also wrote the numbers from right to left, until they changed to Arabic numbers.

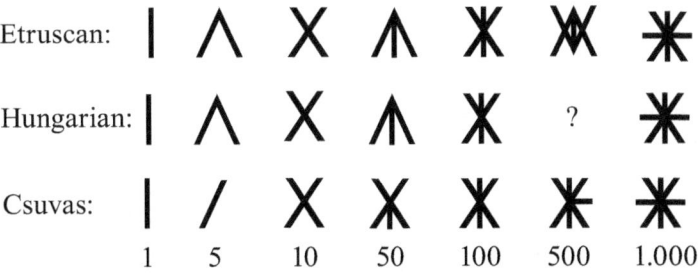

Picture 209. The Turkish-speaking Csuvas are living scattered along the River Volga. The neighbouring Osztyák people use similar numerals.

But parallels can be found in far-off territories as well. A quite interesting find is the tally presented by Georges Ifrah in his book: The Numbers on page 196. This tally - seen below – is the inheritance of the Pueblo Indians in Arizona, US. They still measured the water's height over flooded areas with tallies/scales like this during the 19th century:

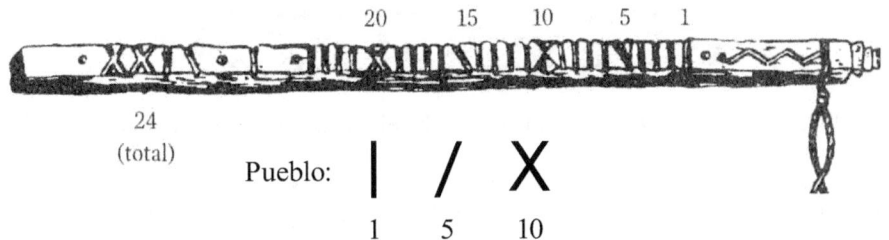

Picture 210. We can still reconstruct the traditions of the pueblos at least for the last 2,000 years

48  Mario Alinei: Etrusco: una forma arcaica di ungherese. Page 8-9, Il Mulino 2005

The pueblo signs 1(I), 5 (/) and 10 (X) render the connection with the numerals of the Picture 196 unquestionable. We can be sure about the connection even without knowing the signs for 100 and 1,000 (the numbers on this tally only go up to 24). Georges Ifrah didn't mention (didn't know?) about the Hungarian-Csuvas connections. However, he assumed that he would have to look into prehistoric times to find the common root of the Etruscan-Pueblo connection. Further (Hungarian-Csuvas) conformity only supports his assumption.

We should take note especially that the number 5 is written here (picture 197), in the Csuvas number-row (196), in the Ruthenian (192) and Vlach(191) samples from Transylvania as in the last presentation of Sebestyén Gyula (189) with the slanted line. One of the lines of the sign Λ is missing.

This is neither Etruscan, Hungarian, Csuvas, nor pueblo and certainly not a Latin numeral sequence. We must see in those names of nationalities the still remaining islands of a once widely-spread and continuous culture.

The big question is where was the origin, the place of birth before the spreading started? There had to be a place where the numerals were born and their spread started.

Answering this question, however, we shouldn't look for political maps, or for the different designs of ceramics as archaeologist like to do. This kind of number-writing was mainly used by shepherds, therefore we must look for their culture or for cultures they were influenced by. We should look for the time at the end of the Ice-Age – 10,000 years ago – when the population of Northern Europe appeared, when agriculture was moving slowly north from Central Europe. The late conquerors probably didn't care much about the numbers carved on the tallies by the shepherds. Nevertheless, we see the modification ('declining') of number-writing under the hands of the late-comers.

Picture 211. Number-Rovás in Braunschweig from 1613. (read as 27)

It seems to be certain that people once recorded the data connected to farming or business with these numerals. Inventory, number of animals, taxes etc.. The registration was always done on wood: one carved the numbers into wooden tallies, or better into wooden plates.

This obviously determined the form of the numerals. They had to be easy to carve.

As we mentioned previously, this method spread over one half of the world and the quipu (see page 111) on the other half. We described quipu earlier in detail; let's stay now with number-Rovás.

Number-Rovás is inseparable from livestock farming. One had to keep record of animals. [49] This fact provides us with further proof for origin of number-Rovás in Middle Europe. Livestock farming is connected to drinking milk and this again with lactase-persistency (the enzyme necessary for milk-digestion). The following picture presents the place of the central appearance and circular spreading of this enzyme 8-9,000 years ago. (Thomas Mark, genetic, University College, London):

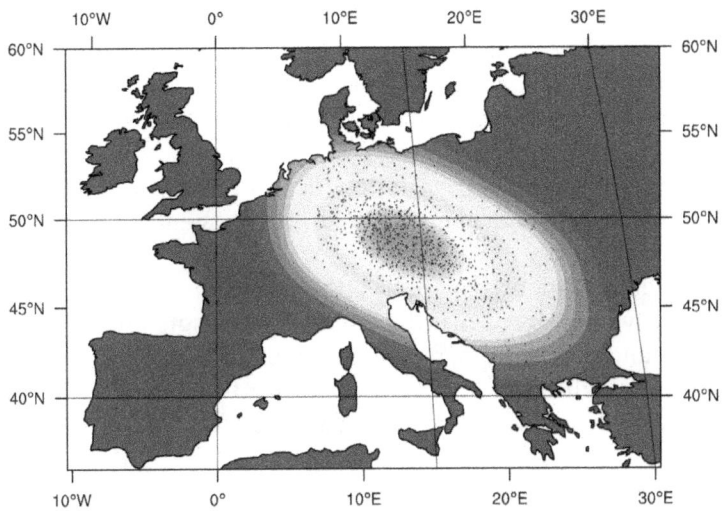

Picture 212. The spread of the lactase persistence shows the propagation of agriculture and livestock-farming.

The picture above seems to represent the spreading-map of the previously discussed numerals and number-Rovás. The number-Rovás finally arrived on the territories of today's England, Scandinavia, Germany, Italy etc... and really into the Far East.

---

49 Number-Rovás also "shepherd's Rovás".

# 16. THE LATE MODIFICATIONS OF THE NUMERALS ENGRAVED IN WOOD

It is understandable that we don't have any really old artefacts of "number-rovás"[50] because of the perishable nature of wood.[51] The oldest finds are from the 17th century. Thus, it is impossible to explore all the changes happening in thousands of years based merely on the few physical findings of the last 3 centuries. There are some helpful facts however; one is the very persistent traditionalism of the number-carving people. We have seen that even after 3,000 years, Hungarian and Etruscan numerals stayed practically the same. (The same can happen in the case of a given language or folk-music.) The important fact is that the system of writing didn't change, not even after the modification of some signs. This is the fact - accompanying us through our time-travel - on which we can rely.

Based on the above, we can follow our story backward by evaluating the differences between the number-rovás findings and arrive at the origin of this type of sign.

We also use the following observations in this evaluation:

1) We proved in detail that the signs of the single numbers may vary between dot and line: written, painted (even on the cheeks of sheep), woven as a dot (knot) and carved into wood as a line (including Chinese numerals constructed from bamboo-sticks where, because of the nature of the material, the dots were necessarily changed into lines). By pushing signs into clay, dots and lines are equally possible, as happened in Mesopotamia during Sumerian times.

2) As its name suggests, the basic material used for the number-rovás has been wood and this influenced the forms of the numerals.[52] They had to be easy to carve into wood. The dot occasionally survived in this method of writing as well, being called a "szúrás" (puncture) by the shepherds. These

---

50 Number-rovás = number-signs engraved into (mostly) wood = tally.

51 Artefacts of text engraved on wood, however, survived more often probably due to transcriptions. For example: the so-called "Rovás-stick of Marsigli", which was carved around 1200, according to the names of holy persons carved on it with Szeklers' script. The Italian military engineer found it 1690 in Transylvania, but this was then probably already a copy of the original.

52 While the material needed for knot-writing is always some kind of thread. It needs a culture in which the manufacture of thread is a daily practice.

tells us that the written number including the dot has always been there as a model.

3) The sign for any given number is built from several, sometimes many, number-signs. This is the most important part: the **number-writing principle,** which is followed by the person writing a number. This is number-writing's spiritual content and as mentioned before, there was never any change in this. It may have become a little more complicated as in the case of Sr. Dán Péter, but the essence stayed the same even there. However the ancient number-rovás preserved by him, are a stand-alone rarity, the extreme result of an effort to become different.

4) The values belonging to a sign can differ. It often happened that different values were assigned to the dot-line, and thus to the number-rovás signs, from the others'. The reason for this was often known: to be different, to decorate, to propose innovations or to deceive other people. The same happened in the history of writing. Very often, if a population group (culture) applied (borrowed) the alphabet - made from the ancient signs – from another culture to their own language, and then they changed the sound-value of several signs.

5) It is clearly visible that the signs of the number-rovás are merely "variations on a theme" and therefore originate from one root. Hence, they started spreading from one given place as we proved previously.

We are lucky to be able to study the number-writing of Sr. Dán Péter, thanks to Sebestyén Gyula. This is irreplaceable, since it is a rich alloy of different number-writing specifics out of different epochs.

Furthermore, finds are available from the Tyrol, Switzerland, Scandinavia, and from German and English territories. Let's look first at the numerals 1-19 of one of the characteristic Western-European number groups:

| 1 | 2 | 3 | 4 | 5 | 6 | 7 | 8 | 9 | 10 | 11 | 12 | 13 | 14 | 15 | 16 | 17 | 18 | 19 |

Picture 213. Western-European (written) rovás-number signs found from Austria to Scandinavia. (Georges Ifrah: Numbers, page 195)

Let's first take a look at the numbers 5-9 in the above picture. The numbers stand perpendicularly, but by turning them through 90o and leaving out the guiding line at the bottom, we can see the identity:

Numbers 5-9 with "guiding line":

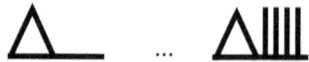

*The same without the "guiding-line":*

| Λ | ΛI | ΛII | ΛIII | ΛIIII |
|---|---|---|---|---|
| 5 | 6 | 7 | 8 | 9 |

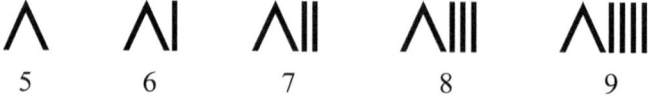

Picture 214. These numbers are written from left to right, which points to a spiritual break, since in earlier times number-writing ran from right to left in every culture (Etruscan, Old-Egyptian and also Hungarian, ...).

We don't see any Roman influence here because there the 9 = IX, but here it is written the original way: ΛIIII. The number 5 is written consequently as Λ and not as the V in Rome. Rome could only spoil it, since the number-rovás number had already been in use for a long time when the Roman culture was born as a mixture of several older cultures.

Now, let's look at the sign for 10:

*Sign of 10 lengthened as a "guiding-line":*

*Sign of 10 without the "guiding line":*

+

10

Picture 215.

This sign is nothing other than the perpendicular sign of '1' crossed by the 'tenfolding' horizontal line. We already discussed writing numbers 10, 20, etc.. using this method (page 49), but it is worth turning back and reading it again.

Thus, the construction of the decimal numbers stayed the same:

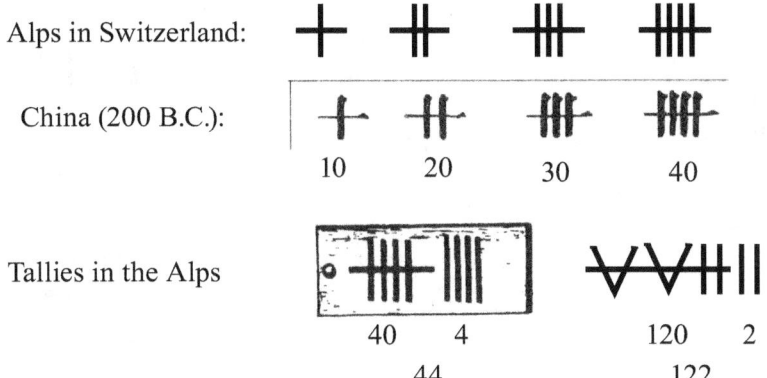

Alps in Switzerland:

China (200 B.C.):

10    20    30    40

Tallies in the Alps

40    4        120    2

44              122

Picture216. The tenfolding line.

The people in Switzerland used a "guiding line" also. The line here tenfolded only the crossed lines, but not the signs written above it. Two examples of this:

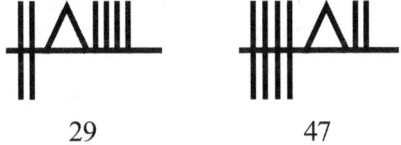

29          47

*let's compare the two numbers above with that below*

149

Picture 217.

The last number is 149 written according to the method of Sr. Dán Péter. We can see that both look alike, but the "guiding line" of Sr.Dán Péter is not just a decoration, it stands for 100. His number is read from right to left, but we can say that Western-European numbers and those of Dán Péter look like "variations on a theme". The differences are only little changes.

Let's see these little changes:

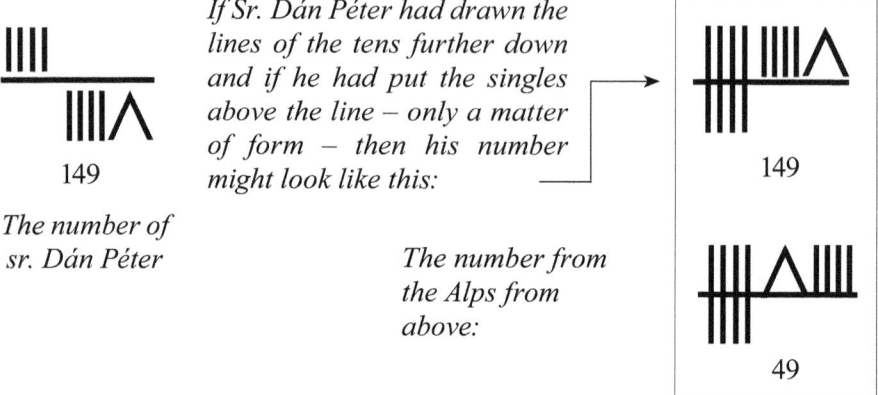

*If Sr. Dán Péter had drawn the lines of the tens further down and if he had put the singles above the line – only a matter of form – then his number might look like this:*

149

*The number of sr. Dán Péter*

*The number from the Alps from above:*

49

*The form of both numbers in the frame is identical. The difference of the 100 turns up because the horizontal line in Sr.Dán Péter's number stands for 100, it is not just a decoration.*

Picture 218. A rearrangement of the signs building a number does not mean a different number-writing method.

We can finally say that the numerals of the number-rovás are for Rovás (carving) specialized forms. These signs are used for the dot-line, or rather: line-line number-writing. This should be clear to everybody who has come this far in the book. Comparing two finds proves this statement.

We see below numbers 1 – 22 recorded in a Year-book during the Renaissance:

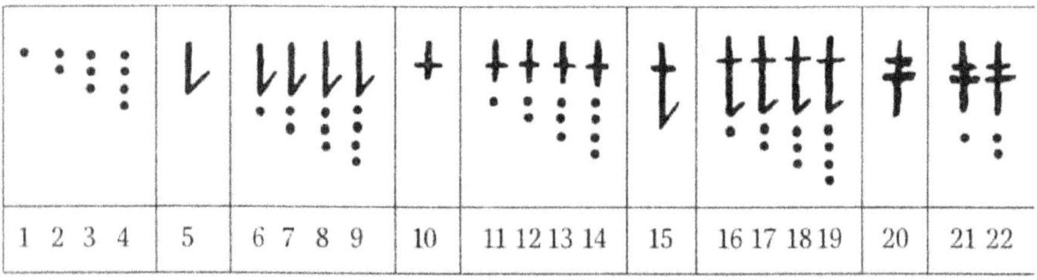

| 1 2 3 4 | 5 | 6 7 8 9 | 10 | 11 12 13 14 | 15 | 16 17 18 19 | 20 | 21 22 |
|---|---|---|---|---|---|---|---|---|

Picture 219. It represents only 22 numbers, but it is merely to demonstrate the signs and number-writing idea. (Georges Ifrah: Numbers, page 195)

We have seen a number-collection also on picture No. 213 (page 179).

Now we are going to compare some of the numbers of picture 213 with the equivalent numbers in picture 219.

See first the two signs for 5:

number-sign from picture 213          nuimber.sign from picture 219

5                    5

*One sees a difference immediately, but the picture is deceptive. Only the sign Λ (5) had been turned through 900 to the right:*

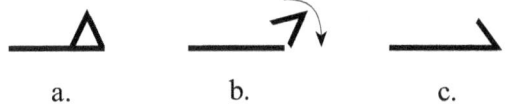

a.                b.                c.

Picture 220. Economy may lie behind this (you have to draw one line fewer), or somebody was joking and his invention became successful.

Let's compare two characteristic numbers from pictures 217 and 219:

number-sign from picture 217          number-sign from picture 219

14:

19:

Picture 221. What we see again, are "variations on a theme", but no major differences. It is a simple rearrangement of the signs

The above pictures stated again that the ancient number-writing of Vízakna, the Etruscans, Egyptians, Chinese, etc... all unfolded from the same ancient number-writing. All the differences between them are just formal: it is a question of arrangement/rearrangement of the signs building a number. This includes the fact that the signs of the ancient dot-line system, especially the perpendicular lines and the long-line, often received different values.

There is something more that we can see looking at the evaluations from above. People self-evidently replaced dots with lines and vice versa by writing singular numbers. The singles are written with lines on picture No. 204 and with dots on picture No. 206, whilst holding on to the same number-writing principle. The same happened in Crete (page 46, picture 56), where people randomly selected dots or lines to present single numbers. Crete is far from England, which means that people generally knew that dots and lines were interchangeable if standing for one.

Finally, we can make an important statement. The starting point is not the over 3,000-year-old – identical with the Etruscan – number-writing, but the much older system used in Western Europe and by Sr. Dán Péter. With this, we arrive at a time 5-6,000 years ago or earlier and not in Asia, but in the middle of Europe. Thus, the oldest known number-writing system, which has survived for the longest time (to the beginning of the 20th century in the salt mines of Transylvania) is European as well.

# 17. ABOUT THE SIGN X (10)

We mentioned the sign X (10) previously when discussing line-line number-writing. In one variant of it, its perpendicular line | (1) is crossed at the middle by the "tenfolding" horizontal line:

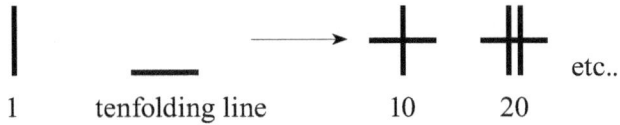

| 1 | tenfolding line | 10 | 20 | etc.. |

*Is it possible that in the sign for 10 the + was turned to give the sign X?*

Picture 222. This process is not possible, only in the reverse direction.

There can't be a + sign in Rovás number-rovás only change from an X to a + (as happened occasionally in the Etruscan writing, never the reverse.

The reason for this is a special feature of the number-rovás.

One method in number-rovás, is a "clearing-procedure". It was equivalent with today's credit cards. The number (number of sheep or value of debt) was carved onto a lath or stick which was then cut in half lengthwise. One half of it stayed with the shepherd or borrower and the other half with the owner of the herd or with the lender. To settle the accounts the two halves of the stick were assembled and became the basis of the settlement. It couldn't be falsified.

Let's look at a tally containing two pieces:

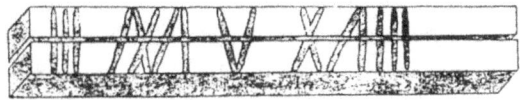

Picture 223. paired Rovás from East Siberia. (Sebestyén Gyula)

We have seen one split tally in picture 211 (page 176) and two in picture 206 page 172), but now here is a Hungarian sample:

Picture 224. Number-rovás from Kovászna, Transylania (Sebestyén Gyula)

As can be seen, the lath has been cut into at the top on the left and at the bottom on the right, thus both halves can easily be adjusted together again and the carved lines will match. Error can be excluded.

Picture 225. The split tally being taken apart.

There was also another procedure, called 'paired-Rovás'.

Two laths were put together and fixed with a pin, and the numbers were carved onto both in one procedure. I emphasize, these are two separate laths and thus the name 'paired Rovás'. They were occasionally even fixed with a nail during the carving procedure.

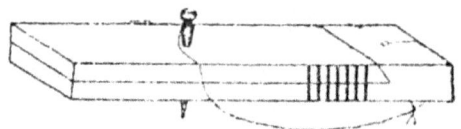

Picture 226, A 'paired tally' (with a nail well-fixed adjustment) from Hódmezővásárhely (Sebestyén Gyula: Rovás és rovásírás). It is evident that the sign + is not usable because the horizontal line would lie over the split and disappear.

There is one more outstanding question.

Every line or dot has its meaning in the process of creating a number in the Rovás system, but is the horizontal line, the "guiding-line" in picture 200 (page 179) merely a   decoration ( _∧||||_  = 8)? An unnecessary line would have been as unthinkable in the old as in today's number-writing. The reason for this line, in my opinion, could be the split in the paired- and split-tallies. This split looks like a line and has no value. The signs stand visibly above and below the achieved split. Therefore, this line could be simply the preservation of a custom.

The split tally spread widely and it was even made out of bone, ceramic, clay and paper. Plinius, a Roman scientist (23 – 79), wrote about split tallies from wood. Marco Polo (1254 -1324) told about its use in China, naming it chi-chin. The German name of the split-Rovás is 'Kerbholz', in Hebrew it is called: 'teomin' (twins) and the Ancient-Greeks named it 'symbolon' (together, doing together).

# 18. THE SIGN Λ OR V (5)

The unnecessary horizontal line in numbers carved into wooden plates is the remnant of the cut in the split-rovás as we've read on the previous page. However, this line may well have had a "tenfolder" role, according to the ancient custom. See below a number carved into wood from the Alps region. We can explain several important things by looking at it:

100 100 50 20  2

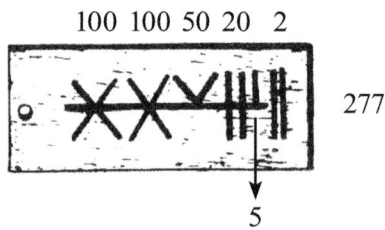

277

5

Picture 227. (Georges Ifrah: Numbers, page 195

The "tenfolding" role of the horizontal line is clear, because by cutting with it, the 10 (**X**) becomes 100 and the one becomes 10. As it can be seen, it does not cut the two single lines standing for two.

It is evident in this case that the signs for 50 and for 5 were created by cutting the lower half of the signs **X** and **I** under the horizontal line. That is, the upper half of the signs for 10 and one are used to write 50 and 5. This is however a total absurdity in number-rovás and a practice which differs from the main trend. Since the main point in the split-rovás is that the upper half of every number will be seen on the lower part - the other half - of the split tally. Thus, the above signs of 50 (**V**) and 5 (**I**) only appear to belong to the number-rovás system (despite V having the same form as 5 in number-writing). After all, these two signs are from the Alps and this kind of number-writing is a dead end variant of the real number-rovás.

We see the same connection between the sign **X** and the value of 5 even in China. This is an unsolved question, but I feel obliged to speak of it.

First of all, I should mention that in China number-writing went in many different directions over thousands of years. The winner was finally the oldest variant with a little modification and they still write their numbers that way (see mainly on page 334).

The reason for keeping this oldest method of writing is probably due to the fact that they wrote with the line-line variant of the dot-line numeral system by putting down bamboo sticks, and counted this way, using suitable rules. This certainly helped to preserve the old way of calculating.

The sign of 5 changed suddenly in one variant to

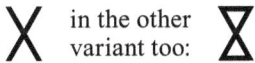

Picture 228. Number 5 in one of the Chinese number-writing systems (around 3000 years ago). Nobody knows why the two closing lines became important.

This X-sign and its 'bow-tie' variant is unexpected in China and it remained a periodic 'stranger', like a 'cuckoo's egg' in a nest. However, all other signs and the number-writing method remained in the most ancient state:

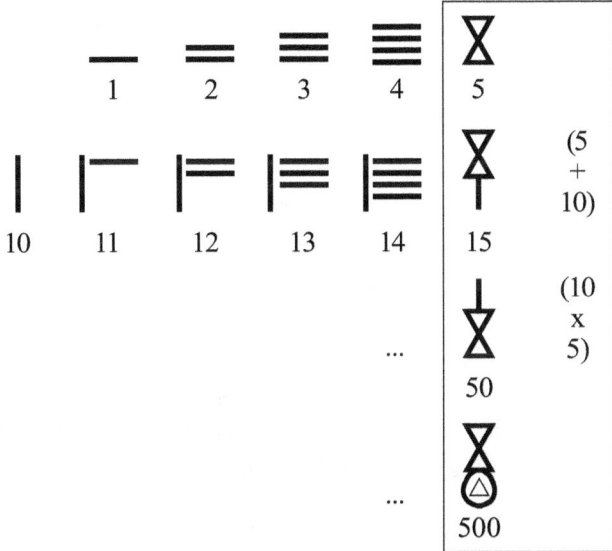

Picture 229

We can see from this that they sometimes found it necessary to introduce signs with bisecting values and chose one identical with X, probably it was not a random choice. Still in old times, many witty variants of these two signs (X and X) were created as can be seen in the following picture:

X X ⚥ ⚥ Ħ Ȟ Χ 8 ∞

*Interestingly, at the same time they also used the
oldest variant of 5 as shown below*

Picture 230. The first line: 2,000-year-old variants of the sign for 5. The old
sign in the second row proves that 5 was written only much later with an
X in China

It is not surprising that the value of the sign **X** is 5 here. As mentioned earlier, it often happened that a value belonging to a sign was changed. Let's see the variety of values associated with the sign **X**:

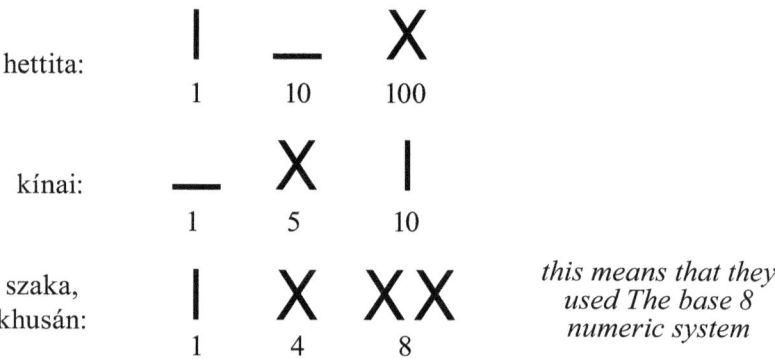

|  | | | |
|---|---|---|---|
| hettita: | 1 | 10 | 100 |
| kínai: | 1 | 5 | 10 |
| szaka, khusán: | 1 | 4 | 8 | *this means that they used The base 8 numeric system* |

Picture 231. The base-numbers are 1, 4, 8 and not 1, 5, 10 in the base 8
numeric system. The Saka and Kushan people used the sign **X**  to bisect
the base-number, thus giving it the value of 4.

It is quite interesting, and fits well with the above samples, that in Egypt the sign Λ stood for 10, but in the Seleucid empire[53] the same sign meant the usual 5 as it did in Kathra (Bengal) turned through 900: **>** (e.g., ⫶⫶**>** = 9).

Let's see two more examples next page:

---

53 The Seleucid Empire included Syria, Mesopotamia, Persia, Asia Minor from 312DC
until its total decay at 64DC. They combined several cultures.

| | | |
|---|---|---|
| Seleukidan: | **丨**<br>1 | **>**<br>5 |
| Egyptian: | **丨**<br>1 | **∧**<br>10 |

Picture 232

Based on the above, it is probable that the signs of the number-rovás once reached China, and influenced their writing for a certain time in a certain area of their territory so that they used the sign **X** (10) from the number-rovás system as the bisecting place value of 5.

# 19. SIGNS FIT FOR NUMBER-ROVÁS

We saw previously how much the original number-writing using the ancient dot-line signs was influenced by the number-rovás even in far-away territories. Its signs (mostly the X and Λ) were wedged between the "traditional", almost timeless, ancient number-signs.

We don't have enough finds to set straight the complete puzzle of writing history, but we have enough to draw the larger steps and connections. Similarly, even if we don't see the wires from further away, but see at least two poles, we know the wires must have stretched between the two. Moreover, we can be convinced of the past existence of more, by now, destroyed poles.

Luckily, we see not only two poles between the ends, but enough to draw the route of the "wire" in big steps fairly well.

Let's repeat again the connection between the two kinds of number-writing. The local values in most number-writing systems are 10 and its multiples:

$$1 \qquad 10 \qquad 100 \qquad 1.000$$

The number-rovás contains also the halves of the local values:

$$1 \quad \mathbf{5} \quad 10 \quad \mathbf{50} \quad 100 \quad \mathbf{500} \quad 1.000$$

These halved local values probably came out of human logic. People had to carve fewer signs onto wood this way, while there was no big difference by drawing the numbers. For example: writing 7 the ancient way: **IIIIIII** but using the carving method: **IIΛ**, it is clearly fewer signs. However, one needed more signs in the case of some numbers:

*44 written with several different variants using the traditional method:*

 etc.

*44 carved on wood:*

*to read from right to left*

Picture 233. While carving the numbers, one doesn't have any possibility to vary the arrangement of signs set for that number.

As we can see, some groups of signs need more work to be engraved, but there is still a good reason to use these signs. Fast and easy engraving onto a tally is only possible if the knife used on the stick overhangs the stick on both sides and the carvings run open to their upper and lower ends.

Picture 234. The carved line runs necessarily open to both edges on the stick.

These lines running openly out at both ends make only the following forms possible (we can further complicate these signs, but then their perspicuity will largely suffer):

$$\lbrace \mathsf{I\ \backslash\ X\ \wedge\ \wedge\ X\ X\ X\ X}\rbrace$$

*Some lines can be left out or new lines can be drawn, like the variants of the "bow-tie" form below:*

$$\lbrace \mathsf{X\ X\ X\ W\ N}\rbrace$$

Picture 235. These are the basic signs possible when applying the condition of carving signs open on their upper and lower ends.

A horizontal line can be ruled out. See again picture 234! A good example of this is also picture 237 on the right. It is almost impossible to cut a short horizontal line into an often very long (occasionally 1.5 m long) but narrow tally. In case of the split- or paired-rovás the horizontal line disappears when splitting the stick.

Let's give an example: If there is a tally from another planet (for example: around Sirius), then the number-signs carved with a knife on wood can't be any different from the signs in picture 235. The only question there could be is which value is assigned to which sign, and this is variable also on our planet.

Therefore, it is not surprising that the Etruscans chose their numbers up to 500 from the above sign group:

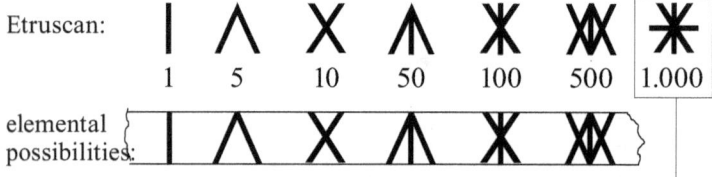

Etruscan:

| I | Λ | X | Λ | X | X | ✳ |
|---|---|---|---|---|---|---|
| 1 | 5 | 10 | 50 | 100 | 500 | 1.000 |

elemental possibilities:

STRANGE

Picture 236. The horizontal line is strange in this system

This concordance between Hungarian and Etruscan number signs is a strong sign of having the same root.[54] Despite sev-

---

54 This agrees with what was said about the ancient writing signs on page 95 in the book "Signs Letters Alphabets". There we stated that 74 different and simple writing-signs can be created from lines ("linear" writing-signs). Not one can be made by turning other signs. At the same time, all old alphabets used only 1/3 of the possible signs and always the same 1/3. Therefore, all of them are the descendants of the same, but much much older alphabet.

Picture 237 Tax-rovás from Komorzán. Sebestyén Gula, pict.20.

eral possibilities, both used the same signs

There is one exception, the Etruscan sign for 500.

Still, we can't say that there was no concordance even with this sign (in fact, it must have been an agreement even if this sign for 500 was in use or a different one from the simplest signs). Well, our finds are only a few hundred years old and we can no longer answer this question.

Finally, look at the possible solutions:

1) A horizontal line can't turn up in any number written using the number rovás system.

It is not even possible if somewhat wider sticks are used. There are some wider wooden sticks in the picture below, but no horizontal lines on them:

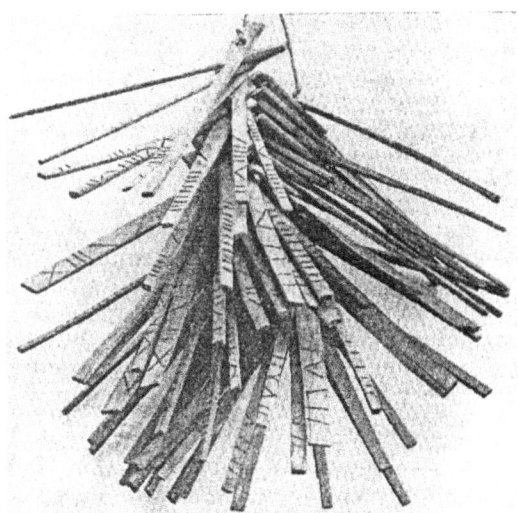

Picture 238. Tax-Rovás tallies from Ruzsinosz (Krassó-Szörény county) 1881. (Sebestyény Gyula, picture 19)

It is easy to see that the width of the tallies in the above picture is quite varied. There are wide laths between the small sticks. The numbers have to be unified, thus, no signs can be written on the wide laths which could not be written on the smallest ones, on which horizontal lines can't be engraved.

2) Neither can the horizontal line be used for the paired or split-rovás. As we discussed earlier, the stick or lath will be cut along its midline after writing on it and any horizontal line would disappear. Splitting not through the middle of the stick would result in different numbers remaining on the two halves of the stick. The remaining horizontal line on the one half would act

as a tenfolder line for the numbers on it but not for the other part.

Therefore, the horizontal line crossing the perpendicular can only be used in the following cases:

a) no split-rovás used,

b) hand-writing

c) the possibility visible in the picture below:

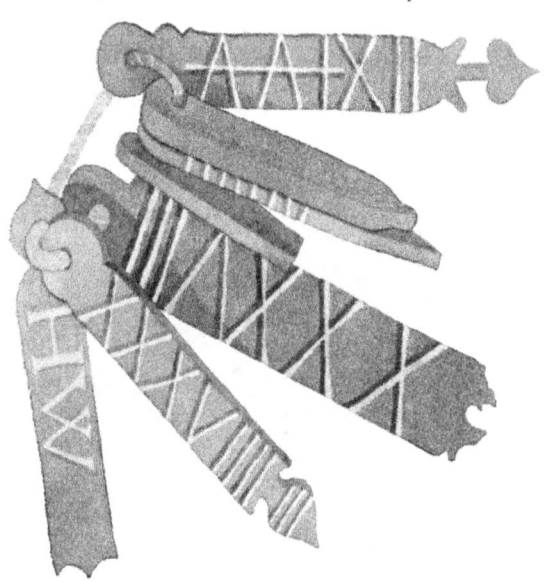

Picture 239. Tallies from Switzerland. The number on the uppermost lath can't be used for split number-rovás, because it has a horizontal line in it determining the number's value: AAI. This line has a tenfolding role, which we introduced in previous chapters. This is not an original picture, only a painted copy, therefore we don't know if the engraver changed his writing directions or the painter's 'free fantasy' lets it run from the right and again from the left.

It is important to see that one can't speak here from general number-rovás numbers, but rather from a fine piece of decorated cabinet-work as the carved laths demonstrate. These all belong rather to the world of xylography, as we see the decorative monogram H W. This is like engravings into stone or metal and this is how the horizontal line might appear in this number, but only after it had been previously written (not carved, but for example painted by shepherds onto the cheeks of animals).

We must also see that the ancient rovás signs were quite durable and were taken up even into the world of Latin signs. In other words, the numbers on the laths in the previous picture were not of Latin origin. Latin numbers don't have the tenfolder-line and 9 should be written as IX instead of IIIIΛ or ΛIIII as on the lath, but the monogram HW is written with Latin letters.

Number-rovás became a real art in wood carving. We see below two pages of a 15th century wooden-calendar from Tyrol. The lower two lines contain only number-signs (see similar picture on page 179).

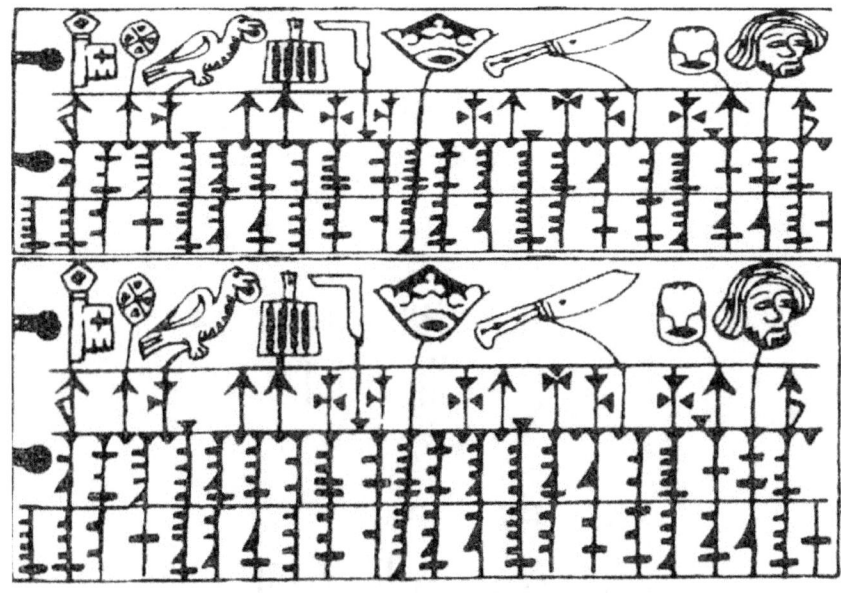

Picture 240. Georges Ifrah: Numbers, page 196. The find is from the Figdorschen Collection, Vienna, no. 800.

This is already the advanced school of woodcarving.

The number signs here are the same as those which were introduced in part 16 on page 178. It is interesting that while the signs used on the tallies from Switzerland have the forms of the ancient number-rovás system, the number signs of the wooden-calendar from Tyrol can only be hand-written or used for xylography. But even these are just little modified variants of the ancient signs.

Handwriting is also a real variant. The next picture shows the number-signs from the Tyrolean calendar, but 1-9 written on paper in Sweden (the sign for 0 is an innovation):

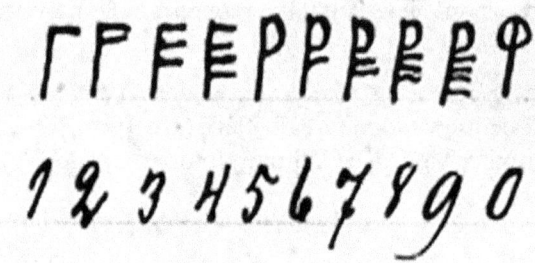

Picture 241. Swedish number-signs written on paper with pen.

Let's lay these signs down. There we see exactly the same ancient number-rovás signs used by the Etruscans, Sr. Dán Péter and by other Hungarian shepherds:

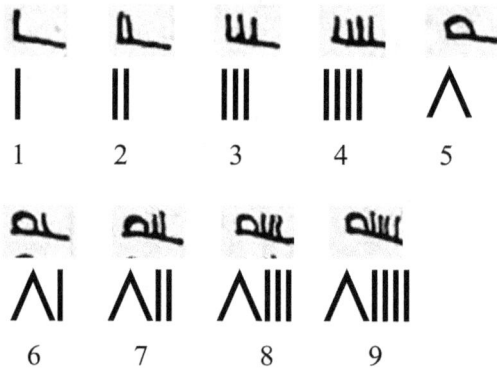

Picture 242. In appearance the handwritten Swedish Signs look different but they are exactly the ancient signs of the number-rovás. The "holding line" here is only a decoration.

Only the sign Λ (5) has been modified, rounded by the handwriting. The same happened in Egypt, where the demotic sign of 10 (Λ) was rounded for the hieroglyph ∩.

This also goes back to the dot-line number-writing and the original dot-line writing which has never been forgotten along with the number-rovás. We can best see this by comparing them with the English written number-signs (seen previously in picture 219, page 182).

Picture 243. The difference here is that the sign (Λ) has been used in its aslant form in the English sign group.

We can see here – as in all other examples in this book – that the origins of these number-signs are the oldest ancient dot-line numerals. Furthermore, based on our previous investigations, we can state that the numerals in the last picture are a mixture: the singles come from the dot-line writing, the Λ from number-rovás and the + taken from the line-line variant of dot-line writing.

Now it is clearly visible that number-rovás originated in middle Europe, since they deteriorate as you move away in every direction, the more so the further you go. However, they stayed practically unchanged with the Etruscans and in the Carpathian Basin (and only at these locations) until the 20th century.

The spreading of the number-Rovás signs matches the map of the lactase-persistency's spreading in Eurasia (see picture on page 177). Keeping any goods on track takes number writing for granted at all times. See for this again the 30-35,000-year-old carved numbers on page 33.

## TABLE IIIV

Interestingly formed number-rovás (from the Internet, author unknown):

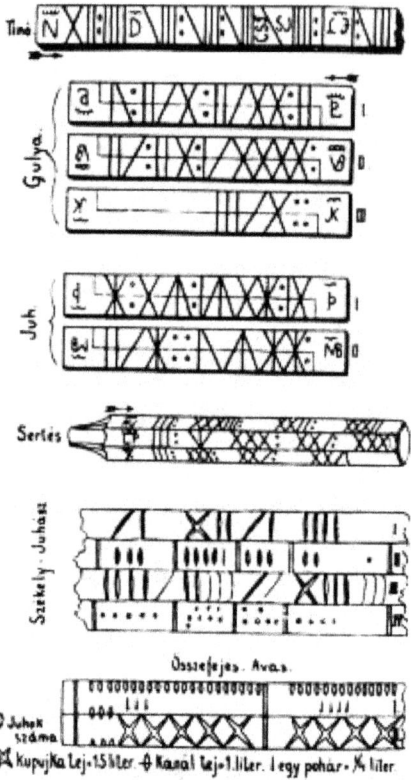

Hungarian Gipsy number-rovás samples from the book Rovás and Rovás-writing by Sebestyén Gyula (Püski Publishers):

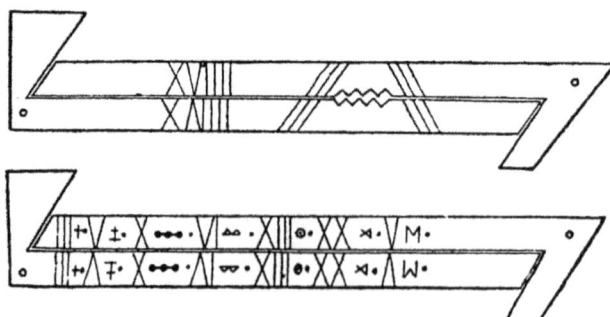

# 20. SINCE WHEN HAVE NUMBER-ROVÁS SIGNS BEEN USED?

As yet, we had only one certain find for the use of number-Rovás: the Etruscans used it 3,000 years ago in the area of Toscana and these signs are identical with the Hungarian signs. These number-rovás signs are probably much older than 3,000 years, for the Etruscans are not supposed to have started using them only at the time of their moving there.

I found conclusive evidence for the early existence of number-rovás in the book The Living Goddesses, by Marija Gimbutas on page 45. This is a necklace, buried with a 30-year-old woman, containing 70 perforated deer teeth. Gimbutas mentioned that 46 of the teeth have scratches on them. The find is from the so-called Magdalenian Era, which was dated between 10-20,000 years before, and they were found at Saint-Germain-la-Rivière.

Let's first look at the 46 scratched teeth presented by Gimbutas:

(b)

Picture 244. The drawing of Gimbutas. She presents the teeth in this size.

We see on all of the teeth exclusively different arrangements of these signs: I, V, X, Ж. There are no other signs on the teeth.

These signs however are the still-alive signs of the number-Rovás. The appearance of only these four signs cannot be random.

Well, the signs I, V, and X could be writing-signs as well, but less probably the fourth sign Ж. Are they writing-signs? Then we must deal with them and with all the consequences of such a date 10-20,000 years ago. Those are therefore writing- or number-writing signs. The repeatedly used 4 signs are too few for possible writing-signs. Furthermore, there is only one arrangement to be found in the grouping of these signs: a sub- or super-ordinate relation to each other, which is possible only in number-writing.

The comparison below makes these arrangements clearly visible. We put only 13 signs in this table, because there are many repetitions among the signs on the 46 teeth. For example, we find the sign I on seven, the sign II on five and sign III on seven teeth independently but we didn't want to show repetitions.

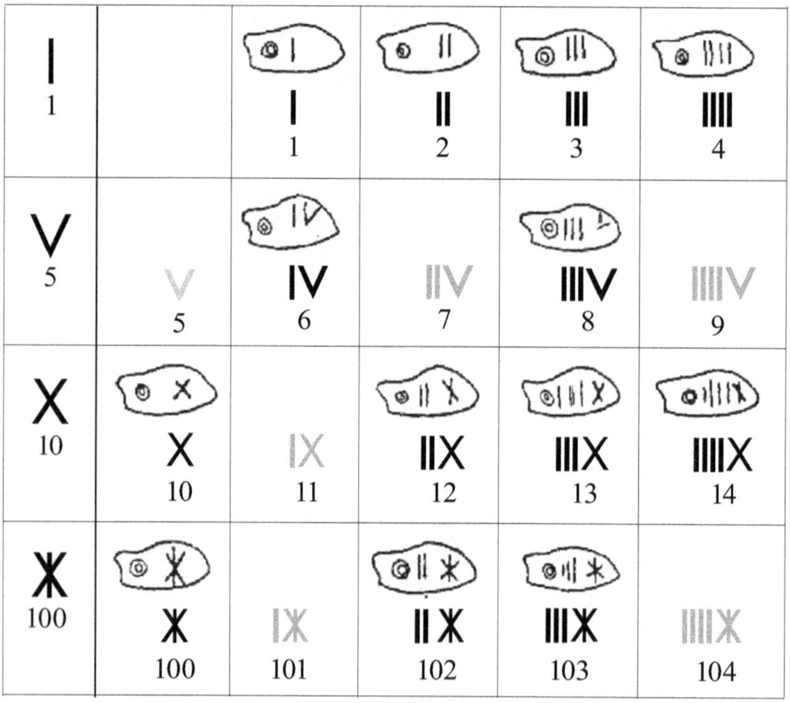

| I<br>1 | | ⊙I<br>I<br>1 | ⊙II<br>II<br>2 | ⊙III<br>III<br>3 | ⊙IIII<br>IIII<br>4 |
|---|---|---|---|---|---|
| V<br>5 | V<br>5 | ⊙IV<br>IV<br>6 | IIV<br>7 | ⊙III▷<br>IIIV<br>8 | IIIIV<br>9 |
| X<br>10 | ⊙X<br>X<br>10 | IX<br>11 | ⊙IIX<br>IIX<br>12 | ⊙IIIX<br>IIIX<br>13 | ⊙IIIIX<br>IIIIX<br>14 |
| Ж<br>100 | ⊙Ж<br>Ж<br>100 | IЖ<br>101 | ⊙IIЖ<br>IIЖ<br>102 | ⊙IIIЖ<br>IIIЖ<br>103 | IIIIЖ<br>104 |

Picture 245. Different number-signs scratched onto teeth were stringed to build a necklace.

The question comes up: if we see a sign for five (V), shouldn't there also be a sign for fifty? The presence of 5 means that this sign-collection is also bisecting that is, having a sign for the halves of its local values. We did not find a sign for fifty on the teeth, because there was no number which needed it. The sign fifty is needed for writing numbers between 50 and 99 and 150 and 199. We would recognize it immediately – knowing the other signs – if they were written on a find.

Therefore, the scratches, presented by Marija Gimbutas as things of interest, but not yet evaluated, are signs of number-rovás. Thus, we can state that the beginning of number-rovás – based on findings known today – can be dated to 20-10,000 years earlier.

Luckily, this is not the only find from this era. The "field marshal's baton" seen below is around 15,000 years old and can be seen in the Archaeological Museum of the city of Santander in Spain. It looks like a horse's head. It is carved out of reindeer shovel, has a length of 16.5 cm and it is covered with several drawings.

Let's look at the left upper corner of the horse's head. There are drawings of a regular row of signs, engraved very carefully, contrary to the easily scratched drawings. These signs are built from straight lines as opposed to the other lines on the right end of the figure. Therefore those signs must have been very important:

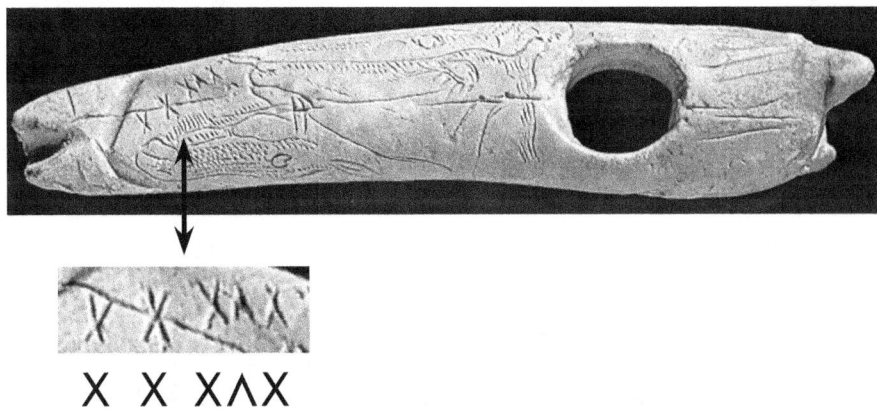

Picture 246. A 15,000 years old "field marshal's baton" and its carefully scratched signs

The signs on the finds of Marija Gimbutas and on the "field marshal's baton" are identical and are both dated for the time 20-10,000BC. We don't think that those carvings were made during the last days of that era, therefore, the age of these finds could be 15,000 years or older.

We do have a younger, only 5,000 years old proof for the existence of number-rovás as well:

The number-signs on it:

Picture 247. a 5,000-year-old pottery found at Lomnic, Transylvania.
(Makkay János: Finds of Tataria, table 26.)

Let's return now to the teeth with numbers.

Why did somebody scratch numbers onto the reindeer teeth found by Marija Gimbutas? Why onto 46 teeth? What did the numbers stand for? Why are the numbered teeth stringed onto a necklace? Why are some numbers repeated several times?

We don't have answers for these questions, but there are other numerated teeth from this Era and this fact points to some importance of numerated teeth being stringed into a necklace.

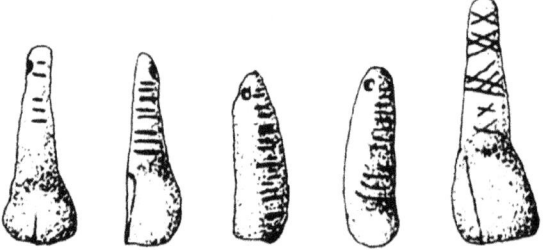

Picture 248. perforated teeth with numbers on them.(Goyet, Belgium, Magdalenian culture)

These teeth are perforated and numbered in the same way as Marija Ginbutas' finds. The numbers are carved using parallel lines, but the signs on the last tooth are partially those of the number-rovás,[55] using X.

---

55 See also page 210

NOTE

The signs on one of the teeth found by Marija Gimbutas are symmetrical in a way that the two outside signs being tilted inwards:

/X\

We have seen examples of symmetrical signs also on the "field marshal's baton" (picture 233), one tooth found by Maria Gimbutas, at Tordos Vincsa (page 102) and also at Tepe Tahya (Page 102):

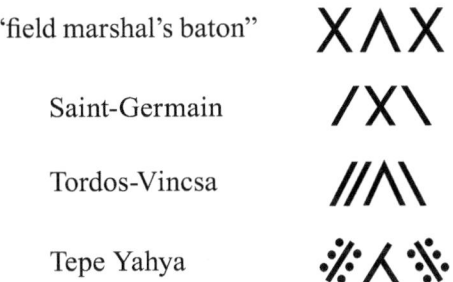

Picture 249. Symmetrically written number signs.

The signs being tilted inward at both ends are unusual. This symmetrical tilting must have had a special meaning.

# PART Λ

## RECKONING

## 1. THE QUESTION OF CONTINUITY

We can rightly assume that number-rovás spread over the last 10,000 years from its main territory of agriculture – most probably from Middle-Europe – in all directions. This happened with such powerful progress that it influenced the much older dot-line number writing. We can see many results of these influences in China (see page 191) as well as in Western-Europe. The American continent shows an interesting mixed picture: the Mayans used the dot-line method to which the quipus also belong and while there are finds of this method on the southern part of the continent (see page 119), the Pueblos from Arizona however carved the signs of the number-Rovás into tallies. I don't know how this might have happened. The historians have avoided this question so far.

At the same time, the signs of the number-Rovás did not have any influence on the oldest ancient dot-line signs in Transylvania (salt mines of Vízakna and Akna-Suhatag), in Egypt, in Sumerian-land, in Aramaic writing or in South-America. The best examples for this are the dot-line number signs which were still in use at the beginning of the 20th century in the Transylvanian salt-mines.

1) We have evidence from around 30,000 years ago for the existence of dot-line number-writing. Below, we repeat some of these finds as perfect evidence of the continuity to reiterate our claims:

Pont d'Arc,
30.000 years ago

Lascaux
17.000 years ago

La Pasiega
15.000 years ago

Tepe Yahya
6.000 years ago

Egypt
6.000 years ago

Transylvania,
20th century

Picture 250.

Dating the oldest find of dot-line number writing to more than 30,000 years ago does not mean that writing numbers with such signs started at that time. The popping up of those finds happened randomly, so we would do better to think of the number-signs on the walls from 30,000 years ago as "snapshots somewhere on the way". Halfway? Well, number-writing could have a 60,000-year-old past. We don't know how long the period was which ended 30,000 years ago. However, we can state, looking at it logically, that dot-line writing is probably much older than 30,000 years.

2) There are two kinds of number-rovás possible. One contains only perpendicular lines and we call this simple number-rovás. The evidence of its existence goes back 37,000 years (see page 35). The other one is the so-called Etruscan-Hungarian number-Rovás using slanted lines as well, as discussed in detail earlier. Slanted lines were introduced in order to be able to mark the position-values.

The number-rovás using merely perpendicular lines did not die out when the Etruscan-Hungarian system was introduced. This most simple and archaic number-rovás – its oldest known find is 37,000 years old – was still used at the beginning of the 20th century, and it stayed continuously in use along with the other system. We can see below four such examples which defied the influence of the number-rovás signs and stayed unchanged in use

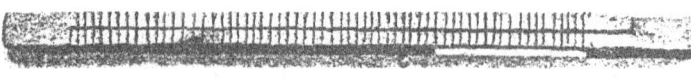

**19th century.**
Freight-rovás
from Csallóköz
(Hungary)

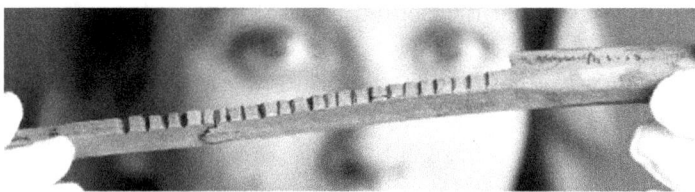

**14th century.**
Number-rovás
from
England

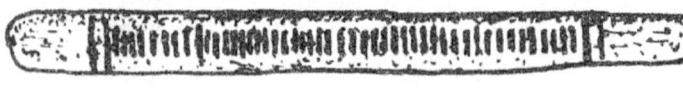

**30.000 years**
old number-
rovás, Saint
Marcel.

**35.000 years** old
number-rovás,
Lebombo

Picture 251.. There are no traces of the Etruscan-Hungarian stile of number-rovás on these tallies.

# 2. THE LARGE COMPARATIVE TABLE

The dates for the finds seen on the previous page cover 35,000 years. Their essential character is identical: a long stick covered with perpendicularly engraved lines identified without doubt as numerals. The question must be asked whether this simple number writing system – signs engraved first into bones, more of them later into wood – have been used continuously throughout this time?

The question is not pointless, for we have several finds from this long period (see page 35). Placing them among the previously presented samples, one should seriously think about the unbroken use of these signs. There still remain, however, notable time gaps. We could only silence every doubt by presenting a similar find for at least every 1-1,500 years of human history. However, I can't present such a large number of finds, here and now. I think that many similar finds are hiding in the world's museums, but probably not enough to provide incontrovertible evidence for the whole of the last 15,000 years. People started using more wood instead of bone for tallies and wood falls into decay fast, while bone can be very durable in favourable surroundings.

The continuity over 35,000 years of the simple number-rovás therefore can't be proven just by the finds. We need something more.

Luckily, there is a way. Number writing had three different methods: simple number-Rovás, rovás with signs turning in different directions (Etruscan-Hungarian) and the mostly written dot/line and line/line number-writing. Let's compare the finds of all three number-writing methods in one table arranged in chronological order. We will cover the known time of human literacy by also adding the most important finds of writing signs to the table. We are going to examine our observation about the continuity of number writing. Is it true or not?

We can see this table on the next page.

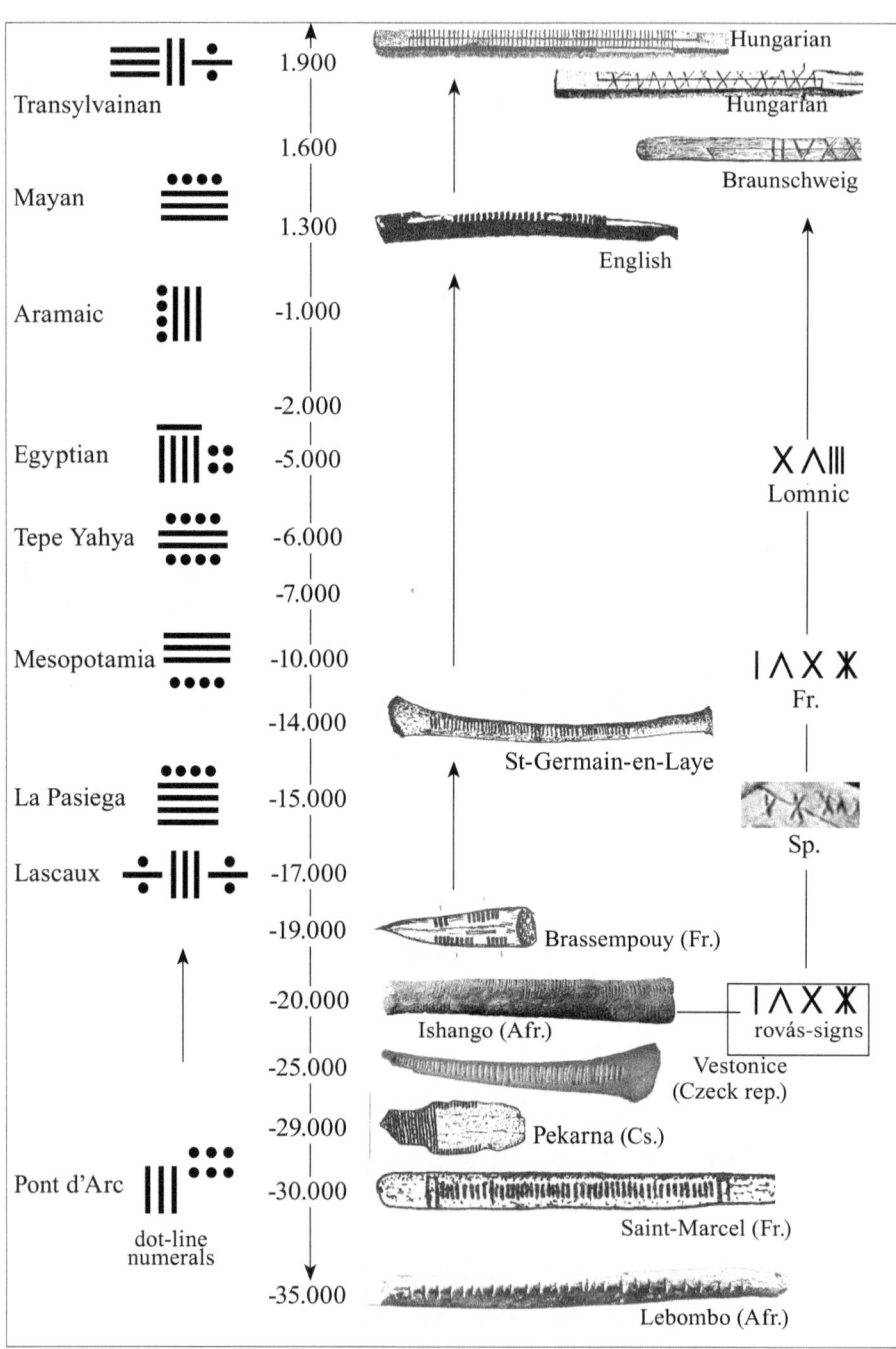

Picture 252. THE LARGE COMPARING TABLE

⊙ ◑ ⊕ I + ⊣ Ж Υ ∧ ↑    székely

I + ⧧ ⊲ ∧ ◇ ⋛⊃ ⱽ    Kenyereér

I + ⊲ ⊙ Ж ⋈ ⋀ ∧ ↑    Carpathian basin

i t   I ⧧ ⊲ ○ ◑ Ж ∨ ⋀ ⊟    latin

⊙ ⊕ I + ⧧ ⊲ Υ ∧ ◇ ⋀    Old-Greek

○ ◑ ⊕ I + ⊲ Ж Υ ∧ ↑    Egypt. hieratic.

I + ⧧ ∧ ⊲ Ж / I Υ ⋈    Alvao

I + ⧧ † ⊲ ∈ ⋌ / I Υ Χ    Tordos-Vincsa

◇ ◇    Lortet

⟨ Χ /    Les Eyzles

I + ∧ Χ ↑ ⊨    Czech Rep.

○ ◑ ⊕ I + ⧧ ↑    Mas d'Azil

Ⅷ / Ụ I    Pech Merle

⊕ I ⧧ ∧ ⟨ Χ /    Grotte Cosquer

⊕ ◑ I + ⧧ □ ⊞    Pont d'Arc

I present just a few characteristic samples for the time-period of the last 10.000 years. Finds are more numerous for this time.

The most important conclusion of this comparison is that writing and number-writing have surely existed continuously alongside each other for the last 30,000 years. Number-rovás with the Etruscan-Hungarian number signs were added to around 15-20,000 years ago (we don't have any earlier finds).

Moreover, the table is somewhat reticent about numbers: we can establish the continuity of dot-line number-writing almost day after day for the last 7-8,000 years, but we have just one sample to suggest that Egyptian number-writing stayed unchanged through the several thousand years of the ancient Egyptian culture. I didn't even mention the variant line-line number-writing ( •••  = Ⅲ ), because the table would have become too dense and the continuity is clearly visible without it.

The continuity of most writing signs is evident also. You can't invent such a row of writing-signs twice. (See more detail about writing in my book Signs Letters

Notes to the table:

The Chinese too would have been in the table from ancient times up until today, had I presented the line-line variant of number writing in it. Number-rovás, in its original form, was widely used even in Russia and tallies were used till the 19th century. Not long ago, engraving numbers onto wood – using tallies – was everydays business in Switzerland, in the northern part of Europe, for example in Sweden and even in England.

Luckily for us, we have direct and perfect evidence of writing signs and their meaning's long-lasting continuity. We can see this on the pictures below which show the same script:

| 21.000 years old script (Grotte Cosquer) | 14.000 years old script (Les Eyzles) |
|---|---|

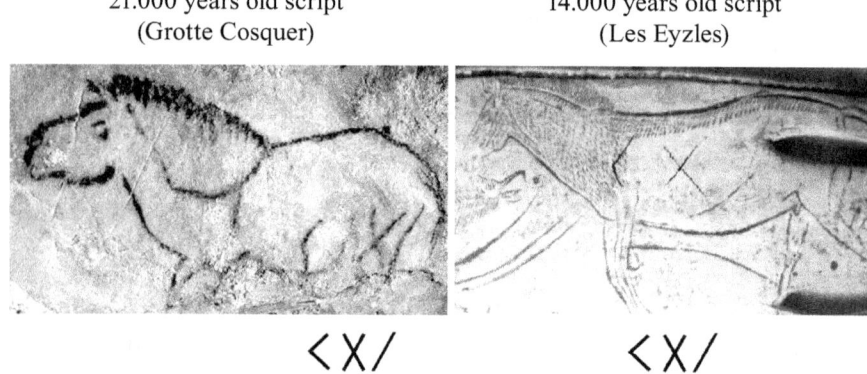

Picture 253. The same three writing signs were used on both pictures.

The two pictures were made 7,000 years apart but carry the same script and both on horses (read more about this in the book Signs, Letters, Alphabets). It is important that in both cases the script is written on a horse, which proves to us the very long-lasting continuity of this triple-sign-group's meaning.

We can apply the same time scale shown by these scripts written on horses to all old writings and number-writings. This can be supported by the four thousand years of continuous literacy in ancient Egypt and China, furthermore we shouldn't forget that our large case, known as "Latin", writing signs are already 3,000 years old and hopefully we won't have to change them to other (e.g. Arabic?) signs in the next few thousand years. Thus, we may correctly think that the different groups of our finds presented with time-gaps in the LARGE TABLE above, were in reality used continuously in time. [56]

---

56 Large time-gaps are only in the case of number-rovás, probably due to the fast decay of wooden material. We can see much older numbers written on ceramics.

The previous picture is an example of durability in time. The people painting the pictures in the Grotte Cosquer knew more than simply the three signs of the picture. Thinking this would be foolish.[57] There is a second picture with writing-signs, in total seven signs, in this cave:

Picture 254.

Seven writing signs from the given time-period survived in the same place (these signs are known even today). This many signs are already one third of a civilized (Latin) writing-sign set! (See in book Signs, Letters Alphabets)

Anyway, who would imagine that the different sign groups seen in the Large comparing table's four columns (picture 239) were not continuously used? Stating that the continuity was broken and the signs forgotten, would mean that the signs were newly invented every time they sank into oblivion. All four sign groups of the table? The dot-line number-writing, the simple number-rovás, the number-writing with the Etruscan-Hungarian sign-group and the group of writing-signs? Newly invented, maybe several times? No way. This is totally impossible. There is no way back in the history of writing. The same writing signs wouldn't be re-invented exactly the same even once.. This means the continuity of overall literacy must be at least 30,000 years old.

Our confidence in this continuity can further be strengthened: there were no trials of inventions to create different kinds of writing-signs or number-signs in the past. Not finding any remnants of other writing-signs means that there were no different writing- or number-writing signs. Thus we have to accept the at least 30,000 years' continuity of our literacy.

---

57 One can often see trees with engraved monograms on them, but we would never imagine that the person who made the "artwork" knew only the two letters he or she engraved. We take it for granted that people know all the necessary signs if they are writing.

The only change seen in number-writing is the "loosening up" of the strong geometric lines of the number-signs. The signs became more and more ornamental and that mostly in far India.[58]

The change in form happened by fast handwriting, when people wrote several parallel lines without lifting their pen. The lines became connected this way. For example: writing 3 lines (≡) without lifting the pen becomes (Ƹ). This is our 3, but much more decorated is the three of Bengal (ৼ). We can read more about the decoration of signs in the next chapter. Only a little change of form happened over time to our writing signs. Almost no change happened in Europe. East and south from us, the old signs were kept but their numbers were extended with new signs (China, Old Egypt, Sumerian, …) Finally, we can state that radiating out from Central Europe in every direction the original writing signs changed more and more with distance.

The modifications or disturbance of the original clarity may have started around 5-6,000 years ago but more probably 3-4,000 years ago in certain territories, but in some other places only after the end of a given culture. In certain more lucky areas and cultures the signs and systems remained unchanged for 10,000 years, right up until the 19th or 20th century.

---

58  The only exception is the script (third row from below) in the Large comparing table, from 20,000 years ago, found in Pech Merle. The letters of this script are beautifully designed and this can only turn up in a highly developed culture. See more about it in the book *"Signs Letters Alphabets"*.

# PART IΛ

OUR PRESENT-DAY NUMERALS

# 1. THE ORIGIN AND HISTORY OF OUR PRESENT-DAY NUMERALS

We can't treat the dot-line and line-line number-writing introduced in detail in the previous chapters as something strange, "primitive", belonging to a forgotten past time and neither can we say that we outgrew it long ago and found a new, revolutionary and better system, taken over from the Arabs and Hindus. There is nothing really new in the numerals we have taken from the Arabs compared with the old ones. Some of them became more ornamental – but most of them not even that. However much we want it, no such development can be seen in them. How it would warm our hearts to see such a thing, because we are brought up to cover up the past. But can we call it a development of the straight line if we draw it a little curved instead of straight? For this is the only change, not a separation from the old system.

It is merely the typical distorting effect of handwriting that awakens in us the false feeling of seeing a novelty: "there has never been such a thing before".

The point is this: sign-forms change easily when they are handwritten. These changes can be inherited too; something we see quite often. Let's see as examples the Egyptian numerals. The numerals 7 and 9 offer a good example of the slow misdrawing of the old line/line signs due to handwriting:

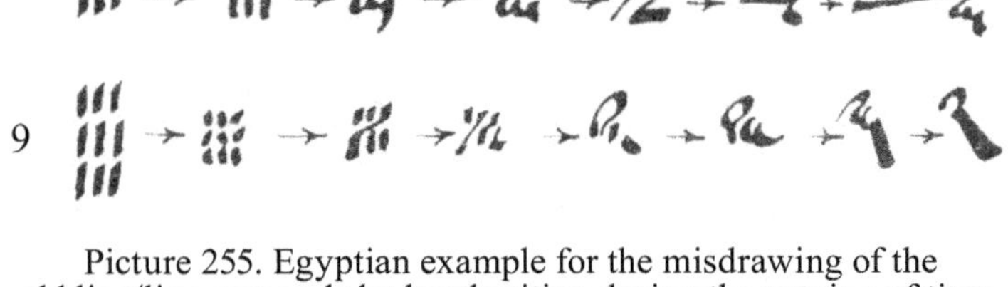

Picture 255. Egyptian example for the misdrawing of the
old line/line numerals by handwriting during the passing of time.

Here, we can see the simplifying or overdrawing effect of handwriting. However, the Egyptian hieroglyph signs didn't change in 5,000 years, since those were engraved in stone and kept their original form.

A fluent writer naturally tries to draw signs - even containing several lines - in one attempt without lifting the writing tool. In this way, he or she connects several horizontal or perpendicular lines with a little curl. To serve this purpose, "fluent writing" (or "joined-up writing") was born.[59]

Our numerals changed in the same way from the ancient forms. The change of the ancient forms started with the Brahmins. The original numerals looked like the ancient ones as we can see in the first row of the following table. These perpendicularly drawn lines changed when handwritten. We can follow four steps of this process in the table below:

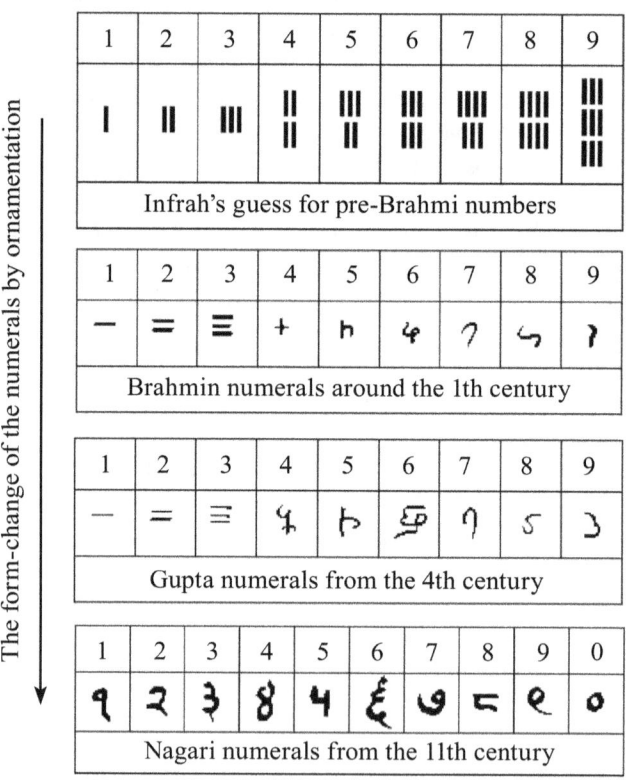

Picture 256. A changing process lasting many centuries

---

[59] The little letters (small case) of our alphabet have existed since 785 AD, invented by a monk named Alkuin.

# 2. THE MISDRAWING OF THE SIGN I

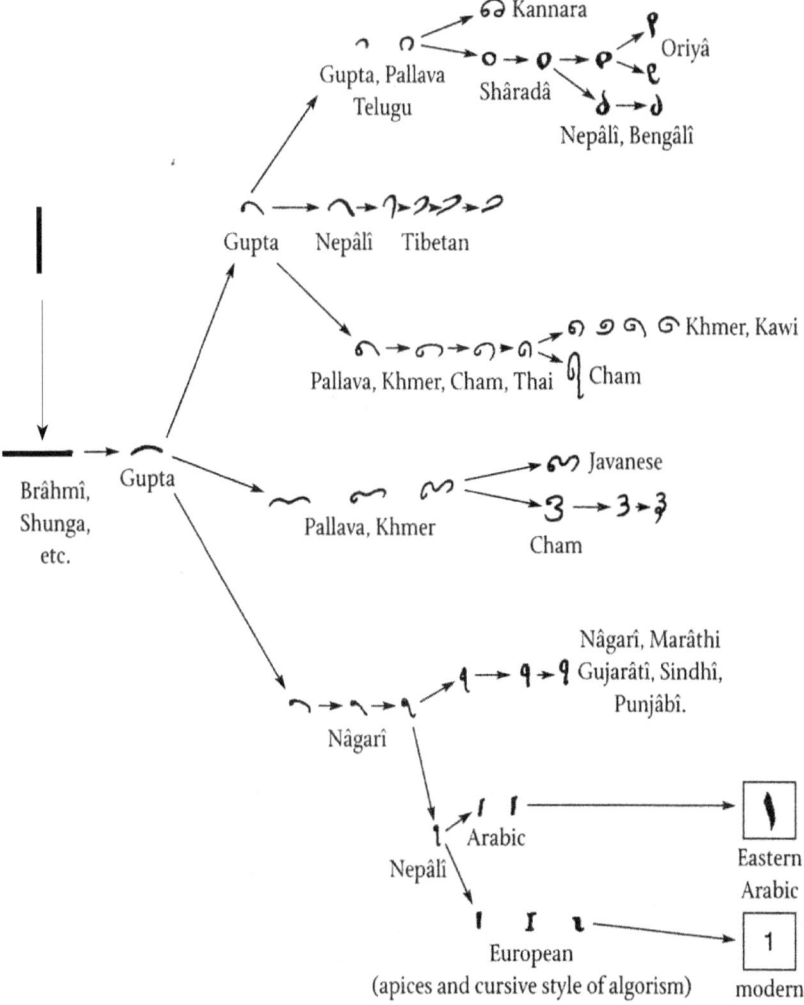

Picture 257. . It doesn't necessarily mean a change, much less an evolution if the perpendicular line is drawn horizontally after a certain time. (From Georges Ifrah: Numbers, page 392)

# 3. THE MISDRAWING OF SIGN ||

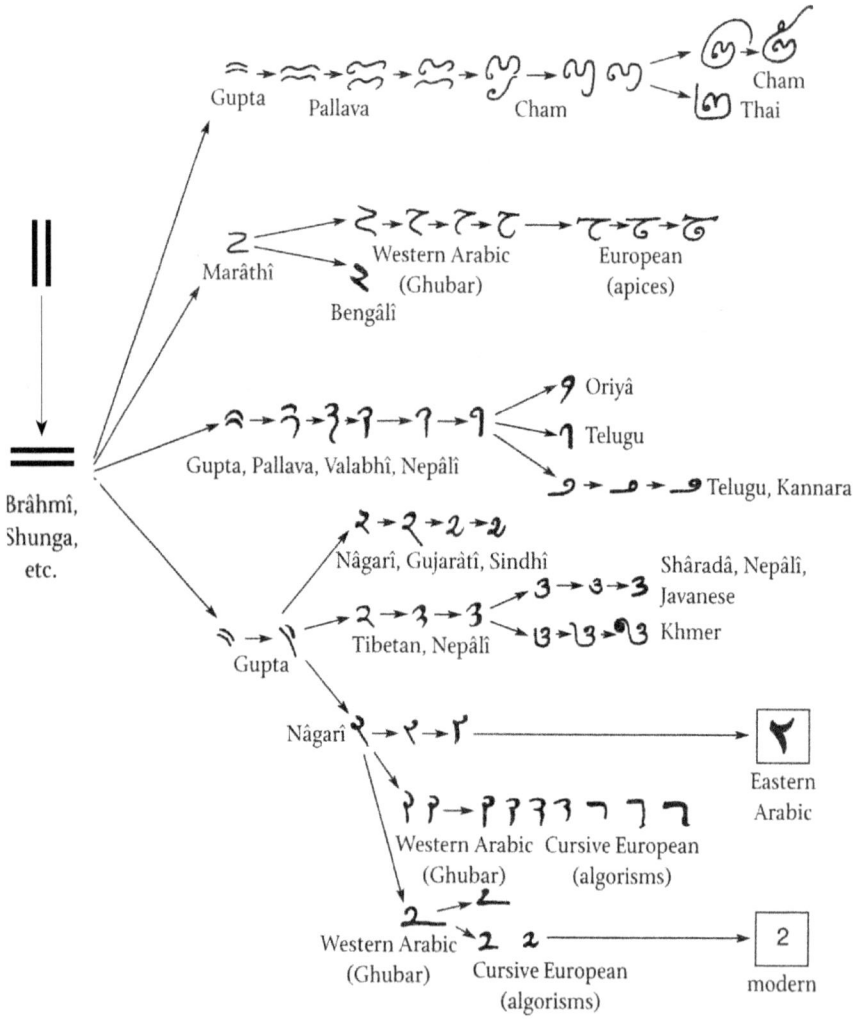

Picture 258. From Georges Ifrah: Numbers, page 339.

One can see the result in many cases, when writers attempt to draw signs containing several parts without lifting their writing-tool. A connecting line will be produced. It can be seen in the previous picture that in the Hindu culture, independently in different places, the two parallel lines became similar to our letter Z. See the signs ⊇, ꓠ, ⟨ on previous page in the middle as examples.

The same modification happened in the territory of the states of Khatra, Nabatea, Palmyra, Phoenicia and in areas where Aramaic numerals were used. Originally in their culture, the horizontal line meant 10 and the perpendiculars stood for the singulars. Two horizontal lines therefore meant 20. (Please look for more details in the part about Phoenician numerals.)

Below, we see the shaping of the sign for the number 20 in the above named territories:

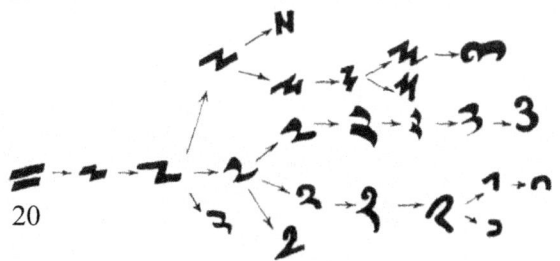

20

Picture 259 from George Ifrah: Numbers page 229

As we can see, the two parallel lines - with time- took on the shape of a Z, if tilted, the shape of an N, but even that of 3. These are the same form-changes where the two horizontal lines went trough seen on Indian territories in the picture 258.

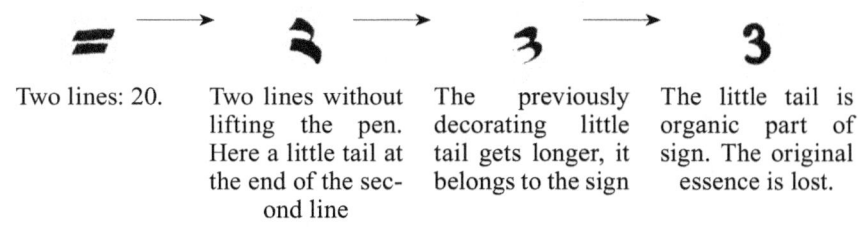

| Two lines: 20. | Two lines without lifting the pen. Here a little tail at the end of the second line | The previously decorating little tail gets longer, it belongs to the sign | The little tail is organic part of sign. The original essence is lost. |

Picture 260

The two horizontal lines, standing for 10 each ꓤ, mean 20, even if they were written with the sign known for us as **3** and, in India shaped by twirling from 3 horizontal lines.

# 4. MISDRAWING THE SIGN OF ‖‖

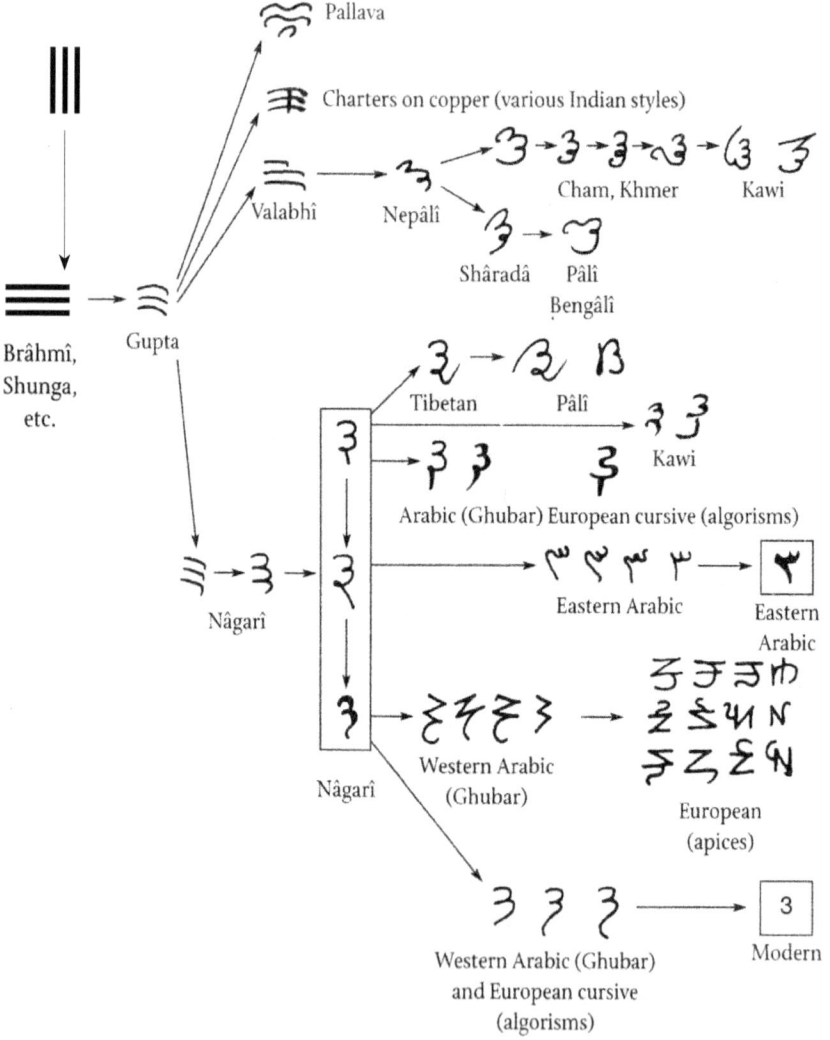

Picture 261. From Georges Ifrah: Numbers, page 393.

# 5. THE MISDRAWING OF THE SIGN ╪ (4)

The writing of number 4 by the sign (╪) instead of the usual IIII, is probably a infiltration from a different system. The sign (╪) stands in most cultures for 5 or 10. It may be a remainder of the base 8 system, where it stood for the number 4.

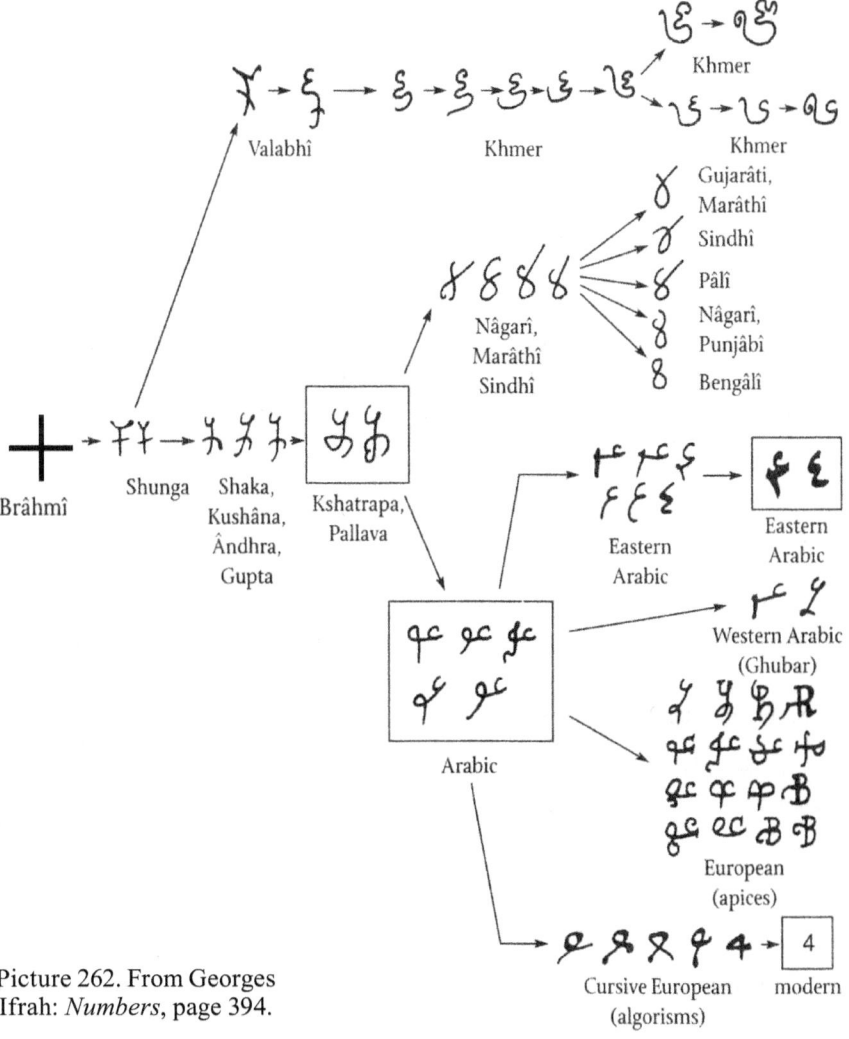

Picture 262. From Georges Ifrah: *Numbers*, page 394.

I couldn't find any hint about how the custom of writing the number four with the sign + reached the Brahmans. The Khmer numerals, while changing a lot above 5, kept the ancient signs out of lines for the numbers 1-4. The Brahman + sign for number 4 – though still ancient – is quite strange.

Comparing the two following rows of signs, we can see that they are almost identical, especially if we looking at them with a graphic artist's eye. The numerals 6, 7, 8, and 9 are of particular note.

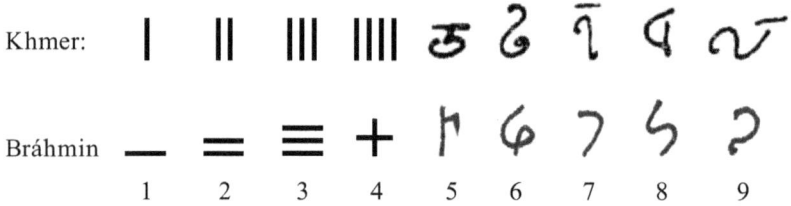

Picture 263. The comparison of the Brahmin and Khmer numerals

Although the sign IIII is much older than the likewise ancient sign +, the above comparison shows clearly that the sign + was first introduced into the Brahmin sign-group by a later sign-exchange. It is worth noting that there are no traces of number-rovás in the introduced Hindu number-presentations; this suggests that it did not reach this continent.

Our numerals, therefore, are not inventions, only the disfigurements of the old clear forms. The Brahmins – as well as other people of the South-East – love curved lines and they let their desires run free. The appearance of today's numerals developed randomly.

We could see in the previous tables that the numerals 1, 2, 3 and 4 among the Arabic and Indian numerals still keep their basic ancient forms, merely the lines —, =, ≡ and + got interwoven when written by hand:

| Brahmin and others | — | = | ≡ | + | |
|---|---|---|---|---|---|
| The joining together of different separate lines when handwritten | I | ⁓ | ⹀ | ✛ | |
| the forms today | 1 | 2 | 3 | 4 | |

Picture 264

There is nothing new in our numerals compared with ancient times. Those only became ornamented. "There is nothing new under the Sun."

# 6. ADORNMENT (CIFRA)

Our Numerals are therefore, not inventions, , they are merely the results of the original, ancient signs' disfigurations. Nobody thought about
the drawings, people only allowed their desire for curved lines to run free,
mainly on the Indian subcontinent. The misdrawing of the signs I  II  III
IIII e.g. resulted in a scribble on the whole sub-continent and hundreds different variants of tortuous figures were created. See more examples for the
numerals 1-9:

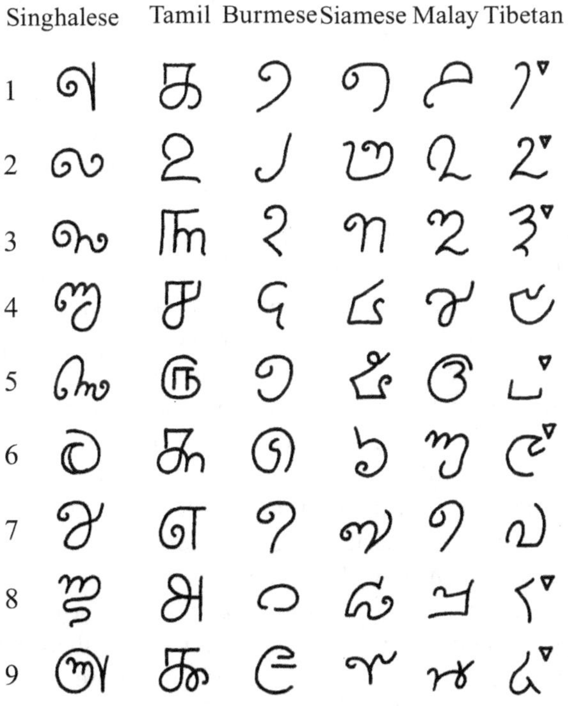

|  | Singhalese | Tamil | Burmese | Siamese | Malay | Tibetan |
|---|---|---|---|---|---|---|
| 1 | ඉ | ௪ | ୨ | ๑ | ۴ | ༡ |
| 2 | ෪ | ௨ | ၂ | ๒ | ۲ | ༢ |
| 3 | ෨ | ௩ | ၃ | ๓ | ۳ | ༣ |
| 4 | ෪ | ௪ | ၄ | ๔ | ۴ | ༤ |
| 5 | ෫ | ௫ | ၅ | ๕ | ๏ | ༥ |
| 6 | ෬ | ௬ | ၆ | ๖ | ๗ | ༦ |
| 7 | ෭ | ௭ | ၇ | ๗ | ๙ | ༧ |
| 8 | ෮ | ௮ | ၈ | ๘ | ๚ | ༨ |
| 9 | ෯ | ௯ | ၉ | ๙ | ๛ | ༩ |

Picture 265. Those look like little pieces of curly threads around a sewing machine. But despite of the disorder, we are able to recognize the decorated variants
- used by us as well – of the archaic signs for 1-3: ━  ═  ≡ .

It happened only that the signs, strongly regulated by mental order, became scribbles without sense. Isn't the peek of stupidity to call these the
result of the "evolution" of number writing?

We are not joking when we call these "from India inherited" numerals "cifra" (scribble). This is indeed the name of those signs.

How did it come about?

Imagine the moment when people accustomed to the ancient dot-line numerals, the majority of humanity, were first confronted with these scribbles from India. They were astounded, seeing, instead of the geometric signs I or III , the scribbles *9* or *(3*. These signs must have looked to them like "cifra" (scribbles).

This must be the reason why the name of these numerals is still "cifra" in many languages. Cifra <czifra> means something over-decorated, over-adorned; too much (superfluous) ornament. Cifra can be: clothing, fur-coat, speech and even a csárda <chaarda> (pub). A dancer can cifráz (add fanciful movements to his or her dance). There is a saying: "it isn't always beautiful, when one is cifra". Earlier, women selling themselves on the corner were called cifra or cafra. Added pieces hanging down from clothes are called cafrang <czafrang>. Cifra is the name of ornamental sounds put between the melodies of a song. "Cifrázza a furujás" <czifraazza a furyaash> (the flautist plays the tunes with variations). The recurring decorations between engravings are also called cifra, because they mean nothing, but are simply decorations.

There are many more examples for "cafrang" <czaphrang>, the deep vowel variant of cifra in the dictionary of Czuczor – Fogarasi, published by the Academy of Science. "Cafrangos" <czaphrangosh> (decorated by cafrang) can be a cloak, horse-blanket, harness, boot, boot-top, but also, figuratively, speech, a greeting, a song and a dance as well. Built from the same word-root (caf) <czaf> is "cafat", a dangling, hanging down piece.

We should know - to understand the deepest meaning of "cifra"- that it is a dialectical variant of "firka". Its root is 'ír' - 'fir' (írkál - firkál = writing – scribbling) and by turning it around cifra <> fircza = e.g.: it is an irkált – firkált = written - scribbled adornment. Numerals were earlier called "cifra" in Hungarian as well.

The real meaning of the word cifra <czifra> is best explained starting from Hungarian because in no other language is there such a large word-cluster built around it as in Hungarian. Therefore, this word must have come from the proto-nostratic language. Thus, we understand its wide distribution:

| | |
|---|---|
| Arabian | Sifr |
| Czech/Slovak | cifra, šifra |
| Danish | ciffer |
| English | cipher |
| German | Ziffer |
| Italian | cifra |
| Norwegian | siffer |
| Persian | Sefr |
| Polish | cyfra |
| Portuguese | cifra |
| Russian | цифра (cifra) |
| Slovenian | cifra |
| Spanish | cifra |
| Swedish | siffra |
| Turkish | sifir (zafir) |
| Urdu | sifer |

# 7. THE FAMILY TREE OF OUR PRESENT NUMERALS

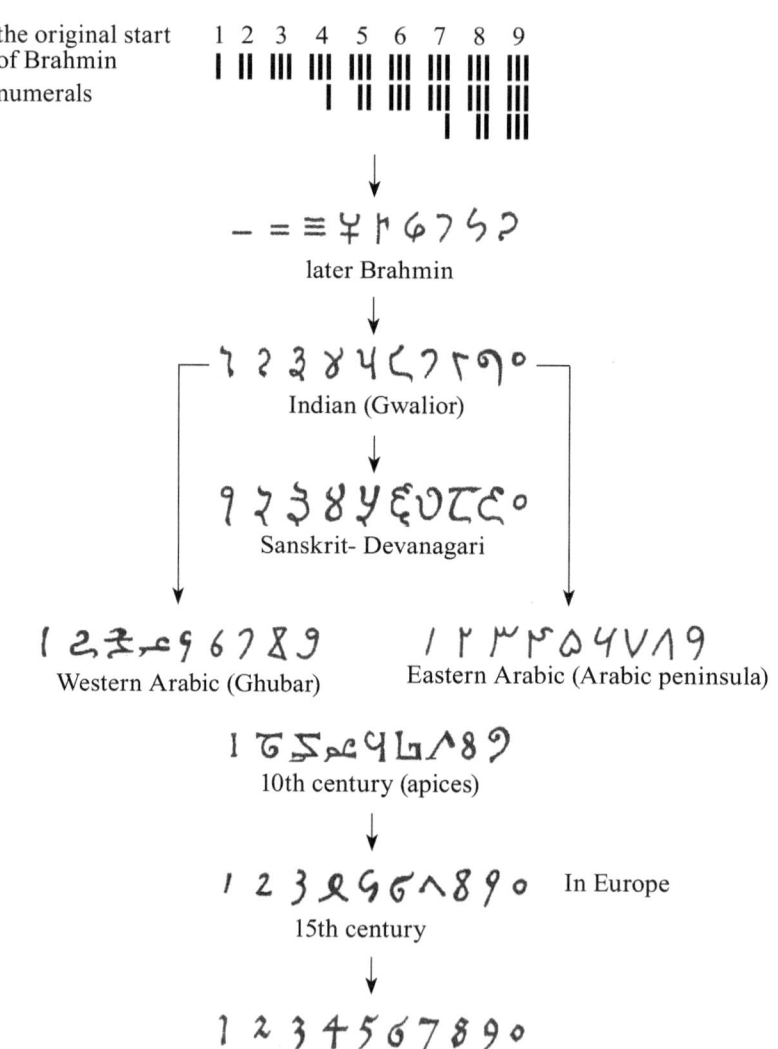

Picture 266. The shaping of our numerals until today. [based on: Filep László-Bereznai Gyula: A számírás története  (The history of number-writing), page 108]

# 8. THE ARABIC NUMERALS IN EUROPA

Well demonstrating the contemporary spiritual level in Western Europe is the fact that the Europeans started to use Arabic and not the Arabs started to write European numerals.

The earliest European script containing Arabic numerals was written 976 in Spain:

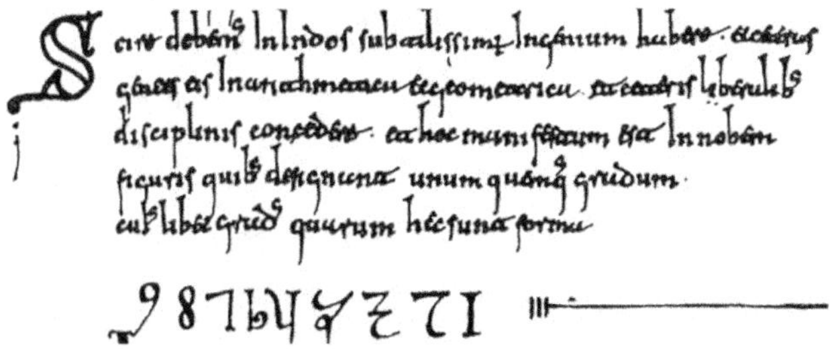

Picture 267. Hindu-Arabic number-signs appeared first in Europe in 976 in the Codex Vigilanus.

It is notable that in the Codex – written in 976 – the text is Latin and is written from left to right, while the numbers are written in the opposite direction. They are "growing" from the right to the left. This has an important cultural-historic significance. To save people from thinking, one often publishes the above numerals in a simply reversed form in scientific as well as in educational works.

EARLY EUROPEAN NUMERALS

Oldest example of our numerals known in any European manuscript. This manuscript was written in Spain in 976

Picture 268. The truth, as in many cases, is just reversed.

We can follow the route of the above signs until the 16th century. The perpendicular columns don't mean any sequence of change. The different variants could have turned up at the same time in different places. The most interesting is the second row from above. Somebody made letters out of the tortuous unidentifiable Arabic signs, but this remained a solitary endeavour:

| | 1 | 2 | 3 | 4 | 5 | 6 | 7 | 8 | 9 |
|---|---|---|---|---|---|---|---|---|---|
| 976 | | | | | | | | | |
| 10th cent. | | | | | | | | | |
| 1077 | | | | | | | | | |
| 11th cent. | | | | | | | | | |
| 11th cent.. | | | | | | | | | |
| 11th cent. | | | | | | | | | |
| 11th cent. | | | | | | | | | |
| 12th cent. | | | | | | | | | |
| 12th cent. | | | | | | | | | |
| 12th cent. | | | | | | | | | |
| 12th cent. | | | | | | | | | |
| 12th cent. | | | | | | | | | |
| 12th cent. | | | | | | | | | |
| ~12th cent. | | | | | | | | | |
| ~12th cent. | | | | | | | | | |
| ? | | | | | | | | | |
| ? | | | | | | | | | |
| ? | | | | | | | | | |
| 15th cent. | | | | | | | | | |
| 15th cent. | | | | | | | | | |
| 16th cent. | | | | | | | | | |

Picture 269. The independent form-changes of the Arabic numerals in Europe. The variants born during 6 centuries. (based on David Eugen Smith: History of Mathematics, page 76. Dover publications, Inc.,New York, 1958)

# 9. OUR NUMERALS' FINAL FORM

The final form of our numerals was designed by Ajtósi[60] Albrecht Dürer, born in Nürnberg in 1471, who was one of the best known representatives of the German Renaissance. The German chalcography reached its peak with his artistic works. His father moved from Hungary (the village Ajtós) to Nürnberg.

Picture 270. The self-portrait of Ajtósi Albrecht Dürer

The Hindu-Arabic numerals, redrafted by Dürer

123456789

Due to their allure, these numeral-forms were fast spreading and being used until today.

---

60 The Hungarian word Ajtós means Türer > Dürer in German ("from the village Ajtós").

# 10. THE EARLIEST ARABIC NUMERALS IN HUNGARY BEFORE DÜRER

The earliest example of the use of Arabic numerals is from 1407. The sign for zero is mysterious):

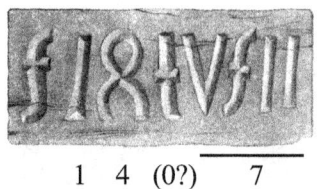

1    4   (0?)    7

Picture 271. A number on the wall of the church in Magyarvalkó –Kalotaszeg (Transylvania) (Filep László – Bereznai Gyula: The history of number-writing, page 140.)

It is interesting that the first half (14) was written with Arabic numerals, the 7 with Latin numerals. The following number, the date of the year, is written above the arms of the nation in the Mathias church in Budapest.

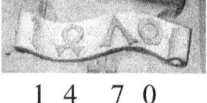

1  4    7  0

Picture 272.
The arms of the nation and the Arabic number written above it

The year 1499 appears on the gold coin of the King Dobzse László (Ulászló the 2th). The number 4 is still written the Arabic way:

1 4 9 9

Picture 273. The nickname of Ulászló was "Dobže, because he said "OK" to everything. The first appearances of coins with Arabic numerals on them: Switzerland – 1424, Austria – 1484, France – 1485, Germany – 1489, England – 1551, Russia – 1654.

The numeral 4 in the year on the contemporary engraving picturing the execution of Dózsa György is definitely Arabic.

1 5 1 4

Picture 274. The sign for 4 is Arabic here as well as on the coin of Ulászló.

# 11. THE COMEDY OF ERRORS

Without the right information, even the smartest minds can make big mistakes. This happened for example to Pierre-Simon Laplace (1749-1827) French mathematician, astrologer and physicist. He knew merely about the - in his time already exclusively used - numerals that those came from the Hindus by mediation of the Arabs. He knew nothing however about the origin of such signs. His widely spread opinion about these numerals became therefore a commonly repeated delusion. His delusion spread through the whole of Europe and distorted the history of number-writing in people's minds to the present day.

Picture 275. Laplace

Let's see the opinion of Laplace:

*"It is India that gave us the ingenious method of expressing all numbers by means of ten symbols, each symbol receiving a value of position as well as an absolute value; a profound and important idea which appears so simple to us now that we ignore its true merit. But its very simplicity and the great ease which it has lent to computations put our arithmetic in the first rank of useful inventions; and we shall appreciate the grandeur of the achievement the more when we remember that it escaped the genius of Archimedes and Apollonius, two of the greatest men produced by antiquity."*

Having read this book so far, we have already learned much about the history of Hindu number-writing– in contradiction to Laplace's totally false belief about it. With our knowledge, it is easy to recognize Laplace's blunders, the indefensibility of his implications and finally, how unfounded was his enthusiasm.

Let's look at his statements:

1) *"It is India that gave us the ingenious method of expressing all numbers"* We agree that these numerals originated from India. We could easily see however, how unsystematically their misdrawing of numerals happened. Thus, their origin is undisputable.

2) *"...expressing all numbers by means of ten symbols ...."* But what is beautiful in this? In ancient times, the system used by everybody – which is still used in China – needed only two signs, a horizontal and a perpendicular line or a dot and line to write every number. Which should make us marvel more?

3) *"...,each symbol receiving a value of position as well as an absolute value ..."* This has happened for at least for 30,000 years. Since number-writing first existed, it verifiably has incorporated a positional value. After so many examples in this book, let's look at two ancient examples and compare them with our present number-writing:

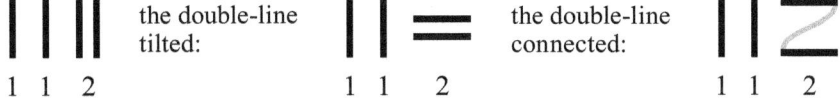

*Isn't this number-writing with position value?*
*It is 112 in both ways, except the numerals*
*are dots or lines as we see below.*

|  |  |  |  |
|--|--|--|--|
| ▌ ▌ ▐▌ | the double-line tilted: | ▌ ▌ ▬ | the double-line connected: |

1  1   2                    1  1   2                    1  1   2

Picture 276. What is new in this? It is the same with the present numerals, we don't even have to write it there.

This has happened for several tens of thousands of years. The start of the position value's use has been lost in the foggy past.

4) *"...appreciate the grandeur of the achievement the more when we remember that it escaped the genius of Archimedes and Apollonius, two of the greatest men produced by antiquity ...."*

It wasn't easy to invent the system we inherited from India? Well, nobody invented anything there. People misdrew and decorated the ancient originally straight lines. Thus, the signs arrived in Europe with random forms. Number-writing using position value is however immeasurably old. The signs merely became distorted by the Indians, and they did not even notice this because of the extreme slowness of the process. But can we speak of "development" if one draws a straight line in a crooked form?

Why should Apollonius and Archimedes have even dreamed about a worse, a not-as-easy to use system than they already had? About a system, in which arithmetic procedures need more time and work? We will see more about this matter in the next chapter IIΛ.

*5) "But its very simplicity and the great ease which it has lent to computations put our arithmetic in the first rank of useful inventions..."* This sounds quite impressive, but it has nothing to do with the truth. The new signs rendered arithmetic computations not far easier, but much more complicated. Practice should decide about the rightness of my statement. I ask the reader to take a piece of paper and a pen to solve the multiplication (or just half of it) below. Please don't do it with a calculator.

$$
\begin{array}{r}
\underline{1.239.458 \times 64.918.473} \\
7436748 \\
4957832 \\
11155122 \\
1239458 \\
9915664 \\
4957832 \\
8676206 \\
\underline{3718374} \\
\hline
80463720707634
\end{array}
$$

Now I ask the dear reader: having done this calculation, did this multiplication not act on you like 30 push-ups done as a punishment? (Would you like to try also the division of these numbers?)

Contrary to this, the same procedure would take only a few seconds in the old dot-line numeral system. You can learn it how to do it in the next chapter. In the following, I just give you a small example of addition in the dot-line numeral system. The ancient number-writing system is based on this. In the case of 6+6 (2x6), you only have **to push the signs of the two numbers together**, like two buttons and two match-sticks, because one can assemble the signs even from those.

|   6   |   6   |   12   |

Well, for dealing with larger numbers you need a handful of seeds or pebbles and a few lines (drawn perhaps even in sand) to mark the position values

# 12. ABOUT THE SPREADING OF THE ARABIC SIGNS

The spread of the Arabic numerals and fall into disuse of the old system was a slow process in Europe. The problems were even not the signs, but the change of the arithmetic processes. One couldn't do the accustomed processes with the new signs. However, old, deeply rooted customs are not easy to change. As well, the new method was more difficult than the old one as seen on the previous page.

Arabic numeral signs, however, play a secondary role in this history, because they are scribbles without any meanings. Consequently, any other scribbles could be used in their place. The mere sight of the signs of the ancient dot-line numerals, on the other hand, provides the number's value. It is not enough just to see the numbers written using Arabic signs, they have to be read, which an additional step is. Look at the following picture:

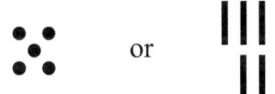 or

5

*Just by glancing at it, the value pops up in our mind immediately: the sign shows its own value*

*This sign means nothing: for this one our mind is forced to start a follow up puzzle-solving procedure.*

Picture 277.

The introduction of Arabic signs - and its associated number-writing system– was for long time undecided, but in the end it was victorious.

Its way wasn't easy. Luckily, due to early printing-works we can follow on several prints the duel between the old and the Arabic methods.

In the following pictures, the Arabic numerals on the table represented the new system, but lines, pebbles or little discs on the table, the old method of calculation, are also present. The latest were the requisites of the old calculating tool, the abacus. We will introduce this in detail in the next chapter. For now, just look at the old pictures.

We can see the old calculating method in the pictures below (lines and little discs). Nothing seems to endanger its use yet:

Picture 278. At an exchange table in 1514

Picture 279. In the 1500s, an older master teaches a young apprentice about the use of the counting table.

Picture 280. A young person studies the use of the counting table in ~1500

Pictur 281. A man sitting at the counting table

Picture 282. People are waiting probably at a money-exchange.

Picture 283. At the money-exchange table. (Woodcut, Jörg Breu, Augsburg, 1531.

Picture 284. A counting table from a Lutheran German Catechism (1530)

Picture 285. Now, the Arabic number-writing is becoming quite widespread. This woodcut is a contemporary espousal of the Arabic number-writing. We see below the competition of the old (right) and the new (left) calculation methods, but the goddess in the middle favours the new one, since her garment is covered with Arabic numerals, in order to persuade people better. Look at the desperate face of the man behind the abacus.

Picture 286. There is a turbulent debate about the advantages and inconveniences of the old and new calculating methods.

Picture 287. Comparison of calculation with Arabic signs and pebbles. The man on the left has a pen in his hands, but the one on the right loads little discs onto the drawn lines. Look at how deeply the man on the right has become immersed in his job.

Picture 288. The lines and pebbles, needed for the old calculation method, are already missing from the table, only Arabic numerals are presented. The fight is finished. The victory belonged to the Indian-Arabic method. In many places in the world however the duel has still not been decided. The old method is alive and blooming in many places.

We could see one interesting aspect in all pictures:

The Arabic numerals are written and the procedures are conducted also in written form. The pictures present it however, scattered on a calculating table. Obviously, they wanted to transfer the validity of the calculating table to the procedures using the Arabic numerals.

Appendix:

I want to support my previously written statements by few quotations out of the book *"Von Pythagoras bis Hilbert"* by Egmont Colerus.[61]

First, a couple of short quotations:

*"Pythagoras and his contemporaries often counted with pebbles."*

*"In an age when we are flying, the Chinese use counting-tables with balls on wires."*

Edmont Colerus describes with much vehemence the fight between the users of the abacus and those of the algorithms:

*"Before we deal with this problem, let's talk about the fight of the users of abacus with the users of algorithms which introduced mathematics into the Western Hemisphere. The fight was about the two methods, calculating on lines or with pens on paper, as people were told later. In an extreme case, the abacus, the line or calculating desk is still present as a kind of "calculator", on which children can learn the first steps of calculations by moving balls on wires. The abacus is therefore a table on which the columns of the "tens'-power" are separated by wires. A zero is not necessary here as, for example in the case of 750 or 3009, one or two wires simply won't be occupied. The calculation happens with signs. Originally, that many signs were put into the respective column, as the co-factor of the ten showed in that column. One puts - in case of the above example - seven into the column of hundreds, five into the column of tens and nothing into the column of the singulars. in the case of 3009, three will go to the thousands and nine to the singulars, but the columns of the hundreds and tens stay empty. This method could become puzzling when used for very large numbers. For those, they started to use signs marked by one of the numbers of 1 – 9 for their different values. The abacus came closer with this to the algorithms, and this innovation became quite striking in the third level of use, when not the usual signs, but numbers were written into the columns. We won't go into further details but state here only that both schools of thought had excellent experts as followers. A famous supporter of the abacus was e.g. Gerbert, the later Pope Sylvester II.*

*The followers of the algorithms were victorious. The debates of the two parties (if we may describe them thus) brought many questions to the surface and gave the Western mathematicians much flexibility and skill, which*

61 Egmont Colerus: *"From Pythagoras to Hilbert"* (original: Von Pythagoras bis Hilbert)

*was not lost after the debate was finished. Scientific debates, along with the major questions, brought up many different programmes, covering even the extraction of roots. There was no contempt for arithmetic in the Middle Ages in Europe as there was in Greece. Already at the beginning of the scientific debates arithmetic and geometry were treated equally, following Arabic customs. In some ways, they gave arithmetic a logical priority."*

Probably due to the absence of the necessary knowledge, the basic fact of every statement about the history of calculation is that the existence of an algorithm (application-rules) didn't exist earlier. Man propagated it as a world-novelty. However this is not accurate. Calculations with written numerals had always followed an algorithm and what we are using today is just one of several possibilities. Furthermore, it is not the fastest method - it was a bad choice. Let's look again at page 234. Turning pages forward, a different algorithm (on page 279) will give us the results with less work and time. There are other algorithms with the capacity for faster calculation."

# PART IIΛ

## THE CALCULATOR

## 1. THE ANCIENT CALCULATING METHOD

There is an ancient calculation tool or method, which has accompanied the ancient dot-line number system for many thousands of years and uses only dots and lines for its calculations. How does it work?

As we have already learned, dot-line number-writing has been based from the beginning on a number-writing using local, or position, values. Giving a value of ten to the line and a value of one to the dot, then the reading of the following number will be 14 as we have seen several times in this book:

Here therefore, the line and only the line stands for the position value, in this case for 10. No change will happen if the dots are put on the line:

As a drawing, it may look a little confused, but as a theorem, it is well suited for putting four pebbles, shells, apricot-stones or other small objects on a line just drawn in the sand - by the rules of position value. Therefore, this arrangement makes it possible to write numbers everywhere with the help of only a few small objects, but without writing tools. See the number 14 above, "written" with kernels:

Picture 289. Kernels on a drawn line: number 14

The line could even be a little stick and the dots pierced beads being threaded onto the stick. The same principle works also using the quipu (knot-writing) system and even if one counts one's beads (but here one only counts the sequence).

We explained here the position value in the decimal system. The question is how to create the other position values? The solution emerges in every dot-line system automatically, but one of its variants, the knot-writing (quipu) is probably the best system to use to understand the principle of the position value's representation.

For this purpose, we repeat again the example taken from the Incas, in which one could see the number 3,643 in knots. The local values follow each other

| 3 | 6 | 4 | 3 |

Picture 290. The local values follow each other from the left in decreasing amounts.

Let's put the local values below each other: 3,643:

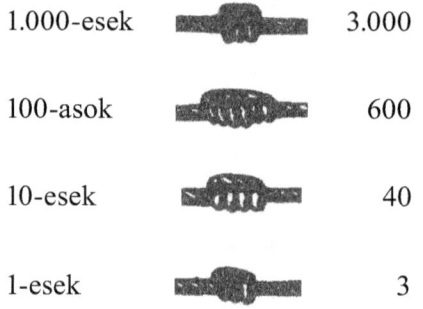

| 1.000-esek | | 3.000 |
| 100-asok | | 600 |
| 10-esek | | 40 |
| 1-esek | | 3 |

Picture 291. We don't even have to add the local values, because we read the whole number anyway as: three-thousand – six-hundred – forty – three

After the knots are tied on the cord, they stay there for good. They can't be moved. This is the point of knot-writing and that makes it perfect for storing numbers, because the knots are fixed on the given spots.

Otherwise, in other counting methods all objects, beads and kernels – representing dots – having been put on a line or pierced and stringed on a cord or stick, can freely be moved. This makes it possible to increase or decrease the numbers belonging to one or other local values. Therefore, one can change the numbers of several local values at once and this again makes it possible to do any mathematical calculation.

Let's mark the local values (in reality, the round numbers) with parallel horizontally drawn lines and put as many peach-kernels on them as to make up the number 3,252:

3 db. 1.000

2 db. 100

5 db. 10

2 db. 1

We just have to read them from the top down: 3 thousand – 2 hundred – fifty – 2. Now add to this number 1,012. It is simple: we take one kernel from our hand and put it on to the thousands, nothing to the hundreds, one to the tens and two to the singles. A few movements and the calculation is finished. See the result:

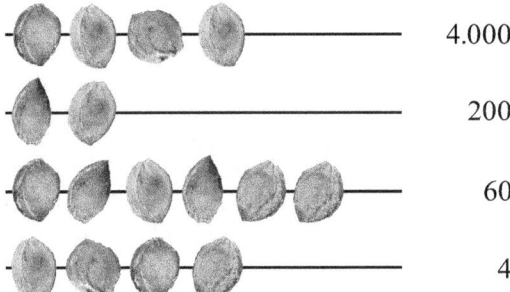

4.000

200

60

4

Now, let's divide the above number by two. We just have to take half of the peach-kernels from every line:

2.000

100

30

2

In the following let's subtract 1,162 from 4,264. This is the form of 4,264:

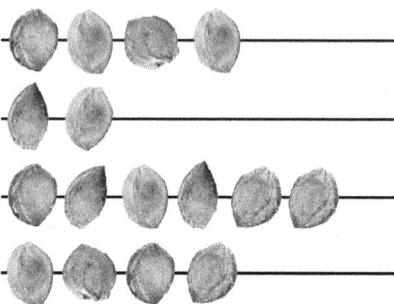

The subtraction of 1,162 means only to take off the table the given numbers of kernels starting from the top. ( 1, 1, 6, 2). The result:

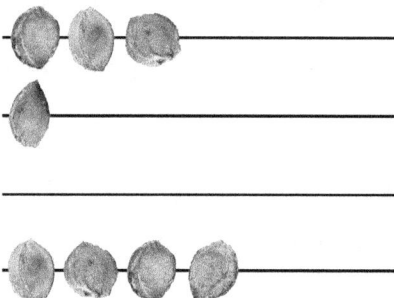

Our readers can clearly see the result at a glance: 3,104. Now we subtract 2,104 and now we see the remainder: 1,000.

Here we can see again that marking zero is here absolutely unnecessary. The empty line of the local value makes it obvious.

Doing mathematical calculations this way is therefore simple and fast. While subtracting 16,199 from 84,727 can be done in a few seconds, the same calculation done with Arabic numbers takes twenty times longer. Please do this subtraction using the modern method and write the result after the ? mark:

$$84727$$
$$-16199$$
$$?:$$

This way, readers can experience for themselves that subtraction compared to the old method seems to be more complicated. The same is true with all other calculations.

The counting method used in dot-line number-writing therefore doesn't need continuous mental arithmetic. It connects -without a detour – the sight and the value of the number. We don't have to do unnecessary mental work while calculating. Nothing will disturb the mental work we do the calculations for.

We will only need our brains in one particular case: if we add so many peach-kernels to the line of the singulars that the result is 10 or more, then the step to the line of the local value 10 is needed. We see 4 pebbles: below.

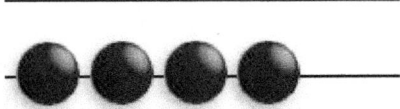

Adding 8 to this number, we have to step to the next value, to the line of the 10. Putting one pebble on this line, there remain 2 more for the singulars.

$4 + 8 = 12.$

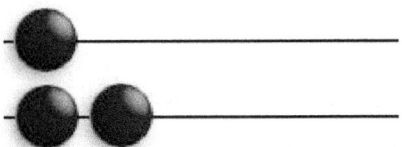

Thinking is needed only for such changes of local values, but even then, one doesn't have to rearrange the whole table, only two lines were touched in this calculation. This is the same procedure we use when we have to give 2 forints to somebody, but have only a 10 forint bill. We change this bill into 10 one forint coins. Now we can pay 2 and retain 8 forints.

Such a calculating tool couldn't survive through the generations, since the pebbles or kernels were put mostly on lines drawn for that occasion, possibly even in sand. There are reports about merchants in Africa who just sit in the sand, draw some lines with their finger in the sand and start calculating with a few pebbles. We know that in ancient Egypt people drew lines in the sand and calculated on them using small stones. Herodotus reported that the Egyptians moved the stones from right to left, but the Greeks from left to right while calculating. One can only find objective proof if the lines of the round numbers (lines of the local values) were engraved into stone and indeed there are some such engravings. We can see below one of these lucky findings. I put the pebbles on it, to see what they looked like during a calculation:

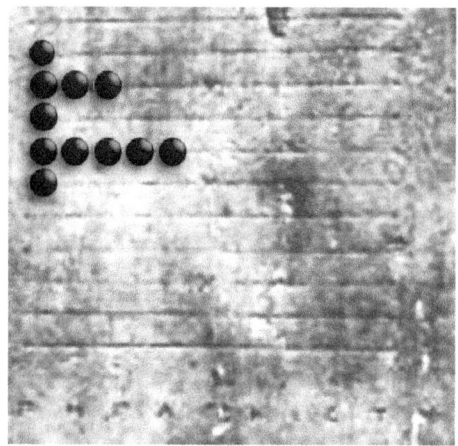

Picture 292. This calculating table is around 2500 years old, found in Salamis (Cyprus). The lines of the local values are engraved. The numbers were evidently little pebbles.

For how long have people calculated by this method?

Presumable since dot-line number-writing has existed. I think, furthermore, that games using lines and pebbles must have the same age, since calculating with pebbles on lines is in reality a game regulated by rules. However, several different rules can be created for pebbles and lines. The histories of games such as "Go", "Backgammon", "Nine Men's Morris" and "Checkers" could go back 10,000 years as well. Their rules are connected to the dot-line calculating system just as the hand is connected with the foot and the eye with the mind.

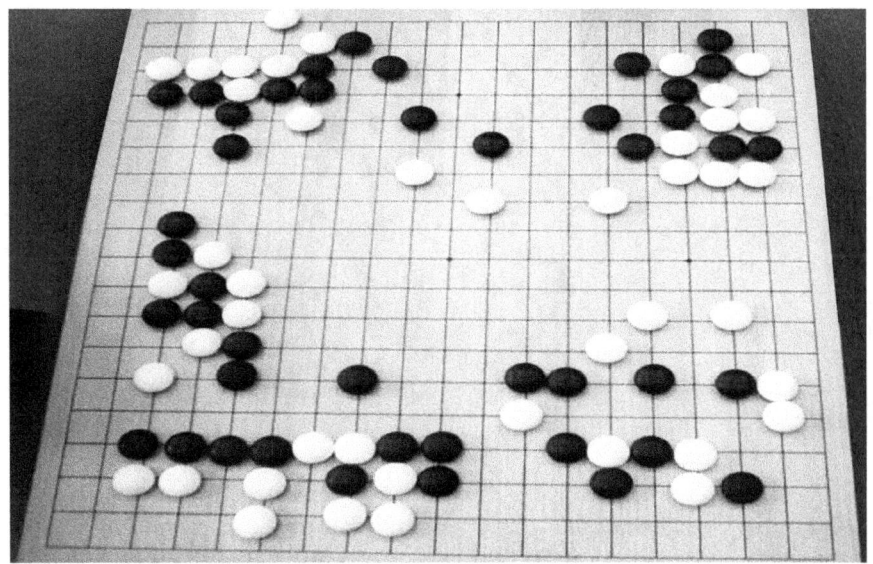

Picture 293. The Go board during a game. (Wikipedia)

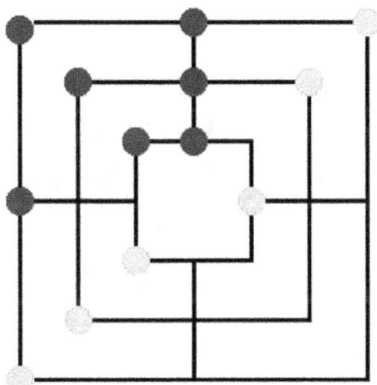

Picture 294 The well-known game Mill "Nine Men's Morris"

There are countless known pebble-line-net games.

People have been playing games involving "putting pebbles on lines, onto the intersections of lines or into the network of cavities" for ten thousand years or more. (Humans have been spiritual creatures as long as they have been humans!)

We don't even need the drawn grid of lines. It is enough to arrange the cavities regularly guided by a network.

We can see games made this way in the next pictures:

Picture 295. Two men are playing a game in Ghana

This is a 5x6 board. The cavities could be the crossings of lines as well and then the board is a quadratic line-net. We can see that one of the players puts clay balls and then the other puts little sticks into the cavities. They could even be white and black pebbles. I drew below a diagram of the actual surface of the game at the time it was photographed:

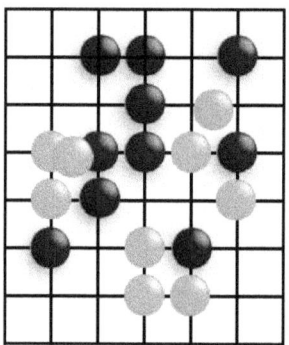

Picture 296. It is evident that the two men in Ghana are playing a variant of the game "Go" for a small board. They use sticks instead of black pebbles.
They may not have found pebbles.

I can't figure out the rules from this picture. The goal could be that one should prevent the other from reaching the side in front of him. He could probably hit and take the pieces of his opponent if they are playing by the rules of Go.

Let's see the relations: this is a 5x6 board, while the boards in the East have mostly 9x9, 13x13 and 19x19 intersections on them.

The board can have an oblong shape, as we can see from the one on the roof of a temple of Thebe.

Picture 297. One can also see such a game-board in Luxor at the temple's entrance. We see pictures of such games painted on the walls of several Egyptian tombs, but they can be found all over Africa and Asia. Game-boards covered with gold have even been found.

This is a game with several pebbles in the holes. (at the cross-points):

Picture 298. This game is probably from Sierra leone

Picture 299 A game somewhere in Africa. We see the row of cavities with pebbles in them

The board below - decorated by beautiful inlay work – is made for different games, but contains three cone-shaped pebbles as well. I present it here to let you see that we shouldn't be afraid thinking about very long time periods.

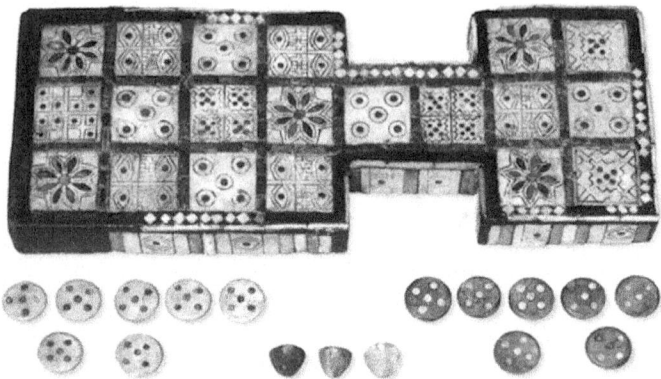

Picture 300. Sumerian table-board. It was found in the oldest layers of the city of UR.

This game and its board is at least 4-5,000 years old. Is it possible that a game with a similar system could have existed 5,000 years ago? Did people have a similarly decorated board for it? What could prevent us from saying that people played similar games 6,000 years ago, or even earlier?

Looking for cultures in the past, we never find beginnings. Every culture of the past – if we find traces of it – was already developed, whole; just popped up? Something working with such simple tools as pebbles however, can certainly be projected back to the dawn of time.

Picture 301. Women play a game with pebbles. There are wooden cavities between them.

# 2. THE STRINGING OF THE BEADS

We can take perforated beads instead of pebbles and kernels and the lines of the local values can be sticks or wires. Ten pierced beads should be stringed on every wire. Parents with children have certainly seen such things. Most toy shops will sell them.

Picture 302 The reader might have such a calculator at home, if she or he has children. Unfortunately, instructions are mostly not given with it. Parents see it in the shop, but from just looking at the mysterious oddity, they won't be able to tell their children how to use it, and this is a big intellectual loss.

Just a few decades ago, a large example of this ancient calculator stood in many classrooms of elementary schools in Hungary.

It is still a beloved calculator around the World.

You can see in the next picture the one widely used in Russia a calculator with beads and wires. It is called 's-choty'. It is placed alongside the newest cash registers, but the cashier uses it for the calculation and only the end-sum will be put into the cash register. The result can be worked out much faster on the 's-choty' than on the cash-register. They save time by doing so. The cash-register will often be checked by using a 's-choty'. Tourists often tell stories about the fantastic speed the cashier can reach with this calculator.

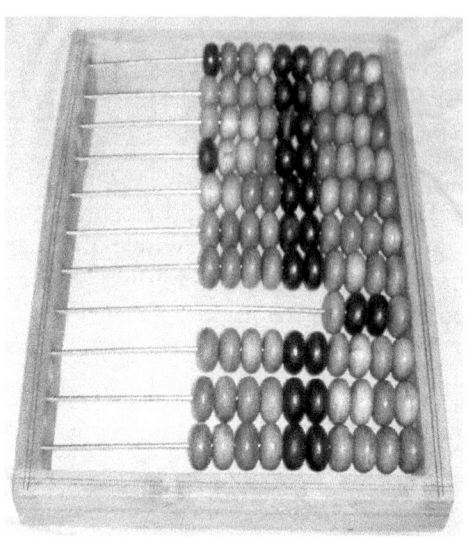

Picture 303 A 's-choty' (счёты). The fourth row from the bottom, with fewer beads, is for counting kopejkas.

We live in the Age of entanglements, many people say. Introducing Arabic numerals added much to this stage. Let's compare the simple calculator above with the one below.

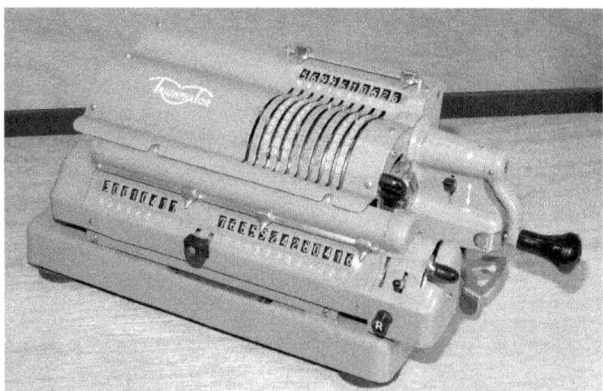

Picture 304. A victory of technical revolution, the calculator with a crank-arm.

This was a very complicated iron construction made out of thousand parts and weighing 10 kg. Its speed was many times less than that of the abacus. Its winder had to be turned like a crushing mill. This machine was

– compared to the abacus – the height of absurdity. It was the over-complication of a very simple but highly ingenious and faster working system and object.

Well, this calculating system, which has survived several tens of thousands of years, is moving on to new territories:

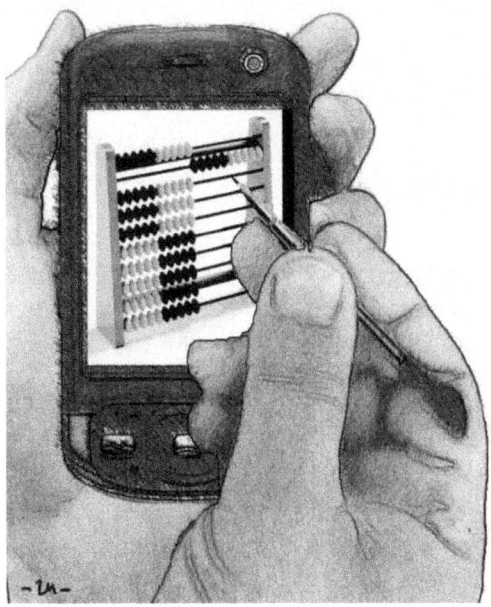

Picture 305. The abacus on a mobile phone.

Halász Géza, the author of this graphic, wrote the following comment under the picture: "I planned this graphic as a joke few years ago, but it has now become a downloadable mobile phone application."

# 3. PRACTICAL ADVICE FOR CALCULATIONS WITH THE ABACUS.

Any calculator, we buy, should have, half of the beads in dark and the other half in a light colour on the same wire. Having all the beads the same colour would make identifying numbers over five more difficult. Having one or two white beads alongside five dark ones, we will know immediately if we are looking at six or seven.

Calculators made for children often have 10 rows of beads. These mean as many local values, or round numbers and are working with up to 1 milliard. Now we only use the upper four rows, the local values of 1,000, 100, 10 and one. With these we are able to calculate with numbers from 0 to 9,999. For larger numbers, we may start with the 100,000 local value and use perhaps 6 rows for our calculations.

First, tilt the frame to the right and all the beads will slide firmly to the right side. Thus, we will count from left to right as we do in writing as well. Had we pushed all the beads to the left, then the calculation would now run unusually from the right as it does when writing with Hungarian rovás. Both directions are right. As mentioned earlier, Herodotus wrote that the Egyptians move the pebbles from right to left when calculating, but the Greeks do it in the reverse direction.

Well, letting all the beads slide to the right will be our zero setting.

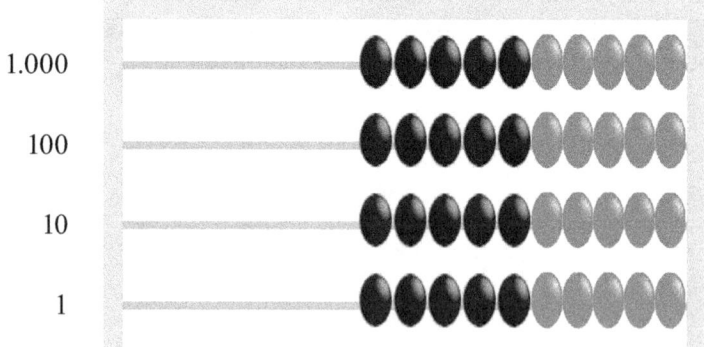

Picture 306 . The zero setting before starting the calculations

Let's see some mathematical procedures. The first number will be 2,001:

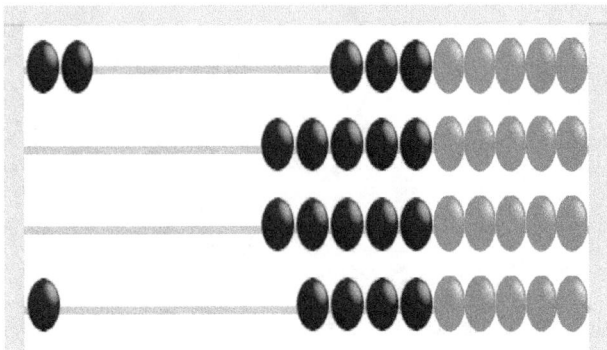

Add to this 483: this many beads, according to the local values, should be pushed to the left.:

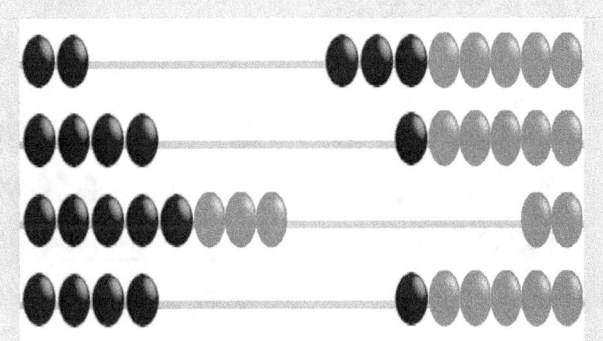

The result of 2,484 should now be divided by 2 (most simple), that is, we have to take away half of the beads in every row. The result is on the left: 1,242:

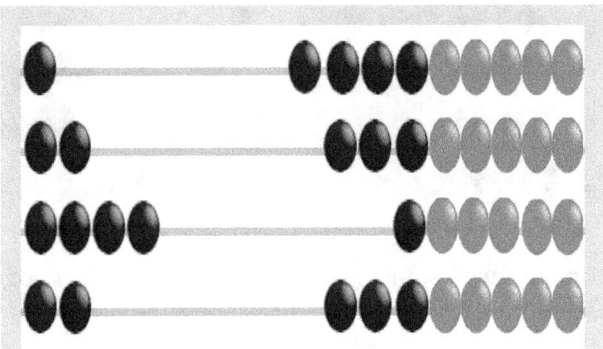

Let's now look at an example requiring a change. We start with 1,007:

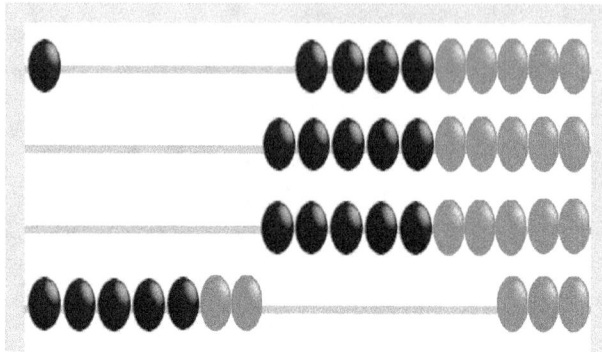

Add 8 to this. This we can do in three steps:

First add 3 to it. Now, there are 10 in the first row and 5 are still waiting.

We have to change 10 up into the row of tens. Here we push one pebble to the left

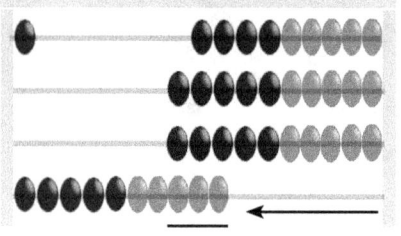

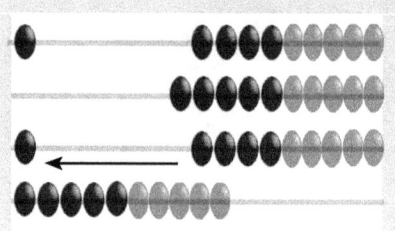

....let's zero the singles' line

...and add to it the 5 still remaining

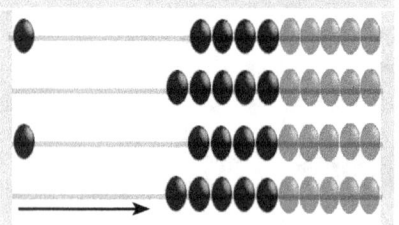

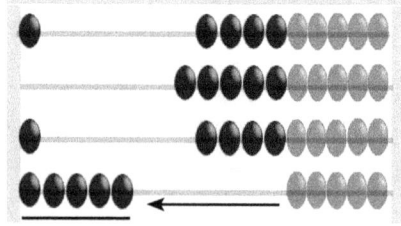

These changes need some practice to become "professional". The first attempts take more time, but it can be learned quite easily and we could become as fast as a Russian cashier.

We can calculate with decimal numbers as well and for this we continue downwards the rows of round numbers: 1/10, 1/100 and so forth. However, for easier reading of the results, we would do better to mark these extensions on the frame:

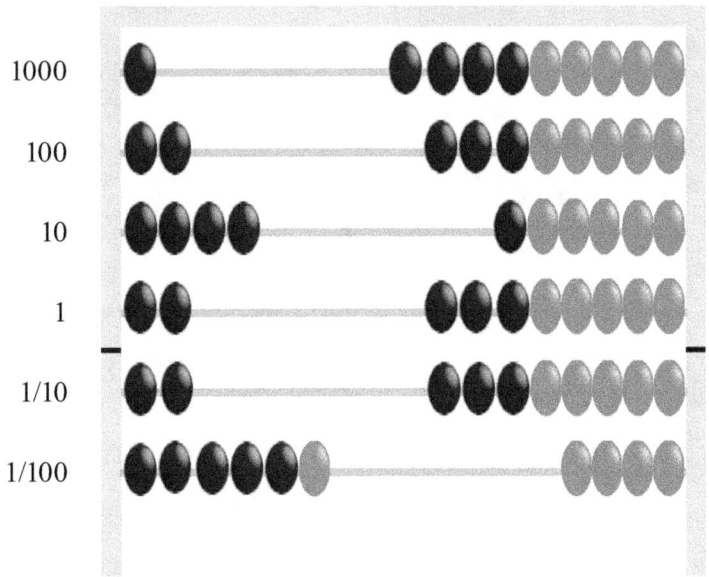

The reading of the number above is: 1242.26.

Divisions often need the change as well.
Let's divide 3 by 2. For this we use the line for 1/10 too:

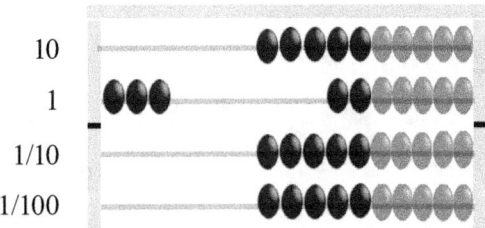

Only paired numbers can be divided into halves, so we therefore change one of the singles to 10 decimals, just as we can change a 10 forint bill to 10 one forint coins. The actions are such that we pull back one of the singles to the right and push 10 beads on the 1/10 line to the left.

(We pushed them into function). No value has been changed. The reading is:

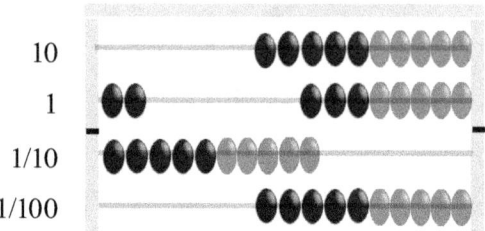

There are paired numbers standing in both active lines. We just have to halve the pebbles in those lines.

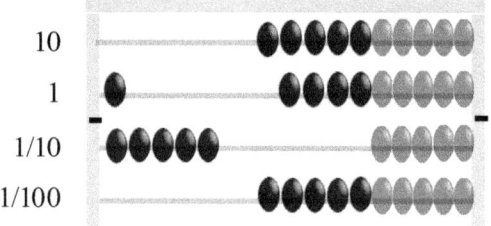

Our result is 1.5, as it should be.

Now let's divide 4 by 3.

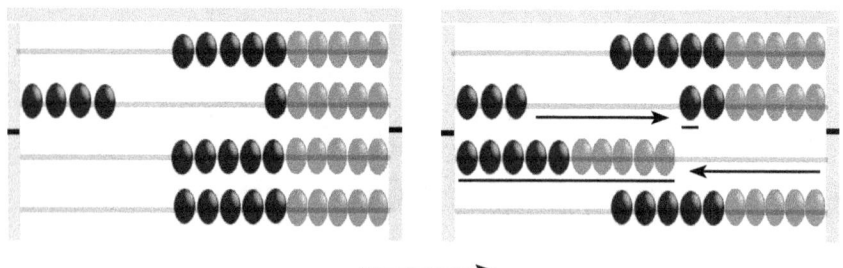

Four can't be divided by 3, therefore we change one pebble of the four: one pebble of the singles must be pulled back to the right and all 10 pebbles on the 1/10 line pushed to the left, "putting them on duty".

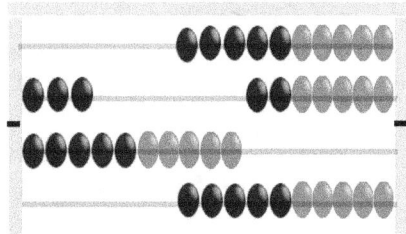

The 10 pebbles on the 1/10 line are however, not divisible by 3,
thus we must change one of them too.

Now, one of the even numbers in the 1/10 line will be changed into ten
pebbles on the 1/100 line:

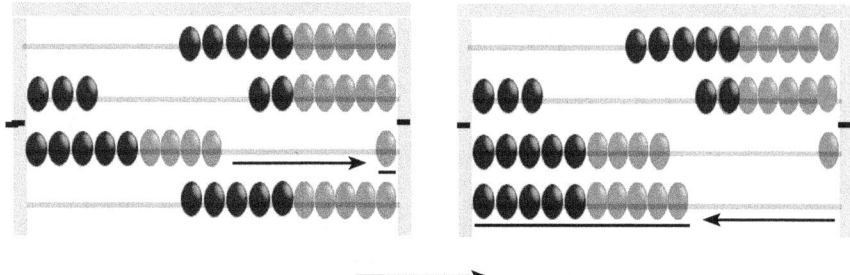

We must now change one pebble in the 1/100 line into 10 pebbles on the
1/1000 line, and then one of those pebbles into 10 on the 1/10,000 line and
so forth by going down the lines.

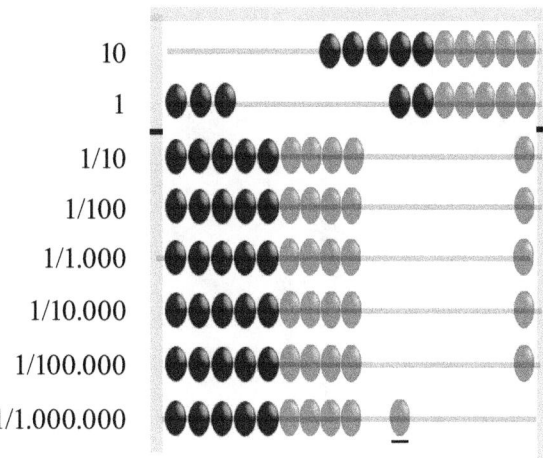

We see now that the sequence can be followed endlessly. Let's stop at the 1/1,000,000 line and take back one third of the pebbles on all lines. The result will be:

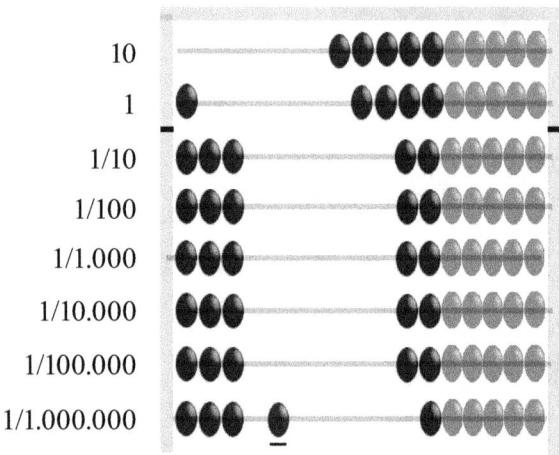

Read the result: 1.333333

Adding and subtracting can easily be learned by children. Division and multiplication need the learning of some special methods and more practice is necessary.

This calculator has a variant which is much faster to handle. We will introduce it in the next part.

# 4. THE BISECTING CALCULATOR

Having reached this far in the book, we have been able to learn about two kinds of number-rows, both based on the decimal system. The first, on which the abacus, introduced in the last chapter, is based, uses round numbers and only these numbers have separate signs if written down:

1      10      100      1.000

In many places, however, the half values of these round numbers have been squeezed between them and have even received their own signs. The Latin row of number-signs is such a system.

1  **5**  10  **50**  100  **500**  1.000

I  V  X  L  C  D  M

It looks as though there were two different ancient cultures. One counted only with round numbers and the other used the half values as well.

This assumption will be further supported by the fact that the same duality exists in calculators with beads and wires. We have mainly seen calculators working with round numbers in the previous presentations. These were used earlier in Hungarian schools, and all over Russia, but people in Ancient Egypt as well as in Mesopotamia used round numbers while calculating with pebbles.

The pebble-calculators had to be somewhat modified for use with numeral systems involving half values.

Looking more closely at the counting tables made for the use of pebbles, we see pebbles even between the lines.

Picture 307. The pebble between the lines marks the local half values

A pebble put between the two lines always stands for five times the value of the lower line. (the number below is 555)

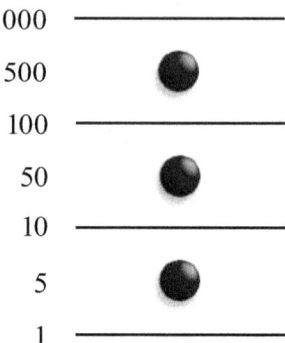

This means that we can only put a maximum of 4 pebbles on the line of the round numbers, because the fifth already belongs to the space between the lines to mark the half value.

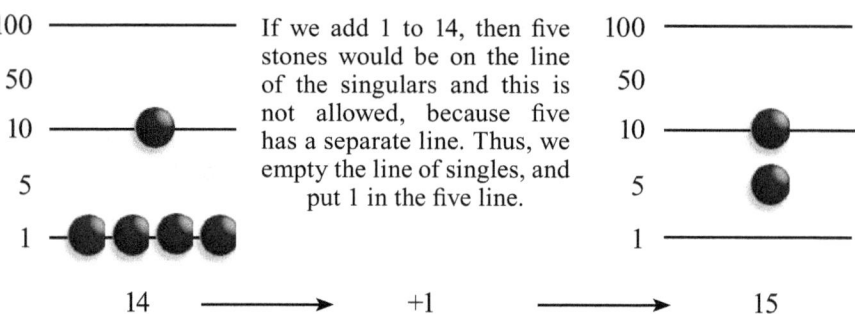

If we add 1 to 14, then five stones would be on the line of the singulars and this is not allowed, because five has a separate line. Thus, we empty the line of singles, and put 1 in the five line.

14 ⟶ +1 ⟶ 15

This rule is valid for all local values. One more example:

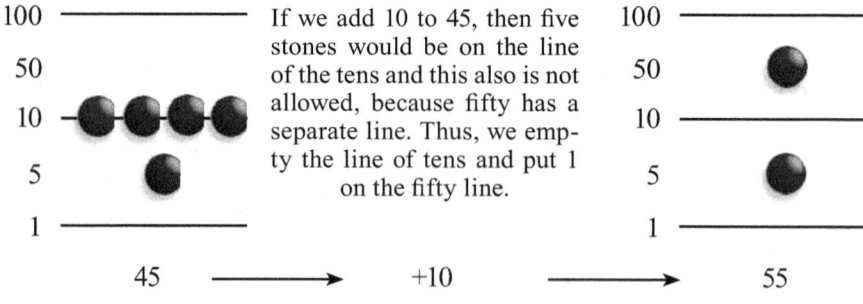

If we add 10 to 45, then five stones would be on the line of the tens and this also is not allowed, because fifty has a separate line. Thus, we empty the line of tens and put 1 on the fifty line.

45 ⟶ +10 ⟶ 55

If we add 3 to 5, then 3 pebbles have to be placed on the line below 5.

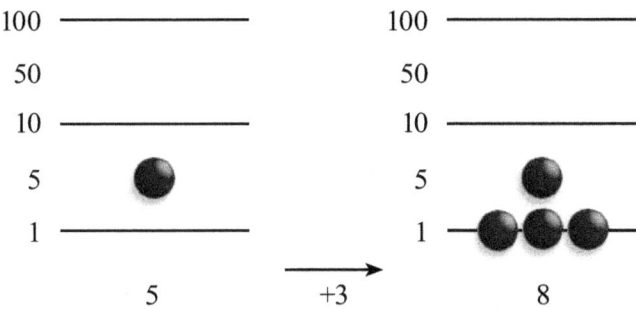

To add 40 to 50, we must put 4 pebbles on the tens line.

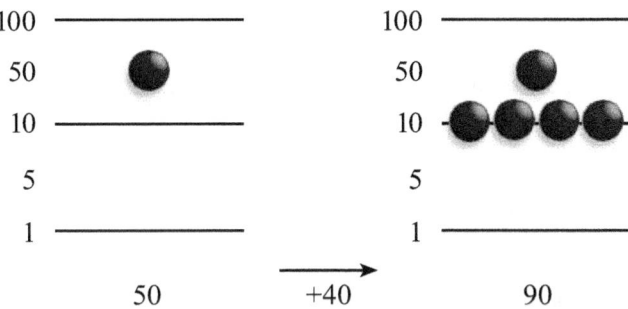

If we want to add 1 to 4, than 1 pebble goes to five and the single's line will be emptied.

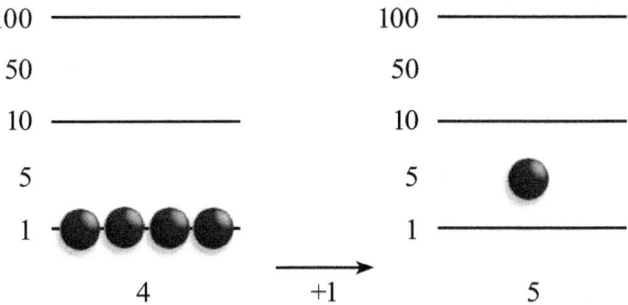

A further example for changing a local half value can be seen when 50 and 50 are added together. Two pebbles would be too many in one line, thus we change them for one pebble on the hundreds line.

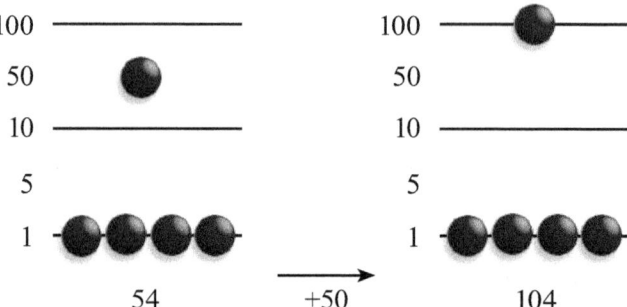

If we were to add 1 to 299, then we would have to follow with a sequence of changes.

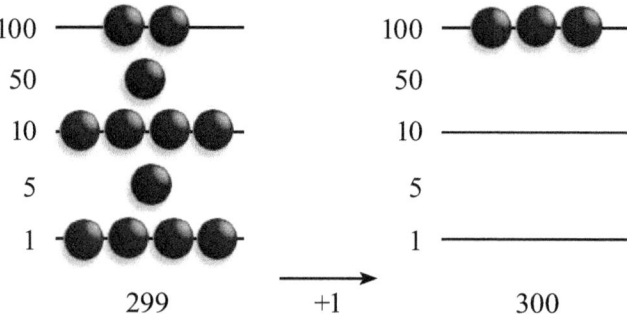

The Mayan number-writing method is in its essence identical with this. It proves the existence of a common ancestor in the very far past. There is only a formal difference. It is reversed: they marked the local half values with lines and put the singles above the lines. I have written the Mayan number in the decimal system here as well, to make the comparison easier:

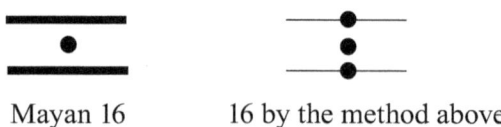

Mayan 16          16 by the method above

Picture 308. The same method used in two different ways. This is a written Mayan number, but the Mayans calculated with pebbles as well, (see page 278).

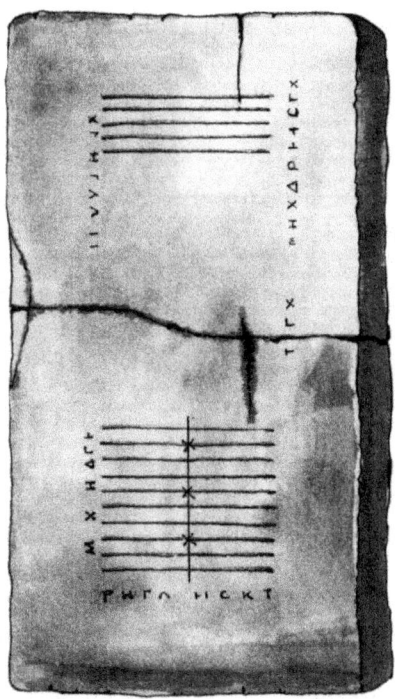

Picture 309. Old calculating stone with local half values. Greek letters show
, which local values person using it had to add to his numbers.

On the table above, even the half values have lines. The local values are
marked on the left side of the lines by Greek letters. The same marking can
be seen on the Roman table, where the half values don't have lines.

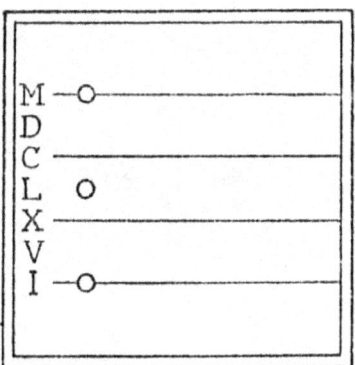

Picture 310. The drawing of a Roman calculating-table. The number set by
the pebbles is 1051.

There now follows an example of addition:

It is practical to use two columns separated by a line. Place the one of the numbers to be added into the left column and the other into the right column.

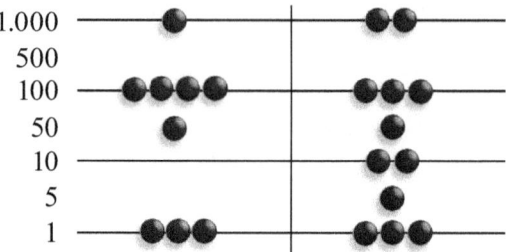

Next put all the pebbles from the left column into the appropriate lines of the right column.

As the first step of the addition, all the pebbles of both columns are on the lines of the right column.

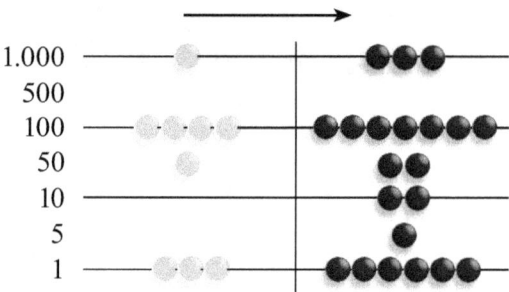

Now is time for clearing, for the changes into higher local values.

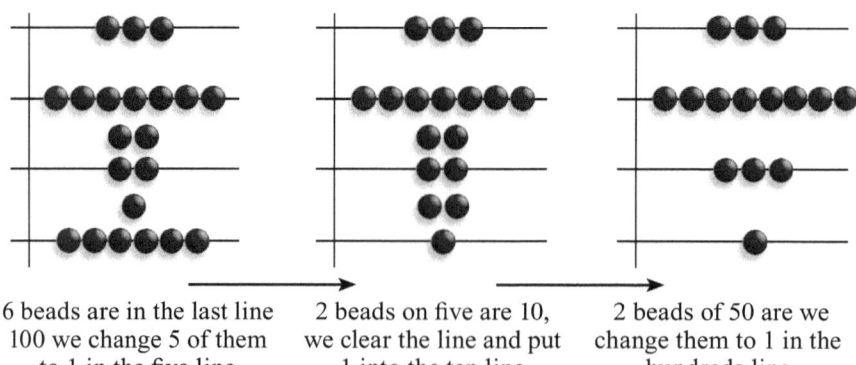

6 beads are in the last line 100 we change 5 of them to 1 in the five line

2 beads on five are 10, we clear the line and put 1 into the ten line

2 beads of 50 are we change them to 1 in the hundreds line

We don't even have to count the beads in the 100 line, they are visibly more than 4. We just have to take five 100-beads away and put one into the place of 500.

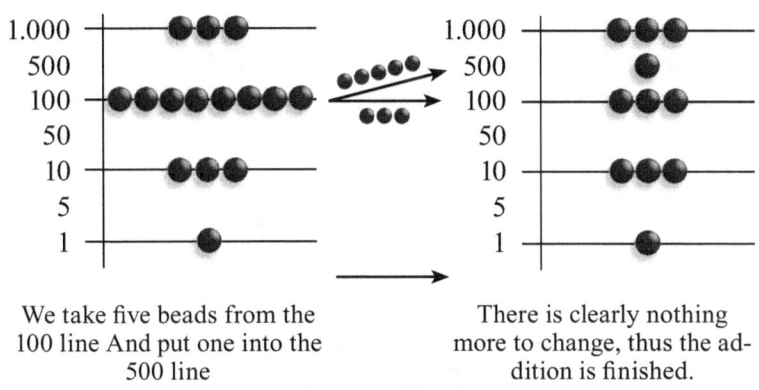

We take five beads from the 100 line And put one into the 500 line

There is clearly nothing more to change, thus the addition is finished.

The result is: 3.831..

Just looking at the pictures above doesn't give a true impression of how quickly such procedures can be done.

# 5. THE FASTEST CALCULATOR

Counting with pebbles, we have to pick up the pebble(s) at each step, putting it somewhere else, onto other lines or off the table. These movements take a lot of time, even the securely taken (grabbed) pebble still needs careful attention.

Picture 311. Grabbing a pebble needs time and careful attention.

In the decimal system of stringed beads however, the hand has to move relatively long distances to pull the beads back and forth. Ten beads are a lot, their handling needs a broader movement. A combination of the two previous methods may eliminate these inconveniences.[62]

The fast calculation speed can first be obtained when the beads can be pushed around by the finger tips and our hand doesn't have to move. However, this is not possible with ten beads on a wire, therefore only the method using the position half values counts. This again needs too many wires.

To solve all these problems, the idea arose not to put the half values in between the wires, but onto separated lengthenings of the wires. Let's look at the next picture:

---

62 I ask the reader not to think of any "evolution". First, we don't know which method came later and we might possibly draw a wrong conclusion. We must follow a logical sequence for the easy explanation of things but this is only a procedure to suit the way our brain functions. The reality does not follow the logical way retroactively invented by man. The urge to show evolution by any means is a mental curse upon European people. For example, the Mayan number writing has been described as a 1-2,000-year-old separate branch, in order to put it into the "evolutionary sequence". However, we saw earlier in this book that Mayan number-writings have been found continuously over the last 15-20,000 years, according to the evidence so far.

Half values

Round values

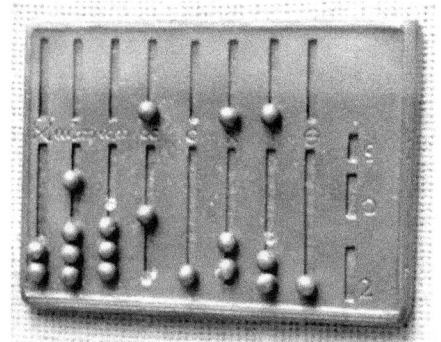

Picture 312. A Roman abacus. It is a fine piece of workmanship made out of bronze.

The beads can slide on this postcard-sized construction, and they can't fall down. It is a beautiful pocket-calculator, like our present electrical ones. There are extra lines for fractions 1/12, 1/24, 1/36 and for 1/48.

A variant of the same system carved out of wood:

Picture 313. This roughly carved counting table is around 2,600 years old. It looks foldable.

Its carving is coarse and the beads are randomly chosen pebbles. It uses exactly the same system as the previous one, but doesn't have the lines for the fractions.

There has been a variant of this system in China, which works with 5 instead of four beads on a wire and has two beads for the local half values making the changes easier to survey:

Picture 314. Chinese calculator with beads.

The Chinese name of this calculator is suanpan. Here there are five beads instead of four on the round values. Furthermore, the half values – carrying two beads – are placed further away from the round values. This fact certainly lessens its speed.

All conditions of speed are fulfilled on the following device:

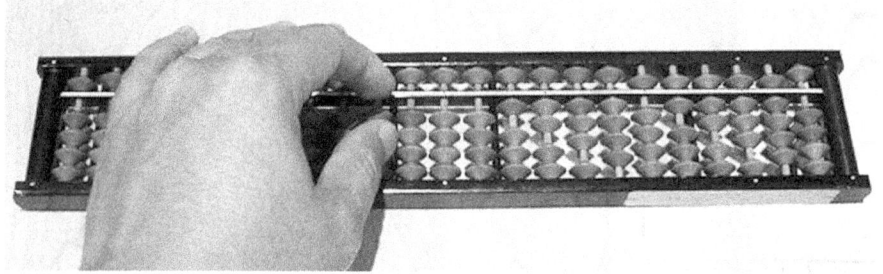

Picture 315. The fastest possible calculator with beads. (from the Internet)

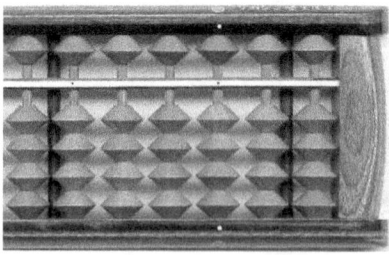

Picture 316. An enlarged part of the above calculator, which is mostly used in Japan.

This calculator is called soroban in Japan.

Using a soroban, our wrist lies on the table and we can calculate merely with the movements of two fingers. It gets faster by moving the bead for round numbers only with our thumb and those for the half values with our index finger. People can reach a fantastic calculating speed using this method. We know from experience that well-trained people in Japan need only 3-4 seconds to extract the root of a 12-digit number using this soroban.

The use of the soroban is identical with the procedures I introduced on page 256, in the chapter about the "bisecting calculator". There are many Internet pages offering descriptions, advice for use, and many films about the use of the soroban. One of these addresses is:

http://www.soroban.com/howto_abacus_eng.html

It is worth having a look at these pages and we may gain much knowledge about it. Less knowledgeable people often say that these bead calculators would be better in a museum.

However, this is not the case in Malaysia, USA, Great Britain, Canada, India, Japan, Russia, Thailand, Sri Lanka, China and New Zealand. In these countries, the use of the soroban is still being taught in schools.

In Japan, pupils in the elementary schools learn calculation basically by using the soroban. Once they have learnt the procedures with this device, the soroban itself is no longer necessary to do them. It is surprising to see the first time: having got the mathematical exercise, they just move their fingers fast over the paper - as if the soroban were there, - playing the beads' movements in the air and write down the results just a few seconds later. One calculation last only a few seconds, or even just one second. This knowledge will help them during their whole life.

Picture 317. A soroban-competition in the Kodály Zoltán
musical school (2001)

Why is it worth using a bead calculator and especially the fast soroban? Because our long-dead ancestors invented the ancient writing-signs, dot-line number-writing and calculation with beads according to suit the way our brain functions. Because of this there is a direct connection between these methods and our mind. Practising this kind of calculation is therefore important not only for counting. It helps to create and improve the direct connection between our mind and the procedures, to cultivate talent and intellectual proficiency. This is the topic of the script: *"How to fight calculating-problems with help of the soroban"* by Vajda Jozsef and his wife Bárdi Magdolna:

*"We build the notion of the natural numbers from the view of their steadily increasing local values. The number-notion becomes more confident due to the manipulative demonstration. The most striking change happens in the territory of verbal calculation. The students perform the procedures self-confidently, precisely and ever faster. While using this instrument, the children's concentration, their persistent attention and memory increase. One should use the same fingers to move the beads as for writing. That way, those fingers will become more skilful, their fine-motoric movement develops faster. Their writing develops faster, better. The learning of soroban use is easy, self-confidence increases due to an early sense of achievement. General capabilities necessary for any work such as punctuality, regularity, task-holding, looking for multiple solutions and a demand on control will become stronger. Practising will become like an easy game and this improves the results."*

The reader can see that the ancient, also called organic, culture of our ancestors was deeply spiritual, even more than our modern cultures. It is more than probable that our early ancestors brought human intellectual life to a high peak at least once in "prehistoric times". Today unfortunately, we are pushing human spiritual functions into the background  And trying to be clever. This is why we call our time the "technical age".

# 6. OTHER BALL-BEARING CALCULATING SYSTEMS

We started the description of ball-bearing calculators on page 246 with the introduction of the quipu. This was the easiest way to understand the system of the abacus, calculation with pebbles. The quipu however was used merely for storage of the numbers. The knots were fixed, tied to a spot.

Countless quipus were produced, because the whole of public administration was based on them. The Inca Empire was huge; it included the territory of today's Columbia, Ecuador, Peru, Bolivia, Argentina and Chile. For the management of this enormous territory, the numbers on the quipus had to be added – subtracted, multiplied and divided - easily and quickly. The Incas had to use a calculator. We know certainly that their counting equipment worked with pebbles or with similar means.

The name of their ball-bearing calculator is *yupana* in Qechua. The Spanish conquistadores forcefully (?) eliminated the use of this device in the whole country probably because they didn't understand its use. No authentic user manual survived this policy. Nobody knows today how the yupana was used, but a researcher could explain some important details by comparing several examples.

The results of these comparisons were quite surprising: Their number system was built on the 1, 1, 2, 3, 5, 8..., i.e., on the Fibonacci-sequence and upon the squares of the numbers 10, 20, and 40. This way even for very high numbers they needed just a few pebbles during the calculations. (Fibonacci sequence: every number is the sum of the two numbers before it.

It's worth remembering these five numbers (1, 1, 2, 3, 5,), the first 5 numbers of the Fibonacci numbers: 1, 1, 2, 3, 5, 8, 13, 21, 34, 55, 89, 144, 233, 377, 610 and so forth after zero. The investment of this number-sequence was a large mathematical achievement. However, the Incas couldn't learn this from Fibonacci, because at that time they had not been "discovered" yet. Thus, the Incas invented this number-sequence by themselves and probably before the Europeans knew it. Interestingly, Leonardo de Pisa Fibonacci (1170-1250) was using ball-bearing devices as the calculators of his time, but preferred the Arabic number system in his book "Liber Abaci" (1202).

The Aztec calculator, named *"nepohueltzitzin"*, was used in Mexico. The balls were replaced by pierced corns stringed on yarn. They used it precisely and quickly. The name is built from pohual (counting) and tzitzin (trifles) meaning: *the person counting on trifles*. The local values increased along the vigesimal numeric system.

The use of this calculator unfortunately ceased after the Spanish occupation in 1521. Some examples of this calculator were found, in which the seeds of the *"nepohueltzitzin"* were replaced by jade or gold. The finds all over showed that these devices were used by *Aztecs, Olmecs and Mayans* as well.

Picture 318. The yupana.

The little kernels or pebbles are not stringed here, but put in the little bowls and then stacked here and there. The placement of these bowls obviously followed the Fibonacci sequence.

# 7. ANCIENT NON BALL-BEARING CALCULATING METHODS

We established in the foregoing chapters that the ball-bearing calculator has been known in ancient times all over the world and it remains an often-used device in most parts of it.

There is only one inconvenience of this device, which is that the result of the calculation is not permanent. It disappears immediately by moving the device. Thus, the result has to be written down. Therefore, number- writing (generally writing) and ball-bearing calculators are inseparable from each other.[63] How we keep the records or what kinds of signs are used to do it does not matter. Therefore, if somewhere there is a use of a ball-bearing calculation method, there are numerals and number-writing as well.

One can only do calculations with written numerals if adequate sequences of procedures (algorithms) are introduced. We all know the algorithms created for the Arabic numerals and are using them today to multiply two numbers. There were other, simpler and sometimes faster methods used in ancient times for this. You can see below a rather special method of multiplication:

It is called the "duplicating method". We show the multiplication of 13 x 17, but it is the same procedure for every number. First we write the products of the multiplier with 1, 2, 4, 8, … under each other (for larger numbers continue the duplications with 16, 32, 64…..)

$$17 \ \text{x} \ \boxed{\begin{array}{l} 1 \\ 2 \\ 4 \\ 8 \end{array}} \ \begin{array}{l} = \quad 17 \\ = \quad 34 \\ = \quad 68 \\ = \ 136 \end{array}$$

For this exercise (17 x 13), we pick the lines 1, 4, and 8 and add their results:

$$17 \ \text{x} \ \boxed{\begin{array}{l} ① \\ 2 \\ ④ \\ ⑧ \end{array}} \ \begin{array}{l} = \quad 17 \\ = \quad 34 \\ = \quad 68 \\ = \ 136 \end{array}$$

---

63 However, it is not so in every case. A money-changer pays out the result shown on his abacus immediately. He will keep records only if the tax-office wants him to do it.

Now, we only add their results: $17 + 68 + 136 = 221$.
$$17 \times 13 = 221$$

Another method runs similarly. It is called the tenfolding-bisecting method. Using this, $17 \times 13$:

$$
\begin{array}{rrll}
13 \;\times & 1 & = & 13 \\
\times & 2 & = & 26 \\
\times & 10 & = & 130 \\
\times & 5 & = & 65
\end{array}
$$

Using this method, the components of 17 must be selected as well, in this case $2 + 10 + 5$:

$$
\begin{array}{rrll}
13 \;\times & 1 & = & 13 \\
\times & ②  & = & 26 \\
\times & ⑩ & = & 130 \\
\times & ⑤ & = & 65
\end{array}
$$

The sum of the results belonging to the chosen components is:
$26 + 130 + 65 = 221$.

There existed a similar procedure for doing divisions easily even with fractions.

Therefore, the algorithms we use for Arabic numerals are not the only possible methods to do basic mathematical procedures and not necessarily the fastest ones.

# PART IIIΛ

## WELL KNOWN ANCIENT NUMBER WRITING METHODS

# 1. THE EGYPTIAN NUMBER-WRITING

Egyptian number writing also proves– and we have already seen many examples of this – that no other number signs existed in ancient times apart from the dot and line or one line standing perpendicular to another. The only differences came through giving different values to the different basic signs (dot, line, perpendicular or long-line) or through the placement of these signs inside a special number. This possibility of various arrangements exists only because all numbers - with the exception of the round numbers - contain not just one, but a whole group of ancient signs.[64]

There were two variants of Egyptian numerals: hieroglyph and hieratic. The two variants were substantially identical from 1-9 and the differences in the larger numbers were merely formal.

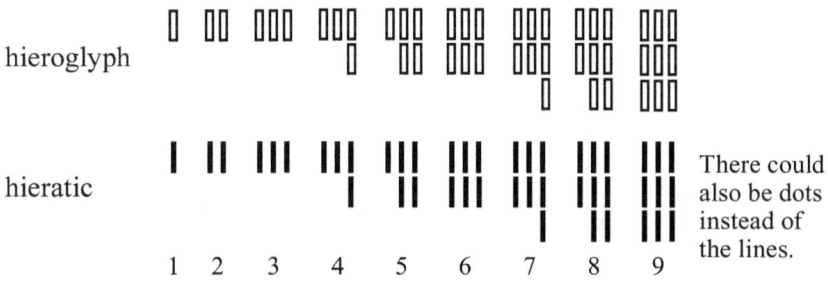

Picture 319. Egyptian hieroglyphic and hieratic number-signs from 1-9.

The hieroglyphic signs are pictured as little narrow slabs, because I wanted to make clear that those were lines engraved in stone. There is no other significance. A line remains the same line, whether it is written on a papyrus or engraved in stone.

64  The word Egypt is a derivative of the Ancient Greek word Aigyptos.

Egyptians used dots as well to mark single numbers, even mixing them into one larger number-sign, if the number became clearer as a result.

The previous picture proves that the hieroglyph and hieratic signs also belong to the ancient number writing system. They were used for thousands of years in the relatively enclosed Egyptian culture and one might expect that the numeral forms would suffer manifold changes during this long period.

The hieroglyphs did not change at all over several thousand years. They were engraved in stone and there was nothing to encourage anybody to change their forms.

Among the hieratic signs – the handwritten form of hieroglyphs –handwriting brought some changes during that time. But compared to the distorting Indian misdrawing of the signs (see page 217-223), these changes served merely to the speed up the handwriting.

In the next few pages, we can study in detail the steadiness of the hieroglyphs and the slight simplifying changes of the hieratic signs in handwriting. These tables don't go into great details to demonstrate every step of the minor changes, but the major steps of change are well demonstrated in them. You can see a few examples below for the changes to the hieratic numbers 5 – 9:

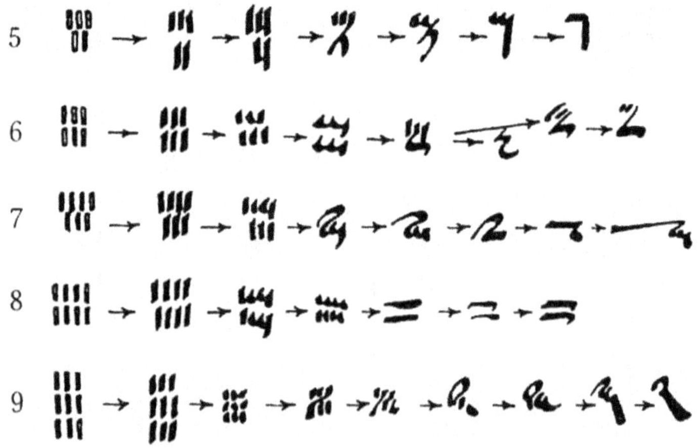

Picture 320. The urge to simplify handwriting is clear.
It is characteristic of handwriting.

These changes are unparalleled and smaller than what happened in the Hindu number writing in a much shorter time.

Picture 321. Geores Ifrah: Numbers, page 172..

Picture 322. Geores Ifrah: Numbers, page 172.

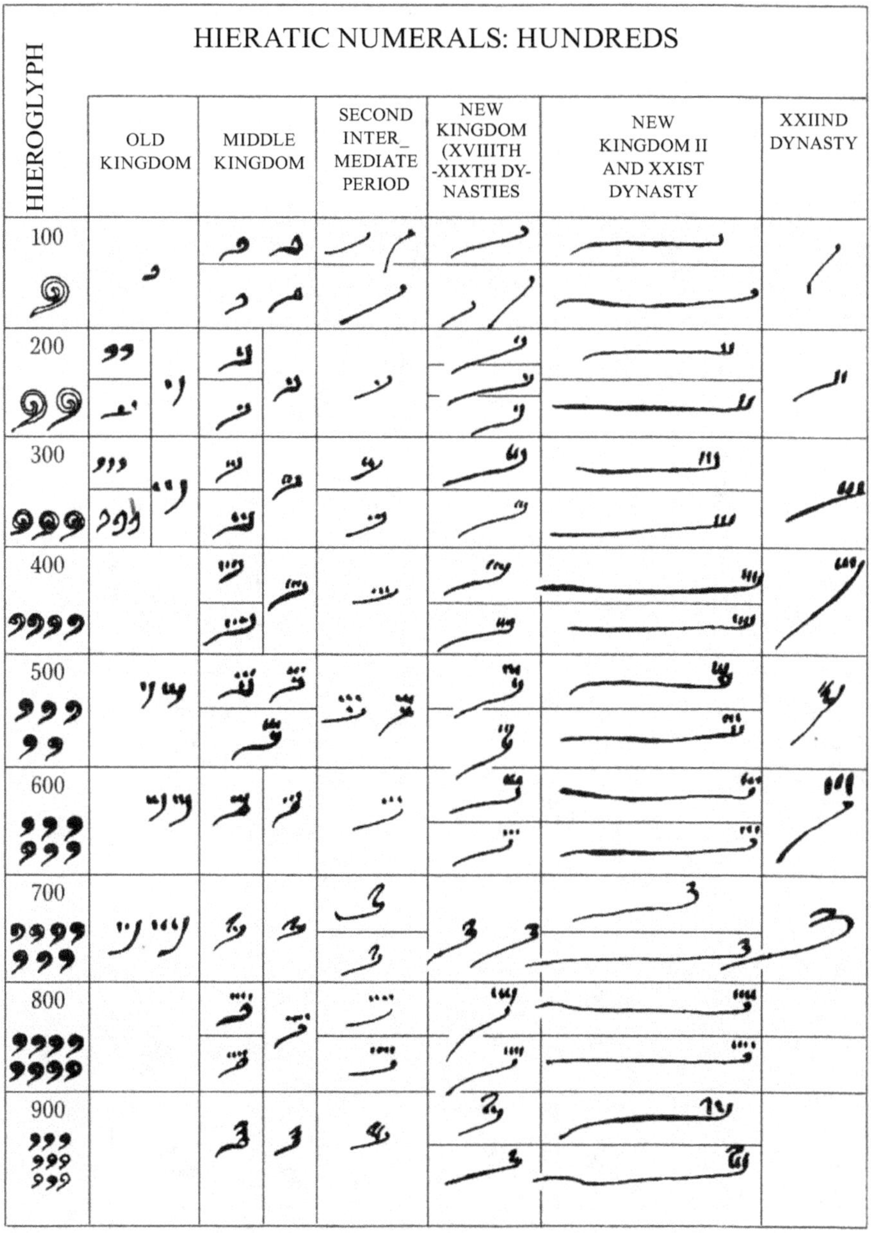

Picture 323. Geores Ifrah: Numbers, page 173

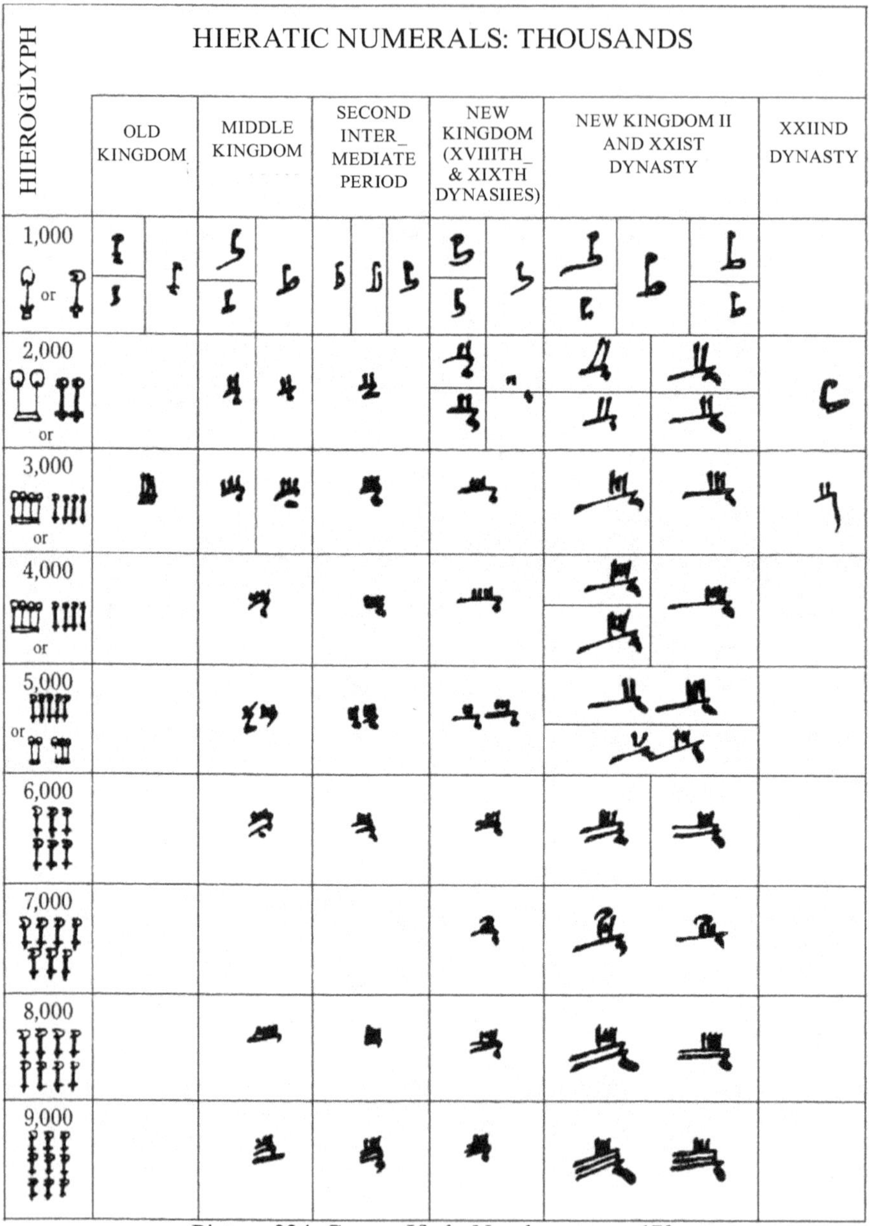
HIEROGLYPH
HIERATIC NUMERALS: THOUSANDS
OLD KINGDOM
MIDDLE KINGDOM
SECOND INTER_ MEDIATE PERIOD
NEW KINGDOM (XVIIITH_ & XIXTH DYNASIIES)
NEW KINGDOM II AND XXIST DYNASTY
XXIIND DYNASTY
1,000
or
2,000
or
3,000
or
4,000
or
5,000
or
6,000
7,000
8,000
9,000

Picture 324. Geores Ifrah: Numbers, page 173

# DOTS AND LINES IN HIERATIC NUMBER WRITING

It is interesting that the Egyptians used alternating dots and lines to mark singular numbers. We can find examples of this in the "Rhind Papyrus", which is a collection of mathematical text-exercises, written in Egypt around 4,000 years ago using hieratic script. This part of the papyrus is quite special, for it shows multiplication done by the "duplicating calculation" method we discussed in the previous chapter. The numbers 1-8 as multipliers can be seen on the right side of the picture (the Egyptians wrote from right to left). The multiplicand is 8 and the products stand on the left side:

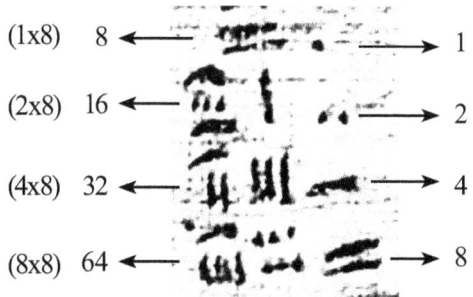

Picture 325. Part of the "Rhind Papyrus"

Let's "print" the numbers, the analysis will be easier:

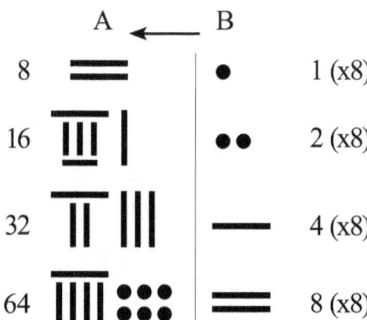

Picture 326.. The example from the Rhind Papyrus in "printed" form.

The numbers 1 and 2 are marked with dots on the side B. The 6 of 64 on the side A, but even here, the 1 of 16 and the 3 of 32 are written with perpendicular lines. On the side B furthermore, the dots of 4 and 8 are jointly drawn as lines.

If we mark the single numbers exclusively with dots, then the numbers from above will look like the following: 1 x 8 = 8, 2 x 8 = 16, 4 x 8n = 32 and so forth.

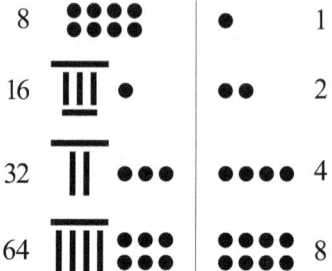

Picture 327. The doubling multiplication from the Rhind Papyrus

We just changed the lines for 1 in 16 and the 3 in 32 into dots, because the signs for 4 and 8 are already the contractions of dots (and not of lines):

The number 16 became mixed. The number itself as seen above:

$$\overline{|||} \bullet$$
16

The simplification of the number |||(6) is quite striking, as follows

| | | |
|---|---|---|
| **The original** | **the 3 lower dots** | **the dots changed** |
| **regular** | **contracted** | **to lines** |

based on these changes, the parts of the number with their values:

the tenfolding linel ⟵ ⎯
6 ⟵ ⎯⎯⎯ ||| • ⟶ the number of tens

Its reading (from right to left) one-ten + six = sixteen.

The same way as in the previous examples (right to left) :

$\overline{\mathsf{TT}}\ \mathsf{III}$  or  $\overline{\mathsf{TT}}$ •••  ⟶  follow the path of the arrows: 3x10+2 = 32

10

2   3

$\overline{\mathsf{IIII}}$ •••
•••  ⟶  64 (6x10+4)

10

4   6

Note: We pronounce 64 today as "hat-von-négy" <hatvonnédj> (six-von-four). What does the word "von" mean? It has to mean "ten". Ten was not just randomly called "von" in the Sumerian language. We are going to explain why we name all numbers from 40 – 99 using the word –von, -van, -ven by looking at the Egyptian numerals further on page 303.

The reader can now read these numbers without help, based on the previous explanations.[65]

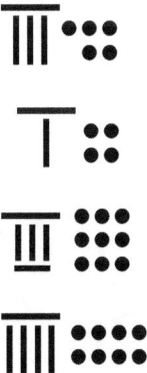

## THE 5 AND THE 7

Contracting one or two sign-rows of the ⠿ (6) and ⠿⠿ (8) won't cause much problem. However, the signs ⠒⠆ (5) and ⠶⠆ (7) make a difference, because the numbers of dots in the rows are not identical. One has to use different methods of contraction. The following solutions were adapted:

There was never any simplification in hieroglyphic writing. The ancient lines or dots were consistently engraved unchanged,

# THE IMPORTANT ROLE OF 9 AND ITS SIGN

Writing the sign for another odd number, 9, happened differently. The dots are again equal in every row: ⦂⦂⦂ but there are three rows.

The number 9 already appears on the oldest known finds (at the time of the Old Kingdom) written with dots, lines and in constricted form:

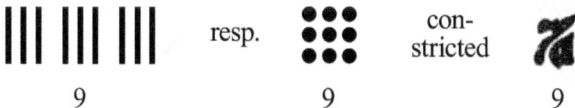

Therefore the two ancient numeral forms of 9 were used alongside its handwritten variant. We don't see a beginning even in that very ancient past. That is, the simplifications applied to the handwritten form were not used at all times and everywhere. One or other of the variants may have been preferred by certain schools or teachers, but in the end, it came to a seemingly unified form. There are however precedents for the use of a very early original variant along with a younger one and this tells us that in one area of the country the modification did not happen in the same way.

Another explanation could be that there were several somewhat differing cultural centres at different places in Egypt.

The reduction of the 3x3 dots sign of the number 9 into 3 lines went along with the marking of this number's important role in calculating circles. One marked this role with a little tail hanged onto the sign and by surrounding this sign with a circle:[66]

The encircled number 9 points to the meaning of the picture below:

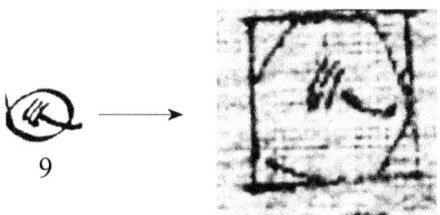

Picture 328. A picture from the Rhind papyrus: The drawing shows the method of calculating the circle's surface.

66  See Borbola J.: Royal Circles (Mathematical text-exercises from the Rhind Papyrus.)

Here, we meet the surface and circumference calculation-method, in which the number 9 plays an important role. The Egyptians received the value of the necessary $\pi$ by using the equation $(8/9)2 = \pi/4$ with an accuracy of 0.02.

Let's draw a quadrate around the circle and divide it by 9 in both directions. Let's connect the corners of every third little quadrate to receive an octagon as seen in the drawing "A" below. The Egyptian scribe put the encircled sign of 9 into the middle of this drawing of the previous picture to help us remember his method of calculating. Here the most important task is to divide by 9.

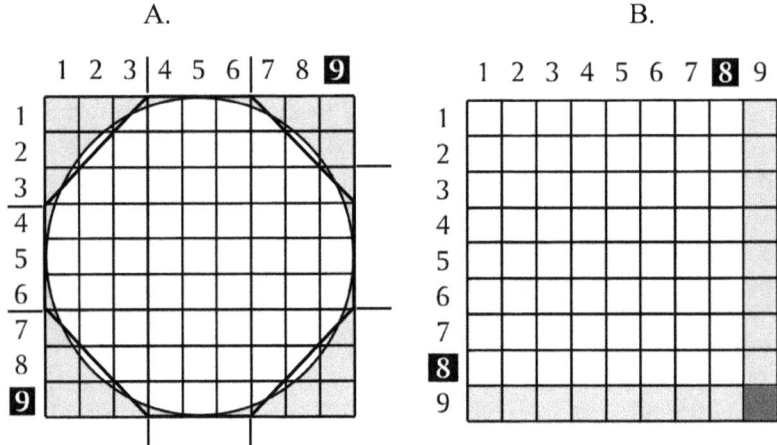

Picture 329. The little quadrates cut out by the octagon (A) and pushed to the sides (B).[67]

The areas of the triangles –cut out by the circle and by the octagon as well – are almost identical. The octagon cut off 18 little quadrates at the corners which, arranged to the sides (painted grey) gave us the drawing B. The remaining 8x8 little quadrates come very close to the circle's area. This is how the number 8 became the second most important number in calculating circles.

We can see however that we did not carry over 18 little quadrates from A to B. The darker one on the right lower corner has been counted twice. Therefore, the remaining 8x8=64 is not the circle's area, but the value is somewhere between 63 and 64.

The right result can be approximated quite well with the numbers 9 and

67 Vogel's diagram of the inscribed octagon of Problem 48 of the RMP. (K.Vogel: *Vorgriechische Mathematik*, Vol.1.)

8. I will show the difference by using our modern method:

The quadrate of 8/9 is: $(8/9)2 = 0.7901...$,
multiplied by 4 = $\qquad$ 3.16049.
Our $\pi$ is $\quad$ 3.14159

That is: $(8/9)2 \approx \pi/4$.
Let's take the radius in our exercise as $9/2 = 4.5$ units. Calculating by our method (circle-area = $r2\pi$): $4.52 \times 3.14159 =$
63.618 units
using $(8/9)2$ however, are $\qquad$ 63.999 units (almost 8x8=64)

The difference is merely 0.4 which is not noticeable in real life. There are, however, larger number-pairs and using those we could approach the value of $\pi$ much more accurately by dividing the circle into more, smaller quadrates. The 9/8 is only the smallest possible number-pair and the Egyptians knew this as well. Nevertheless, the easiest way to calculate is to use the smallest possible number pair and it gave them a sufficient and practical result for everyday use.[68]

---

68 A culture existed in which the value of $\pi$ was calculated by $\sqrt{10} = 3.162277$. This value would have been just as good, just a tiny bit worse, as that of the Egyptians. There is a well-based suspicion that people all over the world knew very well the exact value they had to get close to, but knew an easier way to receive a result sufficient for practical uses.

# THE SIGNS FOR 10, 20 and 30

## *THE SIGN FOR 10*

The sign for 10 is **Λ** in Egypt. This has been a special sign in number-rovás. It even appears on two finds presenting number-rovás 15,000 years ago (see page 223). The value of the sign **Λ** has been mostly 5, but as exceptions, its value was 10 in the Sumerian and Egyptian number-writing. This sign carried an important spiritual content in the past. Its form and value was connected with the so-called "pyramid numbers".

The connection of the form **Λ** and 10:

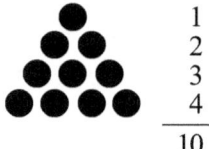

Picture 330  The 10 as a pyramid number

This formation and the **Λ** = 10 is deeply connected in the ancient spiritual world, with the words pír (fire), tűz <tuez> (fire), tíz (10), but even with the names as Zeus and Deus.

We don't have enough finds however to answer the question of why is it so. Is the identity of this **Λ** sign of such spiritual content as the same-looking sign of number-rovás is real or only random? It is possible that the ancient **Λ** sign of number-rovás only later took on this spiritual content. However, this must have happened 10,000 or more years ago. It is possible as well that this sign-form was introduced a second time into the numerals for the reasons we discussed above.

I do not know the answer.

The sign of ten was always written in the rounded form when using hieroglyphs:

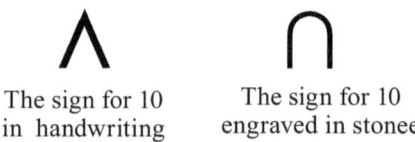

The sign for 10          The sign for 10
in  handwriting      engraved in stonee

Picture 331

# THE SIGN FOR 20

The sign for 20 would be ΛΛ if written the ancient way. It became simplified in the handwritten form:

Λ̄    Λ̂    reduced    ⅄

Picture 332. The sign for 20. To the right: handwritten on papyrus

The horizontal line above the sign means doubling. We know this function of the horizontal line from Roman scripts. When duplicating a writing-sign, the Romans wrote it once and drew a line above it. For example:

Picture 333. We find this solution several times
in old Latin scripts.

The signs were never simplified when using hieroglyph writing.

Λ̄              ∩∩

The sign for 20        twenty engraved
in handwriting         in stone

Picture 334. The sign for 20, handwritten and engraved.

# THE SIGN FOR 30

The sign for 30 – based on the ∩∩∩ hieroglyph signs must have originally been ΛΛΛ. It is however already reduced in the oldest archaeological finds. This means that the reduction must have happened long before the time of the now-known findings. The signs for 30 therefore originate from an epoch before the Ancient Egyptian culture.

Even the oldest known reductions of the numeral 30 did not really change through the known thousands of years of the Egyptian culture. Not even small changes can be seen:

ℵ    ⅄    ⅄    Ⅺ

Pictur 335. It seems that the reduction of the sign 30 happened somewhere before Egyptian times, but it did not change in Egypt throughout its several thousands of years.

# THE SIGNS FOR TENS FROM 40 TO 99

We have established so far:

1) The numerals from 1-9 are ancient, contractions happened only in handwriting.

The sign for 10 (**Λ** or **∩**) is new among the dot-line numerals. Novelties are furthermore, the doubling or tripling of this sign for 20 and 30 (∩∩ resp. ∩∩∩, which in handwriting was **Λ̄** resp. **⅄** ).

The round numbers over 30 again keep their ancient forms. This tells us that the numerals 10, 20 and 30 were introduced later. The assumption that this insertion preceded the era of the oldest Egyptian numerals is supported by the fact that the engraved form (∩ = **Λ**) for the tens stayed unchanged for thousands of years but not in their handwritten form.

The engraved numerals 10 – 90 are arranged the ancient way:

| ∩ | ∩∩ | ∩∩<br>∩ | ∩∩<br>∩∩ | ∩∩∩<br>∩∩ | ∩∩∩<br>∩∩∩ | ∩∩∩∩<br>∩∩∩ | ∩∩∩∩<br>∩∩∩∩ | ∩∩∩<br>∩∩∩<br>∩∩∩ |
|----|----|----|----|----|----|----|----|----|
| 10 | 20 | 30 | 40 | 50 | 60 | 70 | 80 | 90 |

Picture 336. Egyptian numerals 10 – 90..

In handwriting, the numerals for 40-90 follow again the ancient dot-line system as we can see in the table of Georges Ifrah on page 284. This table shows only the interesting simplifications of the numerals 10-40. For practical exercises see the samples from the Rhind-Papyrus on page 287.

Here is the big question: how did they make the tenfold value of a single number readable? As seen before, they often drew a tenfolding line above the number. Even this was often simplified. In the numerals 1-9 both lines were contracted, but in case of a tenfolded value, the dots in the upper line were not contracted. The tenfolding line was hidden this way in the number.

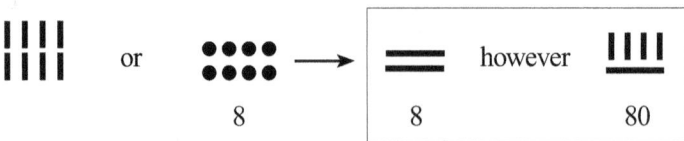

Picture 337. Numerals 8 and 80 written by tenfolding.
The upper line wasn't contracted.

The signs for 9 and 90 are written similarly:

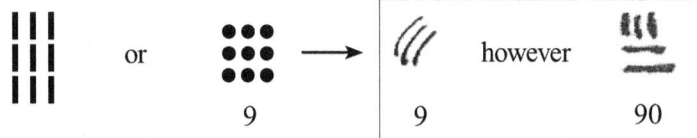

Picture 338. 9 and 90

The numerals 40, 60, 80, and 90 were written the same way:

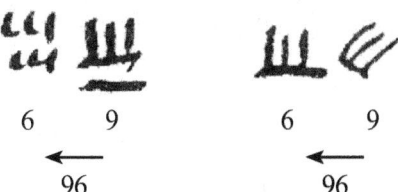

| 40 | 60 | 80 | 90 |

Picture 339. 40, 60, 80, 90.

The numerals 50 and 70 were exceptions as we discussed above when talking about the numerals 5 and 7. Compare these numbers and their tenfolds on pages 283-284.

The Egyptians always knew what the individual sign - included in a numeral – means. This was the reason they could change the way they wrote it. The same hand wrote the numbers below in different parts of the papyrus:

*96 written in two different ways:*

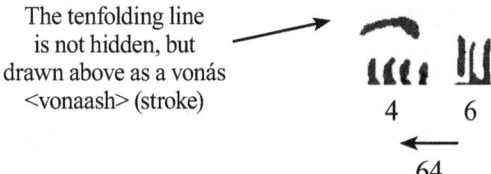

*In 64 however, the tenfolding line is stressed above the numeral:*

The tenfolding line is not hidden, but drawn above as a vonás <vonaash> (stroke)

4    6

64

Picture 340. We call it 6 von 4 (six stroke four), see page 289.
[Von(ás) means stroke in Hungarian.]

## *THE SIGN FOR 100*

As we mentioned earlier, there were four signs in the ancient number-writing: dot, line, a line perpendicular to this and the long line. Thus, as you might expect, the long line represented the number 100 in Egypt.

The engraved form of the long line was like a coiled snail:

100       200       300

*and so forth, continuing in the traditional way
as with the dots:*

800       900

Picture 341. The signs from 100 to 900

The reason for the arrangement above is that the sign itself means one-hundred. Depicting 100 with the long-line would look like this:

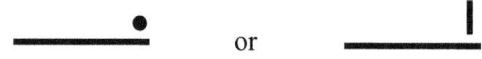

100

*The writing in one compound and ornamenting
started probably from the form with the dot, so
that the writer would not have to lift the writing
tool. The received curve offered itself naturally
for the snail form.*

100                     100

Picture 342. The shaping of the hieroglyph form of the numeral 100

This way of modification is supported by the fact that 100 was written in a quite similar form and led easily to the coil.

Picture 343

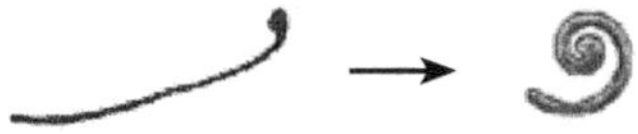

The modification of the sign 100 itself already tells us that the dot-line signs were first and the engraved, hieroglyph signs came later and it couldn't have happened in any other way. However, the engraved 100 sign had a snail-form at the very beginning of the Ancient Egyptian culture just as the **Λ** sign of 10 had the form of **Π**. The suspicion turns up here that the basic signs of hieroglyph writing, the already ornamented signs, were developed somewhere else and certainly much earlier than many people may think.[69]

As mentioned before, the Egyptians marked the singles with dots and lines as well. In the samples below, you may put perpendicular lines for every dot and reverse. It seems that the general use of lines for single numbers over the hundreds started first in the II. New Kingdom, but even later their use alternated.

Let's look at the numerals for the hundreds:

The numerals from 100 to 400:

Picture 344

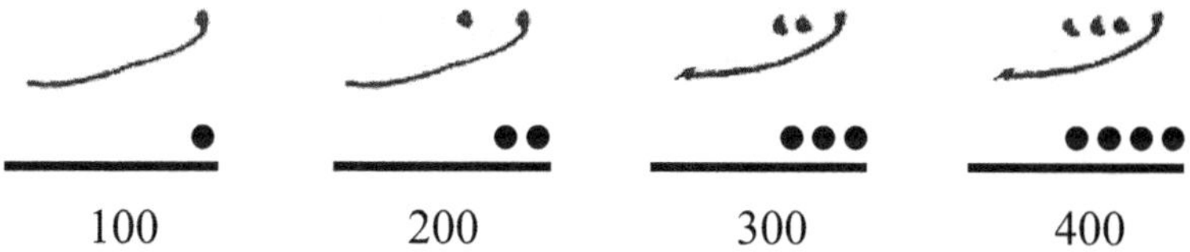

| 100 | 200 | 300 | 400 |

The signs for 500, 600 and 900 retained a much earlier writing method in the old times. Namely, as mentioned before, the perpendicular lines were usually grouped: **II III** (5), **III III** (6), **III IIII** (7). This division in sections was still present in the 100-signs as well at the time of the Old Kingdom:

Picture 345

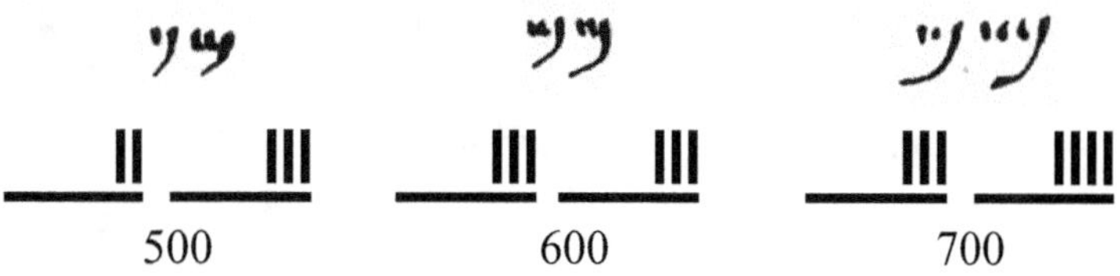

| 500 | 600 | 700 |

69  The number of hieroglyphs, engraved signs, grew to several thousands during the Ancient Egyptian culture.

This kind of division in sections disappeared at the time of the Middle Kingdom and the sections were put beneath each other.

The sectioning of the lines **II III** (5), **III III** (6), **III IIII** (7), **IIII IIII** (8), **III III III** (9) in old times, was followed by the positioning them beneath each other during the Middle Kingdom, which was a general custom in several other ancient cultures. We can see it clearly by looking at the number 500:

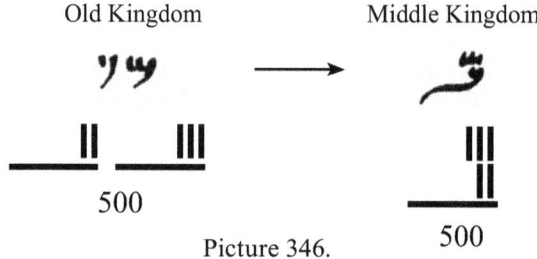

Picture 346.

Writing the numerals 600 and 800 was simplified by not writing the lines for 6 and 8 twice ( **|||** , resp. **||||** ).

The procedure is very interesting. We can see by looking at the numerals 200 and 400 that the dot marking the 1 is connected to the hundred-folding line, it is part of the number and it is standing in one row with the other dots:

Picture 347.

However, writing the numerals 600 and 800, one wrote only one row of the starting number's dots ( **⣿** and **⣿**). They marked the duplication by positioning the dot connected to the hundred-folding line below the row of (3 or 4) dots (or line). Thus, the dot did not belong to the number and told the reader that the singles had to be read twofold:

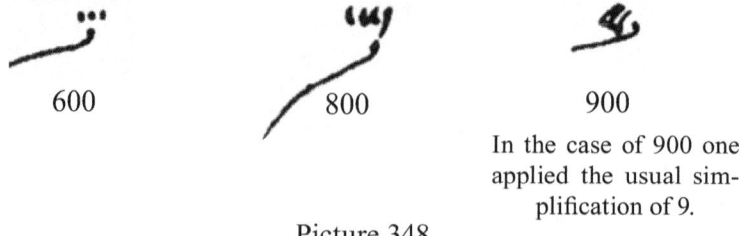

600      800      900

In the case of 900 one applied the usual simplification of 9.

Picture 348.

The shaping of the 700's sign is obscure among the hundreds. The beginning is clear, since it had this form ʸⁿⁿ ʸⁿⁿ in the earliest times of Egypt and its simplification can be easily followed. At the end however the distortion became too much. The original sign "crumbled":

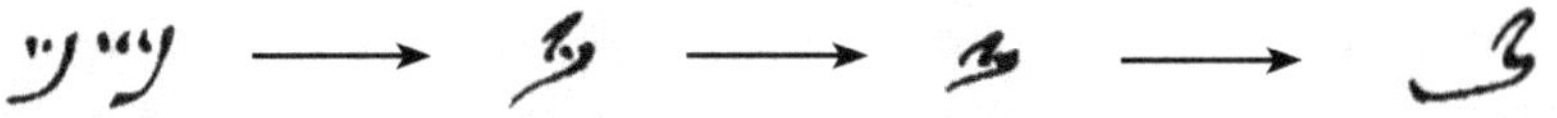

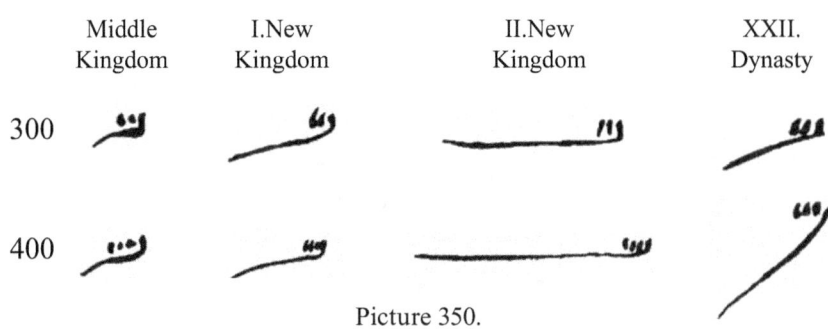

Picture 349. The contraction of the complex sign of 700.

Reaching the last form in the row, it stayed that way till the end.

We may note that the hundred-folding line became very long in time and shortened again toward the end. We don't know what it might have become later, because afterwards Egypt's independent culture came to an end:

|  | Middle Kingdom | I.New Kingdom | II.New Kingdom | XXII. Dynasty |
|---|---|---|---|---|
| 300 | ᴵᴵ | ᴵᴵ | ᴵᴵ | ᴵᴵᴵ |
| 400 | ᴵᴵᴵ | ᴵᴵᴵ | ᴵᴵᴵ | ᴵᴵᴵᴵ |

Picture 350.

The same course of change happened in the case of the other numerals presenting hundreds. It is uncertain what caused these changes. Perhaps a graphologist can give us an answer.

## THE SIGN FOR 1000

The sign for 1,000 is the only one remaining the same, though simplified, even in handwritten form, But only in case of 1,000. From 2,000 on, the writing followed again the archaic method of multiplying it. The sign for 1,000 is probably a clump in a marsh with a water plant having one flower on top:

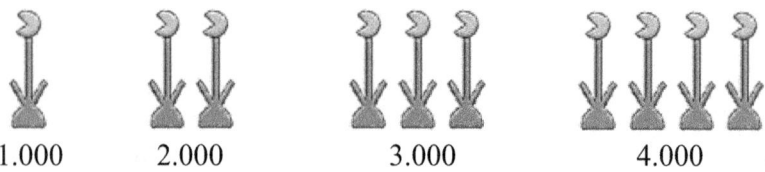

| 1.000 | 2.000 | 3.000 | 4.000 |

*The signs above 5,000 were grouped the same way as the dots or lines marking the singles (see table on page 286). The two parts (2+3) of the 5,000 were mostly engraved side by side:*

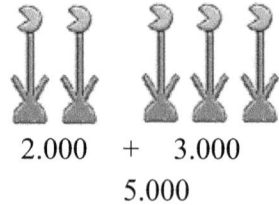

2.000 + 3.000

5.000

Picture 351. The engraved signs for the thousands

The meaning of the sign is that a flower stands for a dot and the number of these dots tell the number of the thousands.

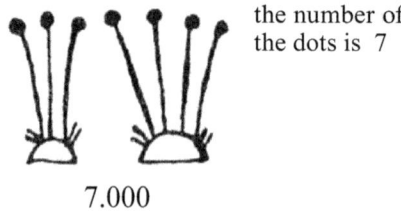

the number of
the dots is 7

7.000

Picture 352. We have to add the dots. The clump at the bottom tells us that the dots mean 1,000.

We might wonder where does the clump come from and what are the stalks of the flowers?

Let's put the hieroglyph sign for 3,000 side by side with its handwritten form:

Picture 353. The hieroglyph and the handwritten forms for 3,000.

It is clearly visible that the perpendicular lines are the equivalents of the stalks and the clump that of the horizontal line with the firmly drawn little tail.

This little tail was always firmly drawn and because of this we may safely assume that it pointed to the value of one thousand and only that pointed to it. This assumption is easy to accept, since the line by itself means 100. This handwritten numeral should be read as 300 without the tail.

This means that there was a sign below the line, which – knowing the signs of the ancient number-writing – could have been only a dot or a line and the Egyptians wrote down this sign connected to the long line in order not to be forced to lift their writing-tool.

According to this, the original sign for 1,000 must have been:

or                                            Picture 354

We found numerals like these elsewhere as well. It is not an Egyptian invention, rather an ancient heritage. The proofs for this are the examples below. We don't know what value these numerals stood for in Lascaux and Akna-Suhatag. The common identity of the sign-creating principle with that of the Egyptian numerals is clear:

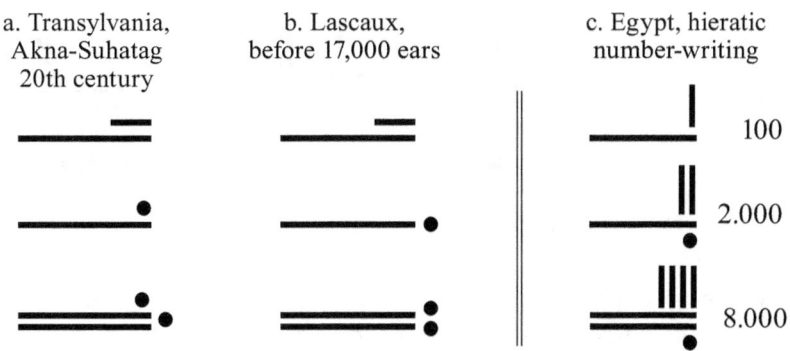

Picture 355. Examples of sign-creation from quite different eras.

Let's look at the thousands now. We will handle the 1,000 later:
2.000, 3.000 and 4.000:

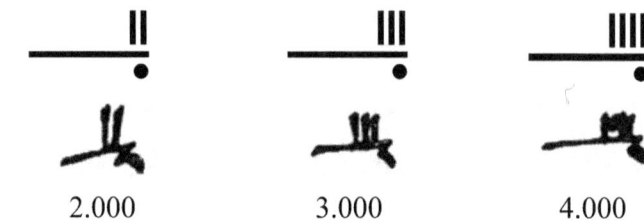

Picture 356.      2.000                    3.000                    4.000

The 5,000 is special in the way that it was written according to the old legacy in 2 + 3 form, like the 5: II III. It is interesting that this custom had never been changed, while the similar writing of 500 (200 + 300) was simplified during the time of the Middle Kingdom. This clinging to the old legacy when writing 5,000 is exceptional

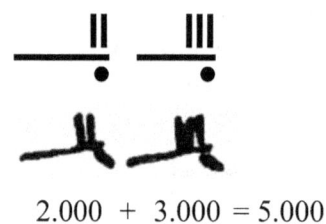

Picture 357.                2.000 + 3.000 = 5.000

We have seen in the case of the tens and hundreds with values of 6, 8 and 9 that the recurring rows of dots ( ••• , •••• and ••• ) were con-

stricted or a doubling line was drawn beneath the first row. We have to understand the numbers 6,000, 8,000 and 9,000

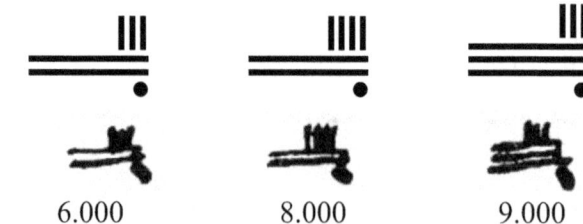

Picture 358.        6.000                    8.000                    9.000

The 5 in 5,000 has been written as usual (see picture 346), but when writing 7,000 they put that form of 7 above the 1,000, which had been shortened a very long time ago:

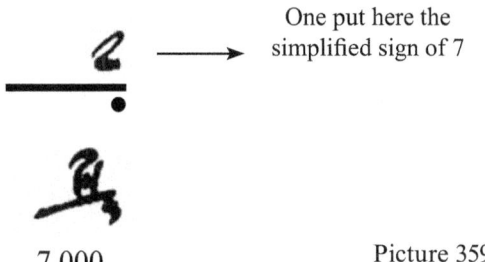

One put here the
simplified sign of 7

7.000                                          Picture 359. .

Let's now look at 1,000, better known as 1x1,000.

It is interesting that its oldest known hand-written (hieratic) sign was identical from the start to the simple form of its hieroglyph sign and did not change throughout history.

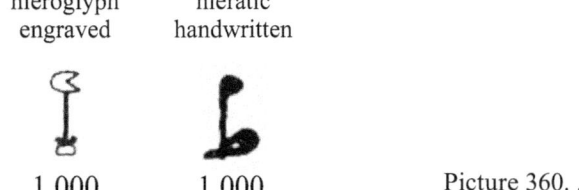

| hieroglyph engraved | hieratic handwritten |
|---|---|
| 1.000 | 1.000 |

Picture 360. .

Here again, we can see that the carved hieroglyph sign for 1,000 must have existed before the earliest known time of Egyptian history, because its handwritten variant could only have been born after that. We see in the case of 7,000 that the simplified sign for 7 ( •••• ) was there in the earliest known times as well. At the same time, most of the other signs had the possible oldest form and were being influenced by the hieroglyphs, which are certainly younger than the signs of ancient dot-line number-writing.

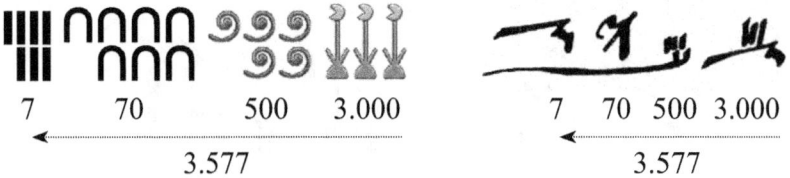

Picture 361. Two different forms of 3,577

## *THE SIGNS FOR 10,000, 100,000 and 1,000,000*

The signs for 10,000, 100,000 and 1,000,000 have nothing in common with the signs of the ancient dot-line numerals. The 10,000 sign is a finger drawn in straight and also in bent form. The sign for 100,000 looks like a chameleon (?) or a fish (?) and the sign for 1,000,000 is a picture of a kneeling man with outstretched arms:

10.000

100.000

1.000.000

Picture 362. The signs for 10,000, 100,000 and 1,000,000

The finger, 10,000, remembers occasionally rather of a sickle. This similarity could have broth people to the bend finger.

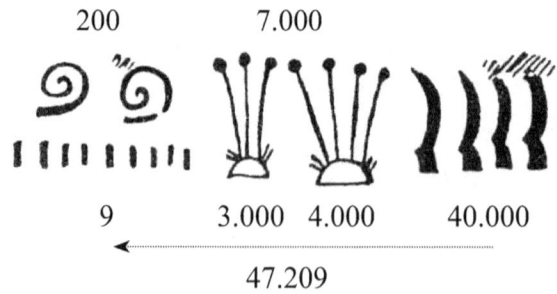

Picture 363. from Georges Ifrah: Numbers, page 166.

At this point, there should be no problem recognizing that there is a number 6,000 in the lower right corner of the following hieroglyph. Look at the six papoose flowers standing in a clump:

Picture 364. 6,000

I think the reader should have no trouble reading these following numbers carved into the wall of the temple in Luxor.

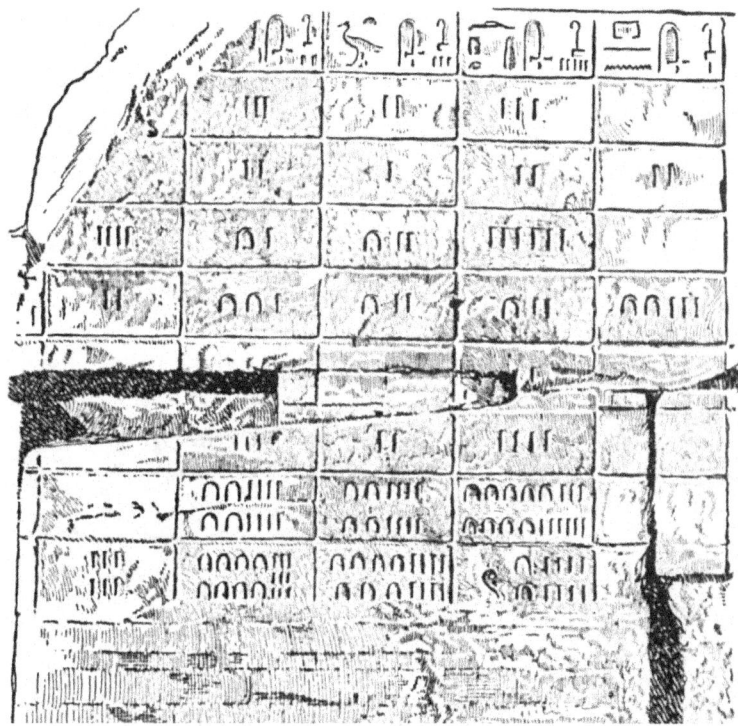

Picture 365. Luxor, numbers carved into the wall of the temple.
(Old copperplate.)

# A FEW WORDS ABOUT FRACTIONS.

If the numerator is 1, then the Egyptians marked it in handwriting with a dot and in hieroglyph-writing with a flat ellipsis ⌒ above the numeral. They always calculated with a numerator of 1, except in the often used 2/3 and ¾. It is not easy to find the original numeral-form of frequently used numerals due to the simplification coming from the hieratic writing:

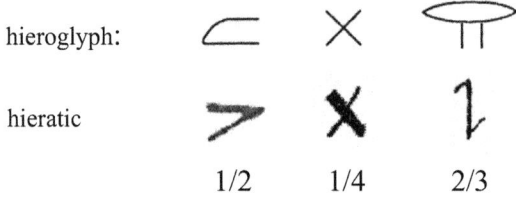

hieroglyph:

hieratic

| 1/2 | 1/4 | 2/3 |

*The basic idea is hidden in the hieroglyph of the ¾ as well:*

3/4

Picture 366. ½, ¼, 2/3, ¾.

The rest of the numerals were built logically:

| 1/9 | 1/10 | 1/20 | 1/90 |

*We introduce the general rule by writing 1/18:*

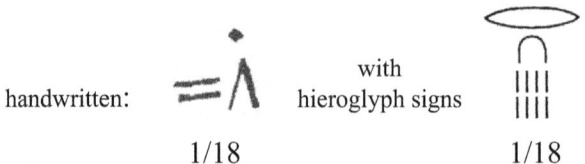

handwritten:     with
hieroglyph signs

1/18                    1/18

Picture 367. Two examples for writing fractions

# EXPLORING DEEP CONNECTIONS

Looking at the forms for the number-signs, we can see a striking difference between the group of 10 (∧), 20 (∧∧), 30 (∧∧∧) and the rest of the numerals. The sign ∧ is known, but making the three numerals by doubling and tripling this sign is unknown outside Egypt. The insertion of this group of numerals among the ancient signs has a close connection with the Hungarian numerals as well as the name and functions of the numerals 8 and 9.

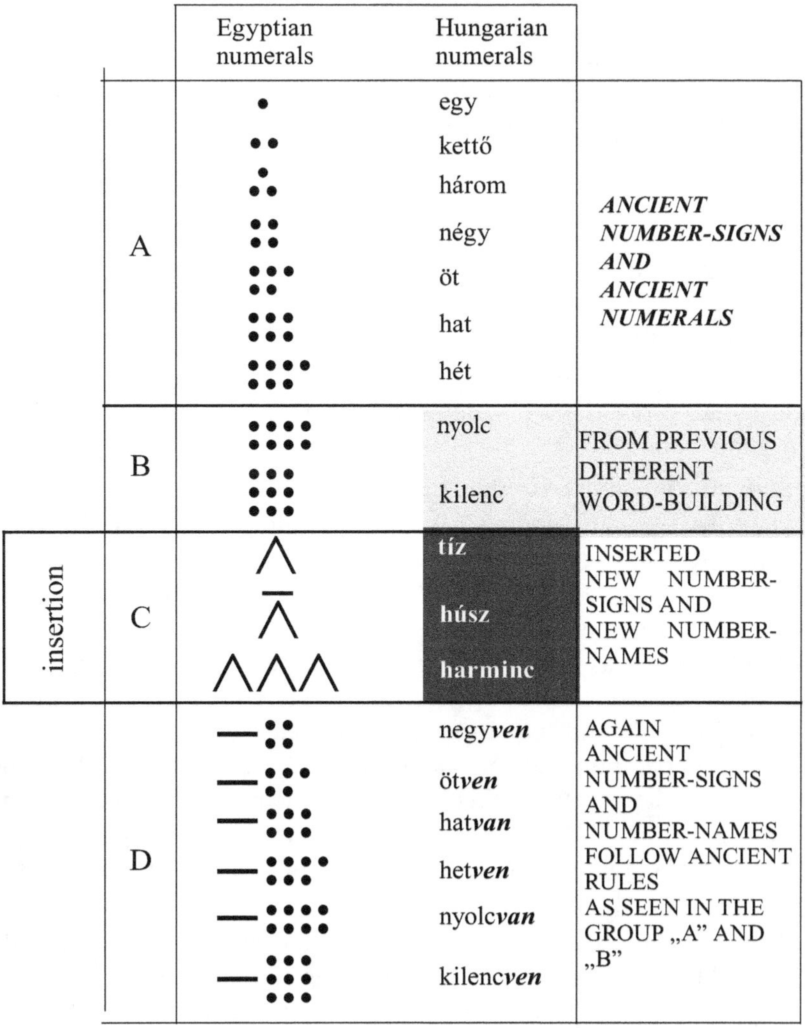

| | | Egyptian numerals | Hungarian numerals | |
|---|---|---|---|---|
| | A | • | egy | ***ANCIENT NUMBER-SIGNS AND ANCIENT NUMERALS*** |
| | | • • | kettő | |
| | | • •• | három | |
| | | •• •• | négy | |
| | | ••• •• • | öt | |
| | | ••• ••• | hat | |
| | | •••• ••• | hét | |
| | B | •••• •••• | nyolc | FROM PREVIOUS DIFFERENT WORD-BUILDING |
| | | ••• ••• | kilenc | |
| insertion | C | ∧ | tíz | INSERTED NEW NUMBER-SIGNS AND NEW NUMBER-NAMES |
| | | ∧̄ | húsz | |
| | | ∧∧∧ | harminc | |
| | D | — •• •• | negyven | AGAIN ANCIENT NUMBER-SIGNS AND NUMBER-NAMES FOLLOW ANCIENT RULES AS SEEN IN THE GROUP „A" AND „B" |
| | | — ••• •• | ötven | |
| | | — •• ••• | hatvan | |
| | | — •• •••• | hetven | |
| | | — •••• ••• | nyolcvan | |
| | | — ••• ••• | kilencven | |

Picture 368.

We can state the following about Egyptian numerals:

1) Numerals 1-9 (groups A and B) as well as those from 40-90 (group D) are written uniformly in the ancient way.

2) The three signs of group C (10, 20, 30) are clearly insertions between the ancient groups of A, B and D. Look again at the row of Egyptian numerals. Once therefore, certain people decided to insert these new number-signs in between the uniform ancient numerals.

   By contrast, this group of numbers is treated differently in Hungarian. And look at the number-names 10 and 20, they are followed by the positioning suffix -en, -on (tiz-en-egy = 11, husz-on-három = 23) instead of -von, -ven of the group D (öt-ven hat = 56 , hat-van három = 63). The number 30 is not followed by any suffixes (harminc-egy = 32).

3) There are no parallel examples in other languages that the signs of 10, 20, and 30 are a separate group – using whatever signs - being inserted and acting as a foreign body among the unified ancient number-signs as in Egypt.

4) We can't call it a random occurrence if a similar insertion into the numerals of a different language can be noted, as is the case in Hungarian. To prove the last statement, one must look deeper and search for more connections.

Examining the numerals based on the previous table:

1) The Hungarian numerals in group D, and only in this group, are built uniformly: negy-ven, öt-ven, …kilenc-ven (40, 50, … 90). It is the name of the singular number (4, 5,…9) + the word "ven", "van", which therefore has to mean 10. Similar to Hungarian, "van" or "ven" means 10 in Sumerian as well. (You would not expect the same name to be invented for the same thing twice.) Therefore, négy-ven = négy tíz <nedj tiiz> = 40 and hat-van = hat tíz =60, etc.

   The Egyptian numerals belong to group D. They are built the same way: from the corresponding singular number and a horizontal line added meaning 10. For example: if 10 = von, then (reading from right to left) ▬ ∷ = 4 von = négy-von = with today's pronunciation negy-ven <nɛdjvɛn> = 40.

2) The names for 10, 20 and 30 – tíz, húsz and harminc – are differently built and have nothing to do with the names of the singulars as used in any other language. Even három (3) and 30 (harminc <harmincz>) are not directly connected (see (6) below for explanation).

3) The direct meaning of the word "tíz" is uncertain. It could mean fog (tooth or teeth), fok (cap, degree) going by the form of the sign Λ. It looks like a pyramid and "pyr" means "pirit", "tűz" (fire) as well. The idea of "tűz-háromszög" (fire-triangle) is very old and there is the question of the identity of the words "tűz" and "tíz" (fire and 10). It is possible that the semantic range of the two words was somehow connected in early times.

4) The Hungarian number-name "húsz" (twenty) does not mean directly a number, neither do our other number-names. We should call it rather "kettő-tíz" <kɛttœ-tiiz> (two-ten), if handling as a number-name. This has been done in other languages: Zwanzig in German was built from zwei and twenty in English from two. In Hungarian however, the words "húsz" and "kettő" (20 and 2) have morphologically nothing in common.

In Egyptian, the sign for 20 differs from that for 10 (Λ) only in the horizontal line above it. There is no other difference and the words 'tíz" and "húsz" are morphologically not related. Thus the question may arise whether the word "húsz" could mean the naming of that "húz"-ott (drawn, pulled) line above the sign for 10: Ā? As words, "húsz" and "húz" sound quite similar and we also find support for this in the etymological dictionary, which tells us that they could cover the same meaning in number-writing:

The Etymological Dictionary by Czuczor-Fogarasi explains HÚSZ as follows: *"The cardinal number HÚSZ means twice ten and Hungarian linguists refer etymologically to the verb "húz" as being most natural, because the reckoner **draws a line** on arriving at the second ten while counting."*

Consequently, the word "húsz" could express "húzás" <huuzaash> (pulling) as a number-name. However, Egyptian number-writing offers a perfect explanation for it.

5) Borbola János poses the question in his book *Királykörök* <Kiraaykœrœk> (*Royal circles)* on pages 209-210 as well: might the name for the number 20 have become "húsz", because we "húz" (pull) a line above the 10 (Λ) while writing it. Furthermore, for the numbers 40 and above we "von" (draw) the tenfolding line after the singular number: "negy-ven", etc. We "tenfold" the singular numbers with the suffix "-ven, -van". We can explain the unique forms of our numerals perfectly by accepting the role of this dual expression of drawing a line: húza-vona (drawing-pulling). All these can only be explained by the form of the Egyptian numerals.

312

6) Listening to the official explanation, the word "harminc" (30) is the result of adding three words, however this statement isn't right.

The root morpheme of the word *"három"*, earlier *"harm"*, is *"har"* which means pointing up, protruding, emerging... Look at the finger of our hand: the third is the longest, the most protruding and we call it har, harmadik (third). The word *"harminc"* as explained by Czuczor-Fogarasi: *"Harmonc is identical with harminc and triplets was its earlier meaning".* Elsewhere: *"We call 'harmonc' an amount which contains three of a certain unit"* The word "harmonc" by itself therefore means merely the triplet of something. We can only say that the word *harminc* tells us to write the sign **Λ** three times in written **ΛΛΛ** or in engraved (hieroglyph) ∩∩∩ form.

The above reasoning can even be proved etymologically.

The ending -onc, -enc is known, but the "n" in it is a late insertion. For example vadóc<>vadonc (darnel, wilding), magóc<>magonc (seedling). The word tömlőc <tœmlœcz> (dungeon) could even be pronounced as tömlönc. The sound 'c' stands here for the previous 's'. For example: gömbőc <gœmbœcz> or with 'o' gombóc <gomboocz> (dumpling) was pronounced in old-Greek as kümbos. Following this path, 'harmonc' means, by the dictionary of Czuczor-Fogarasi, in reality 'hármas' <haarmash> (threefold, triple). If 'hármas' <haarmash> is pronounced with a front-vowel and with an 's', we come to Hermes. Well, *Hermes Trismegistos*[70] said about himself that "... my name is *Hermes Trismegistos*[71] because I owe all three parts of the Universe's knowledge". Therefore, Hermes (hármas) himself is 'harmósz', 'harmonc' (triple), meaning hereby his triple knowledge and, in the case of 'harminc', meaning the triple of ten.

Finally, let's see the numbers nyolc <njolcz> (8) and kilenc <kilɛncz> (9) in group B.

1) There must be a definite reason for such a structural difference of the number-names 'nyolc' (8) and 'kilenc' (9) from all others before them.

2) Also, we should be aware of the identical word-building in 'harminc' and 'kilenc'.

70  Corpus Hermeticum/Tabula Smaragdina

71  'Tris' means triple in most other languages as well. 'Megistos' is the old Greek word megas = magas <magash> (tall) and magasztos means majestic, dignified in Hungarian. His whole name would be in Hungarian: 'Háromszor Háromszor Magasztos. (the second háromszor = tris is superfluous, because the first 'hermes' = 'hármas' already includes 'tris' (in Hungarian). This fact may have been not known to the Latin scribes of later times; otherwise his name wouldn't have become as lengthy.

3) The 8 and 9 number-pair was very important in Old Egypt. They were used in calculations for circles. When doing this, 9 was written in a circle ⟨𝒬⟩ , to tell us that they were calculating the circle with this method. The question pops up: isn't 'kilenc' identical with the word 'körönc' <kœrœkncz>, köröz <kœrœz> (circling)? Borbola János writes in his book *'The Royal Circles'* (page 212) that 'kelence' means an encircled apiary or a circular beehive. Therefore the words 'kilenc', 'kelenc' and 'körönc' sound alike and mean all things connected with the circle. As Borbola János says on the same page, about the number 'kilenc': *"It can't be just accidental that this one number includes the idea of the circle as well as the forgotten key to the ancient calculation of the circle's surface."*

4) The meaning of the word 'nyolc' can only be rendered probable. The dictionary by Czuczor-Fogarasi says that it was pronounced earlier as 'nyult' and before that as 'nyolt', but its explanation about the length of the different fingers is not convincing. The theory of Borbola János however should be seriously considered. He pointed to the fact that when calculating the circle's area by the method of '9' in Ancient Egypt, the number 8 had to grow (be stretched) by its quadrate. The expression nyú, nyő – nyújt <njuuyt> (stretches), növest <nœvɛst> (lets grow) are acceptable names of the function of 8 in the procedure. This is only an assumption, but it is a strong one looking at it among all the other etymological explanations. By examining the whole and its parts we must always explain the parts based on and out of the whole.

*"The well of our past is profoundly deep".* (Thomas Mann)

\*

We could look into the very deep well of the past for a short glimpse. We have recognized connections between times separated by large gaps. These connections disappeared long ago into the labyrinths of written history. I do not intend to rewrite history, I have merely been searching for what has stayed unchanged in the human mind. The writing signs, numerals and number writing are perfect tools for finding what has remained unchanged. Therefore, we can't hide the results of the investigations just to "keep quiet". To keep them covered just because these results are not what we believed up to this moment.

A discovery is always a surprise. We call it a discovery, because it is not the result of our previous knowledge.

Finally, I want to show you that the surprising Ancient Egyptian –Hungarian connection in the world of numerals is not isolated. The writing signs inform us of similar connections. Don't ask me how this happened. The historians must answer that question.

Look at the next picture, where we can see nine signs of the demotic alphabet which are identical in form and sound with the corresponding signs of the Hungarian-Székely rovás. These signs are taken from the so-called 'Rosetta Stone' written using multiple languages and alphabets. These nine signs are already 1/3 of an alphabet and the concordances are indisputable, it can't be random.

Egyptian-demotic Székely-Hungarian

/ (r)    / (r)
ʟʟ (g)   ∧ (g)
ℙ (i)    ◁ (a)
ᕐ (n)    Ɔ (n)
◥ (k)    ⱱ (k)
ſ (o)    ℗ (o)
≼ (t)    Ƴ (t)
ꜱ (w)    ꙅ (ü)
ʔ (e)    ʔ (e)

Picture 369. The certain concordances between the signs of the demotic and Hungarian-Székely rovás writing-signs.

Reading:

Egyptians and Székely-Hun people used earlier the two similar calendars:

1) In Egypt the Sun-year lasted three seasons of 4 months each and every month had three times ten days. The 5 left-over days were handled separately as holidays.

2) The archaic moon-calendar was run by the moon-cycle and used for example to calculate holidays, like the date of death and resurrection of Osiris, as we calculate Eastern today.

19 Sun-years are equal to 235 moon-months.

This makes it possible and indeed probable that the well known vowel-elision method of the Székely rovás-writing was used in the Egyptian demotic writing as well. The possibility of using vowel-elision depends strongly on the language. Actually, you can only do it with a couple of languages. It is not a conceptual dispute. Try it. Try it with English.

# 2. A FEW MORE WORDS ABOUT THE PYRAMID FIGURES

The number-sequence in the pyramids can be followed into infinity.

We received the number-sequence of 1, 3, 6, 10, 15, 21, 27 etc. If we order the balls into a quadratic form instead of a triangle, then we receive the figures below:

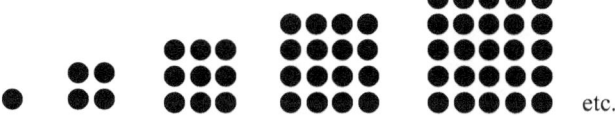

The number-sequence in the quadrate-forms is: 1, 4, 9, 16, 25, 36, 49 etc. Putting them one onto the other in a growing sequence, we receive a spatial pyramid and the sides of it are fire-angles as well. This deepens the belief in the wholeness of the pyramid. The number-sequence received this way is: 1, 5, 14, 30, 55, 91, 140, 204, 285, 385, 506, 650, 819, etc.

Picture 370. An old, spatial - but front-wise looking as a fire-angle – formation, put together from globes. (Musèe historique de Strasbourg)

It was a fascinating observation to recognize the connection between the numbers building a triangle and a quadrate, namely that both contain the pyramid-formation: the one on a flat surface and the other spatially. No wonder so many of our ancestors saw a divine manifestation in the pyramid-numbers.

The pyramid-angle (fire-angle) can be created on a flat surface. We think that studying numbers this way has happened since calculations were done using dots, discs, flat pebbles, in other words since dot-line numerals existed.

# 3. SUMERIAN NUMBER WRITING

It is common knowledge that from the beginning the Sumerians also wrote numbers with their usual wedge-form writing stick into clay. However, on the very oldest finds they had not yet started to use their typical wedge, but rather cylindrical sticks. The find below is around 5,000 years old and was found in Uruk.

Picture 371

This clay-plate is a relic of early Sumerian bookkeeping. We can see on it the summary of the details on the back. According to it, the number of cows and oxen is 54.

The signs were engraved with the same round stick into the clay. The stick was held perpendicularly to print the tens, but for the singulars however, our master tilted the stick and caused a longer imprint starting with a half-circle. Actually, he created a line by tilting the stick. No new numerals were formed this way; he merely applied the ancestral number-writing to clay. Only here – as we also could see elsewhere – the value of the dot (1) and the line (10) were exchanged.

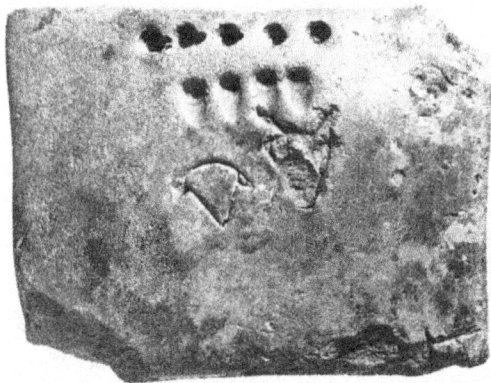

1          10

Picture 372

Knowing these two signs, we can easily read the numbers on the 5,100-year-old clay-table from Uruk below.

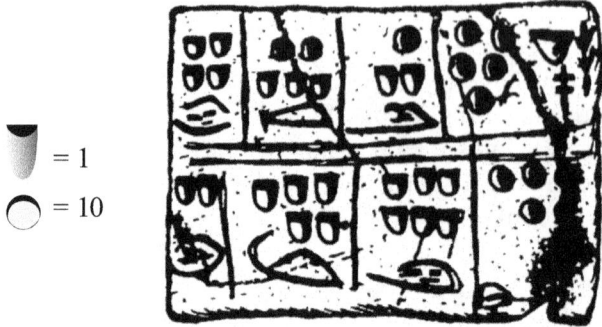

= 1

= 10

Picture 373. The pictures below the numbers probably show the articles being counted.

The signs for 1 and 10 were engraved with the smaller end of a conically formed round stick into the clay:

Picture 374. The smaller end of the stick (here on the left side) was used to print the 1 and 10 into clay, by holding it perpendicularly or slanted.

The thicker end of the stick was used for printing larger place-values. This is the reason for using the same forms, but larger or smaller, for printing different place-values. Below, we can see the signs for 1 and 10, and also the place-values standing above them, which are merely the larger or combined forms of these two signs:

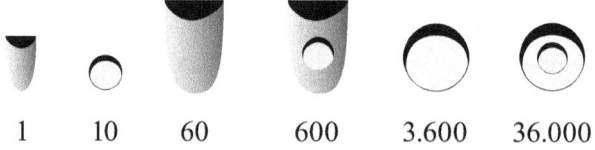

| 1 | 10 | 60 | 600 | 3.600 | 36.000 |

Picture 375. Sumerian place-values.

The Sumerians used the base 60 calculating system. The place-values however are not regular, for example the presence of the 10 points to a start of the decimal system.

Our suggestion about their early use of the decimal system is supported by the larger place-values. First let's look at the round numbers of the base 60 system (1, 60, 60x60, 60x60x60, etc.).

1      60      3.600      216.000

There are place-values between these values in the base 60 system as well (see previous picture). These are supposed to be the so-called "bisecting" place-values seen in the decimal system: the numbers put between 1, 5, 10, 50, 100, 500, 1,000, etc. were always half of the next higher value. However the Sumerian additional values are not bisecting, they are the tenfolded value of the previous round number. It points again to an extension of the decimal system. The place-value between 60 and 3,600 is not half of 3,600, but 10 x 60 and the next is not half of 216,000, but 10 x 3,600 as well, clearly recognized by looking at the signs for these numbers:

60 x 10 = 600        3.600 x 10 = 36.000

Picture 376. The place-values in between were therefore not built based on mathematical means as seen in other cultures.

The system will be clear for everybody, if we divide the round and the in-between place-values. There, it will be more visible that these in-between numbers are the tenfold values of numbers below them.

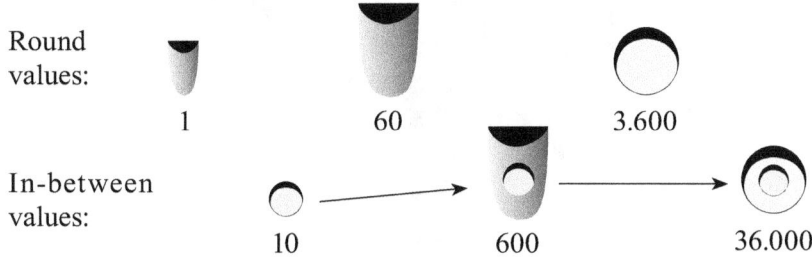

Picture 377. The round place-values are on the upper line, the in-between values are on the lower line.

Based on the above, we can read even the animal's registered numbers in the book-keeping below:

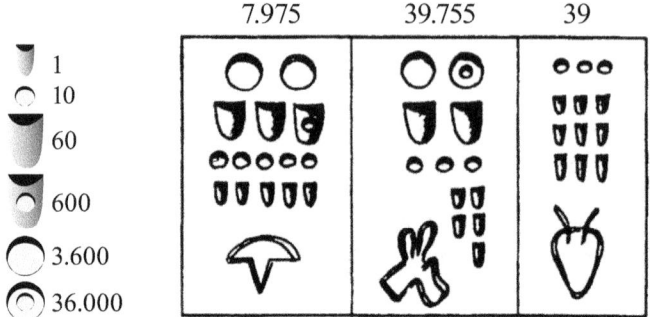

Picture 378. The animals from the left: cow, donkey, ox.

The signs for 1, 60 and 600 can be tilted as well. The signs for 60 can even overlap each other, but not the 600, because the circle of 10 in the middle would disappear:

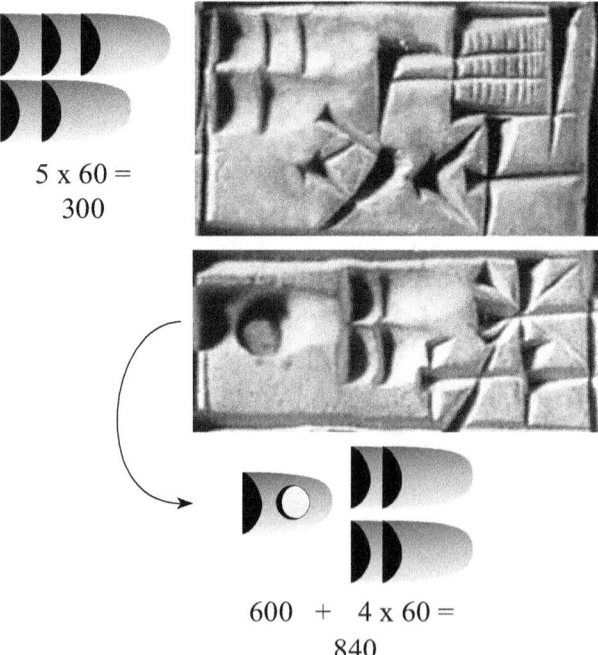

5 x 60 = 300

600 + 4 x 60 = 840

Picture 379. The sign for 600 is definitely marked by the circular form in it.

## Table IIV/I

Early Sumerian clay tables, containing only numbers:

These numbers above can be read easily by applying the number-values learned on the previous pages. The only exemptions are the two half-moons on the last plate, which stand for 1/5. The fractions will be discussed in the lower part of page 330.

322

TABLE IIV/2

Small Sumerian clay tables with text and many numbers:

The registered data are clearly inventory numbers.

The sign for 600 is a good example to present mathematical procedures. It contains two signs: 10 and 60. Their relative position to each other will show whether there is an addition or a multiplication. We see an addition if they are printed side by side and a multiplication if one is inside the other. We can even see two examples below:

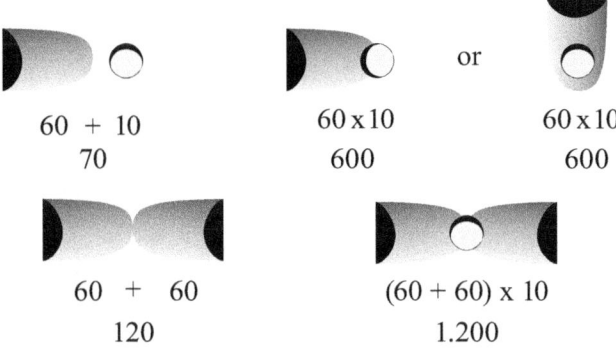

60 + 10          60 x 10          60 x 10          or
   70               600              600

60 + 60          (60 + 60) x 10
  120                1.200

Picture 380.  Marking addition and multiplication

To signal extraction Sumerians used their known sticks in the following way:

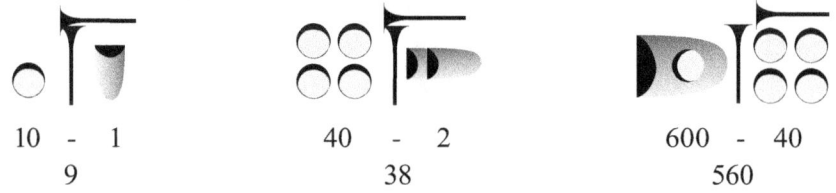

10 - 1          40 - 2          600 - 40
   9               38              560

*If many numbers are used together in a calculation, congestion can occur. However, no mistakes should happen, because the signs are grouped around the part framed by the lines drawn by the writing sticks:*

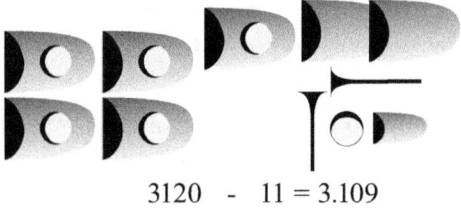

3120 - 11 = 3.109

Picture 381.

All the signs presented so far and printed into clay had the form of a circle or a cone. There are no other forms apart from these two.

It is interesting that number-signs existed not only in the form of a print, but even as small figures moulded from clay. Additionally, the number-set was transformed in order to be able to be printed with their writing-sticks into clay. Therefore they used the same number-set in three different ways:

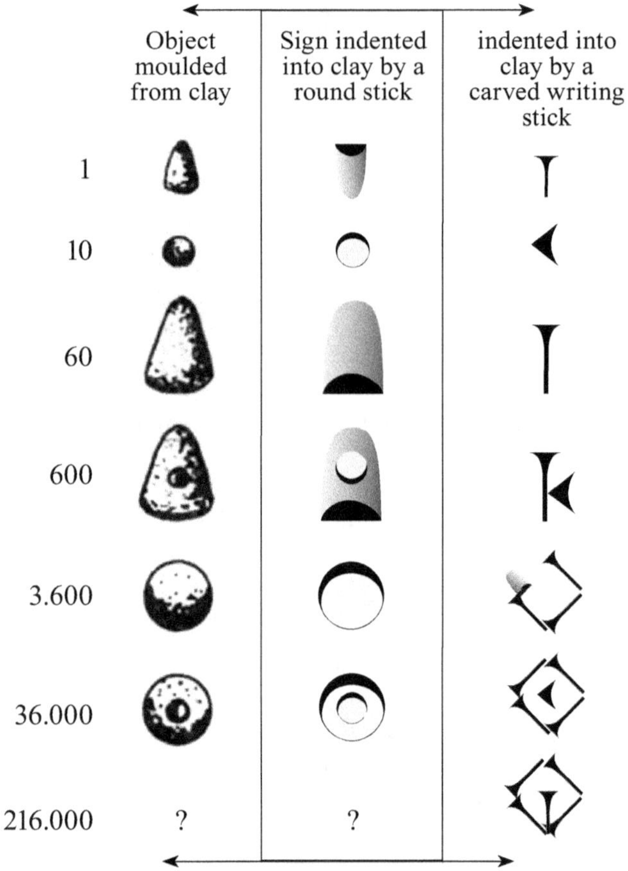

| | Object moulded from clay | Sign indented into clay by a round stick | indented into clay by a carved writing stick |
|---|---|---|---|
| 1 | | | |
| 10 | | | |
| 60 | | | |
| 600 | | | |
| 3.600 | | | |
| 36.000 | | | |
| 216.000 | ? | ? | |

Picture 382. The same numeral-collection using three different designs

The signs indented into clay (middle column) were moulded as objects as well (column on the left). The reverse could also have happened. The small clay figures may have existed first. We don't have enough finds to be able to answer the question of which was first, but this won't lessen the fact of their identity. .

In the column on the right are the writing-stick variants of the middle-column's signs printed into clay. Writing this way, the small and large cones were restricted to a single line and the circle became a quadrate, because no circle could be printed with the end of the carved stick. The small disc could only be printed with the end of the stick containing a small semi-circle: ◀ .

What purpose could the clay-figures representing the numerals have had, as objects in the hand?

The benefit of this method was that one could put the numbers into the pocket or into a small container. It was unnecessary to count them again, because their form already indicated their value. We count our metal coins the same way: we just add the coins with the same forms. Therefore, the coins in our pocket represent a number, the sum of several parts having the same amount, made up of different values. Even a blind person can count them, because the form of a part indicates its value.

This method was a big help to the Sumerians in keeping their records. If they had to record the number of articles changing hands in a business, they just had to assemble the correct clay figures, and put them into a hollow clay-globe, which they closed. The number stayed in there indefinitely and could not be falsified.

We see below two "number-stores". The right picture is an X-ray of an intact globe and we can clearly see the numerals in there:

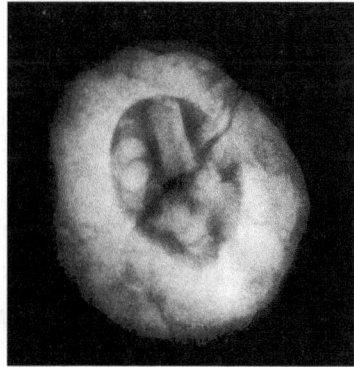

Picture 383. "canned numbers". This method is similar to our saving-box also made from clay, which often has the form of a pig and also has to be broken at the end.

We see a broken saving box in the following picture with its contents before it:

Picture 384. The sum of the numerals results in the number 4,292. (The whole conception of the picture is mine, but it represents the truth.)

Sumerians tried to use the same stick for writing numbers and text. Using different tools would certainly been inconvenient. The imprint of the writing stick is easily visible on the picture below:

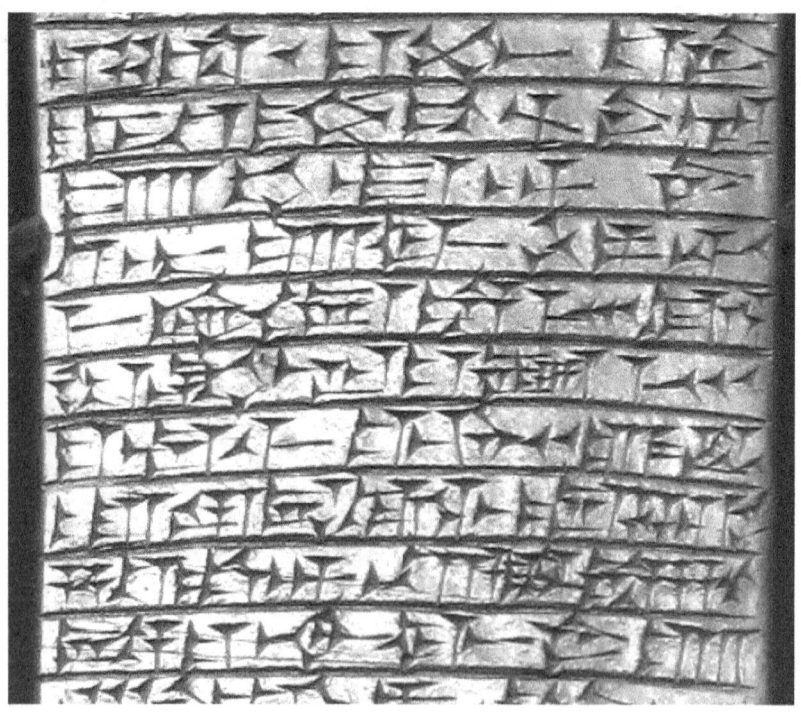

Picture 385. Text is written with cuneiform signs, this time on a gold plate

There are four cuneiform numerals, where ◀ = 10 and Ţ = 1:

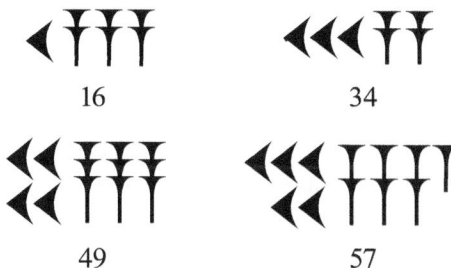

Picture 386. It is easy to read these numbers, when knowing the two signs.

As we could see, the sign for the 1 is identical with that for 60, only much smaller.

Picture 387. Only the difference in size matters, as in the case of the original signs: 1 = small cone, 60 = large cone.

This distinction in size was used mostly in cases which were not totally clear. The signs could have been the same size, if their place-value was easily recognizable. See the following numbers.

The number 60 is again a good example:

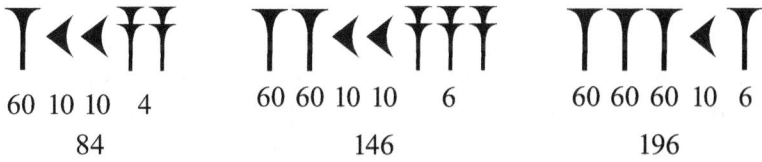

*Now we are able to read and write every cuneiform number written with the stick:*

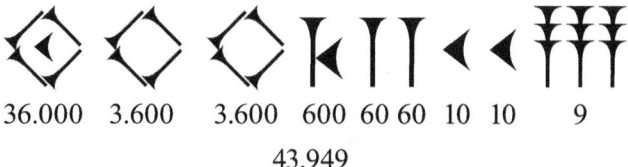

Picture 388.

The sign for 36,000 can further be enhanced by multiplying the sign of 10 ( ◀ ), written inside it.

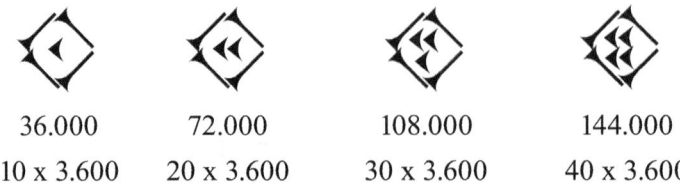

| 36.000 | 72.000 | 108.000 | 144.000 |
|--------|--------|---------|---------|
| 10 x 3.600 | 20 x 3.600 | 30 x 3.600 | 40 x 3.600 |

Picture 389. Even more tens can be printed into the square.

It is important whether the two numbers touch each other or not. If they touch then it is a multiplication. With a space between them it is an addition.

| 600 | 70 |
|-----|-----|
| 60 x 10 | 60 + 10 |

*In the case of the same sign for 1 and 60:*

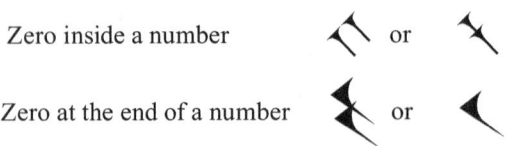

| 2 | 61 |
|---|-----|
| 1 + 1 | 60 + 1 |

Picture 390. The second example must be read by the place-value.

It seems that the clear base 60 numeric system had been introduced. Add to that the marking for the zero in Babylon, for which they had even two different signs. One was the doubled miniature sign of 1 slanted through 450. The other was used at end of the number, if there was no value in the singulars. The sign of 10 was used for the latter , but its lower tail was lengthened::

Zero inside a number     ⩘ or ⩘

Zero at the end of a number     ⩘ or ◀

Picture 391. The two signs for the empty place-value

Using this method however, the value of the number depends on the place where it stands

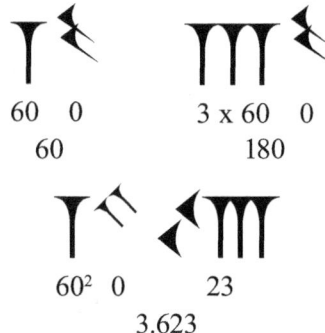

| 60 | 0 | | 3 x 60 | 0 |

60                    180

| $60^2$ | 0 | | 23 |

3.623

Picture 392. The 60 in the example below stands in a higher place than the 60 above it. (This is the reason for marking it 602)

We get a better picture of the system if the place-values are marked. The big difference is that in our decimal system, the round numbers are 100, 101, 102, 103, etc., but in the base 60 system the round numbers are 600, 601, 602, 603, 604, etc.

| local values | | | | |
|---|---|---|---|---|
| $60^4$ | $60^3$ | $60^2$ | $60^1$ | $60^0$ |
| 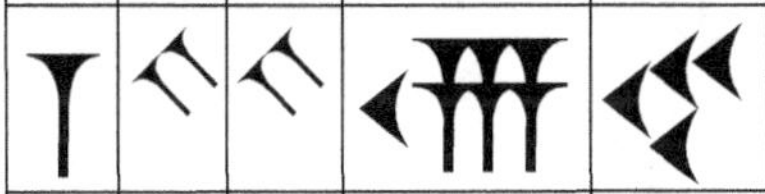 𒁹 | 𒌋 | 𒌋 | 𒐕 | 𒐏 |
| $60^4$ | 0 | 0 | 960 | 40 |

12.961.000

Picture 393.

The marking of the empty place-values allowed the writing of fractions with numerals. See the method:

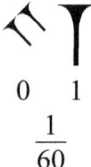

0   1

$\dfrac{1}{60}$

Picture 394.

Further examples make the Babylonian writing of the fraction clearer:

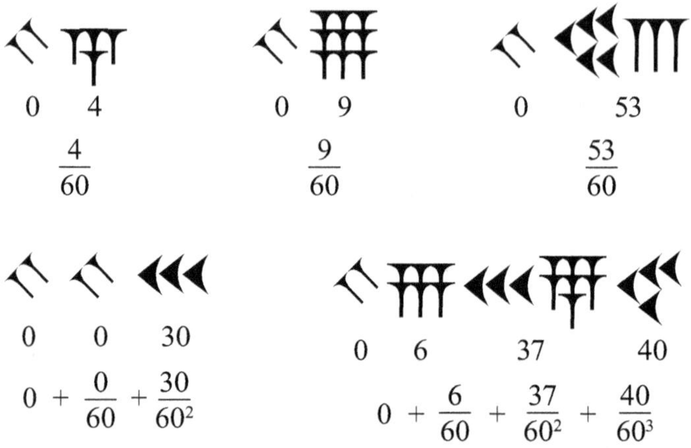

| | | |
|---|---|---|
| 0  4 | 0  9 | 0  53 |
| $\dfrac{4}{60}$ | $\dfrac{9}{60}$ | $\dfrac{53}{60}$ |

| | |
|---|---|
| 0  0  30 | 0  6  37  40 |
| $0 + \dfrac{0}{60} + \dfrac{30}{60^2}$ | $0 + \dfrac{6}{60} + \dfrac{37}{60^2} + \dfrac{40}{60^3}$ |

Picture 394. Fractions written the Babylonian way.

The pre-Elamites had separate signs for fractions (but it is possible that the Sumerians had them as well):

| $\dfrac{1}{120}$ | $\dfrac{1}{60}$ | $\dfrac{1}{30}$ | $\dfrac{1}{10}$ | $\dfrac{1}{5}$ |
|---|---|---|---|---|

Picture 385. Separate signs for the fractions, for printing into clay

A written number (to be read from right to left):

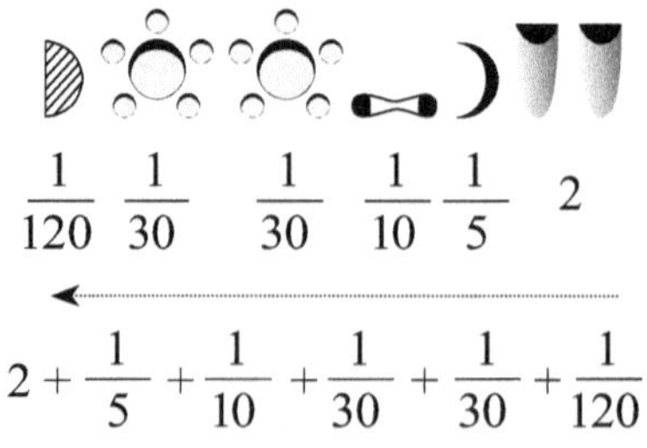

| $\dfrac{1}{120}$ | $\dfrac{1}{30}$ | $\dfrac{1}{30}$ | $\dfrac{1}{10}$ | $\dfrac{1}{5}$ | 2 |
|---|---|---|---|---|---|

$$2 + \frac{1}{5} + \frac{1}{10} + \frac{1}{30} + \frac{1}{30} + \frac{1}{120}$$

Picture 397

The round numbers of the pre-Elamites were also partially different from that of the Sumerians:

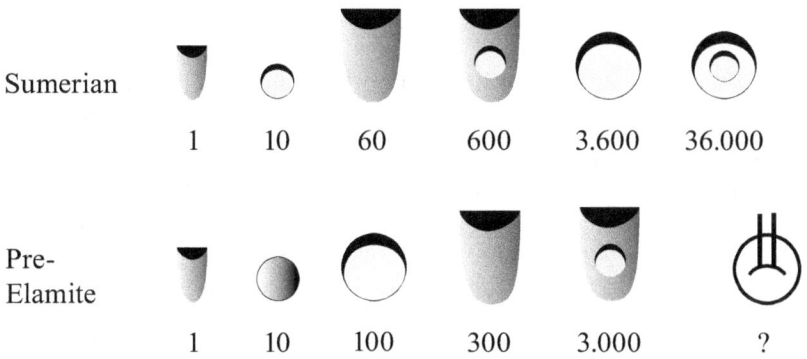

Picture 398. The question-mark may stand for 9,000 or 18,000

They also used their signs moulded from clay, but with a small difference. The sign for 10 was a small globe instead of a disc and that for the 1 was a small cylinder instead of a cone:

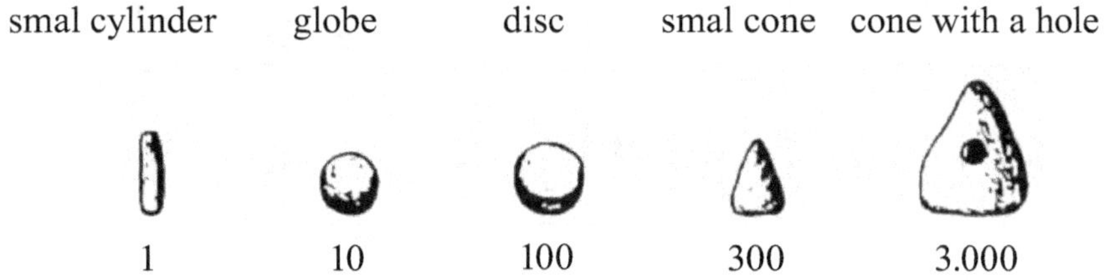

| smal cylinder | globe | disc | smal cone | cone with a hole |
|:---:|:---:|:---:|:---:|:---:|
| 1 | 10 | 100 | 300 | 3.000 |

Picture 399. The clay-numerals of the Elamites. This means that they too used the closed clay-globe for "canned numbers".

This wasn't a clear base 60 system anymore. It was a decimal system with a few parts of the base 60 system (300, 3,000) mixed into it.

Finally, the abacus was extensively used all over Asia Minor as well.

# 4. THE CHINESE NUMBER-WRITING

The Chinese used the dot-line or line-line method in ancient times like everybody else. In time they changed, as happened in every independent culture, but the changes here meant the replacement of some line-line numerals by writing-signs. Of course, there were territorial differences.

We can see a part of this process in the picture below. In the first row are the original numerals, below them, we can see quite clearly the path of the changes down to the 5th row, to *the final official variant:*

Picture 400. A few variants of ancient Chinese numerals. The number-signs for 6, 7, 8, and 9 in the last row demonstrate the changes happening after the change to paint the signs with a brush. Looking at these signs, we can see that they are the same as the signs in the 4th row, only looking a little more "Chinese".

Some signs became more figurative as seen on the next picture. This is the sign for the number 10, assembled from the perpendicular line of 1 and the horizontal tenfolding line::

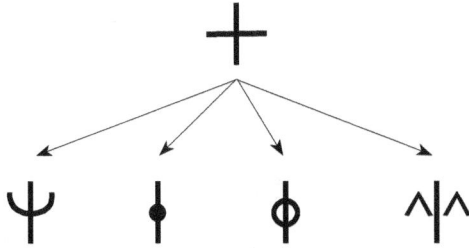

Picture 401. As we can see, this is only a drawing game, trickery every time with the same number.

Let's stay with the + sign. Originally – as we have seen previously – the signs for 10, 20, 30, 40 are the horizontally crossed signs for 1, 2, 3, 4:

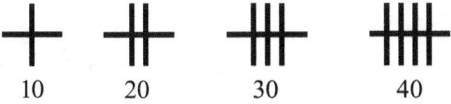

*Chinese people wrote the following:*

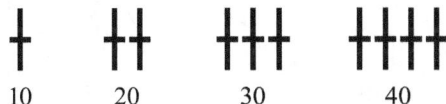

*signs' decorative variants separately:*

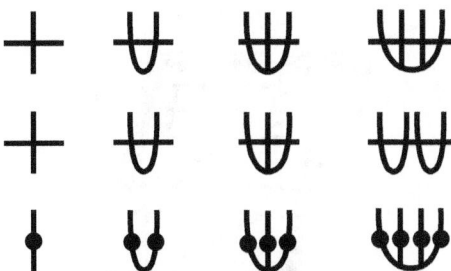

Picture 402. This signs are merely "variations on a theme", modifications of the same basic characters.

However, the Chinese people couldn't forget the original line-line numeral system. They managed their calculations with their characteristic
bamboo sticks, assembling with them the numerals composed from horizontal and perpendicular lines. The numeral system was the same in China,
Japan and Korea. We can see a portable set of sticks below:

Picture 403. Korean stick set, used for calculations, in a portable little bag.

Picture 404. Japanese anecdotal drawing: calculating with sticks on a suitable table.

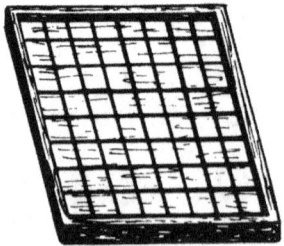

Picture 405. A Chinese table subdivided into partitions used for calculation
with sticks.

Picture 406. A Chinese man doing calculations with sticks in the courtyard.

Nevertheless, they used the calculator with the beads (soroban).

Picture 407.. A Chinese merchant with suan-pan (soroban)

Chinese number-writing went in two different directions. One followed a method which enabled form-variability in hand writing (see later) and the other kept the line-line number-writing intact. The tools used for calculation (bamboo sticks) influenced and preserved the method of using perpendicular and horizontal lines for numerals. However, they just kept the line as a sign and lost the opportunity to create the place-value of the numbers within the forms of the numerals. We will see more about this in Chapter 10, "About the zero (0)" on page 55. I proved in this chapter that the introduction of the sign for 0 (zero) was merely necessary due to the deterioration of the ancient writing-method and it is neither a genial invention, nor a new idea.

In any case, the sign did not tell any information about the place-value for the Chinese people. They tried to make up for this lack in three different ways. 1) They left gaps for "empty" place-values, telling the reader that one is needed, but that it is an empty place value. 2) They marked the empty place with a sign which finally became a circle. 3) They wrote with Chinese signs the words "hundred", "thousand", etc into the empty space.

We should mention, before presenting the possible solutions from above, that if two signs are put beside each other by writing them in sequence, then one of them can be turned in order to mark the end of a number-sign and the start of the next one. For example, the usual form of ⊥⊥ (67) would be quite unambiguous as ⊥〒, just as the form ∥≡ (23) would be better if turned Therefore, the numbers 1 – 5 could be turned and the numbers 6 – 9 could even be written in two different ways:

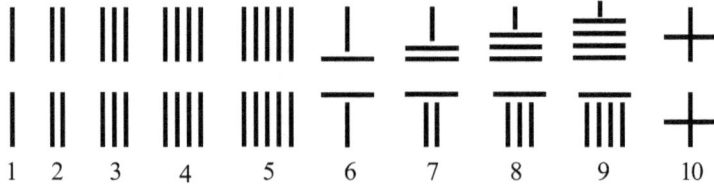

Picture 408. The signs of these two variants could always be chosen freely.

Thus, it became possible in this way to turn the next sign through 900 if both signs have parallel line signs. The horizontal and perpendicular became alternating this way.

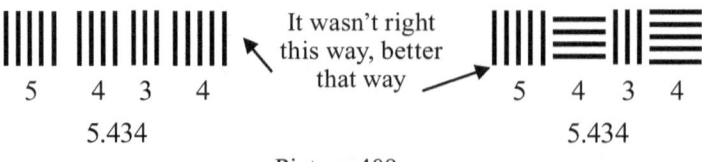

Picture 409.

A number could be written in a more condensed way due to the possibility of turning the numerals:

6.666                6.666

Picture 410. Nice patterns in sequence are created this way.

Let's look now at the possibilities for marking the empty place values: Leaving a gap between the signs for an empty space-value, was one possible solution. Fore example:

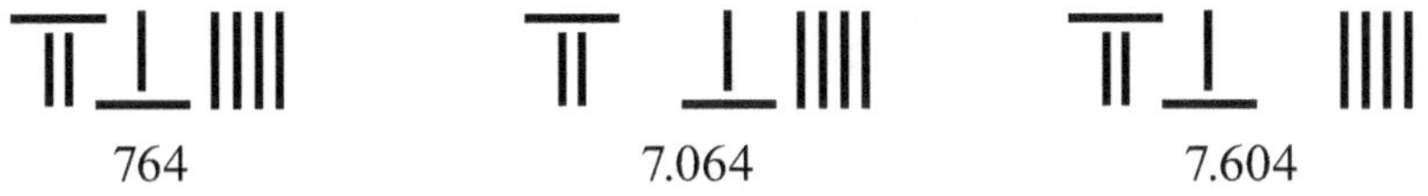

764                7.064                7.604

Picture 411. A gap points to the existence of an empty place-value.

There is a problem when the last number (singular) is missing; no gap can be left there, because there is no following number . One very logical solution in this case was to put the sign for 10 (十) at the end or if two numbers were missing, the sign for 100 (百):

7.640                76.400

100

Picture 412.

The number on the left therefore: 764 x 10 = 7,640. On the right however, 764 x 100 = 76,400. Even the gap can be avoided this way, since the recognition of a gap might be questionable. See the introduced sign of 10,000:

70.640

10.000

Picture 413.

This number is: 7 x 10,000 + 640 = 70,640.

This solution for the missing place-value is usable, but not the best. The best solution is to mark the gap with a sign. The circle (**O**) is a Chinese writing-sign meaning "empty", or a "well".

Let's see some numbers written using this method:

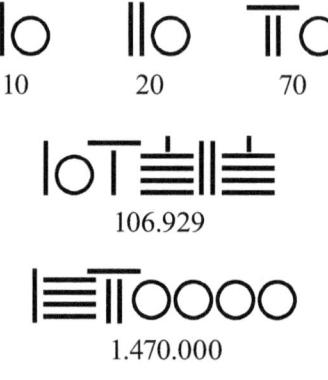

Picture 414.

Even fractions are easy to write after the introduction of this sign. The circle would stand at the beginning, indicating that the number should be read as a fraction with zero integers before the comma:

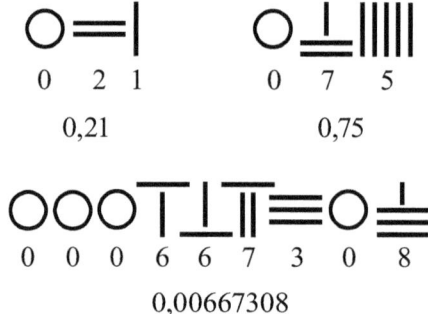

Picture 415. Numbers written with a sign marking the empty place.

Negative numbers were marked by a slanted line crossing the last number:

Picture 416..

This way, one can create a beautiful sequential pattern out of the number signs. This is called the Sanghi method. The following page was created (drawn) by the Japanese mathematician Fujita Sadasuke in 1779. It shows the reading of the number on the top, but having read the pages before, the rest of the numbers can be read easily as well:

46.431

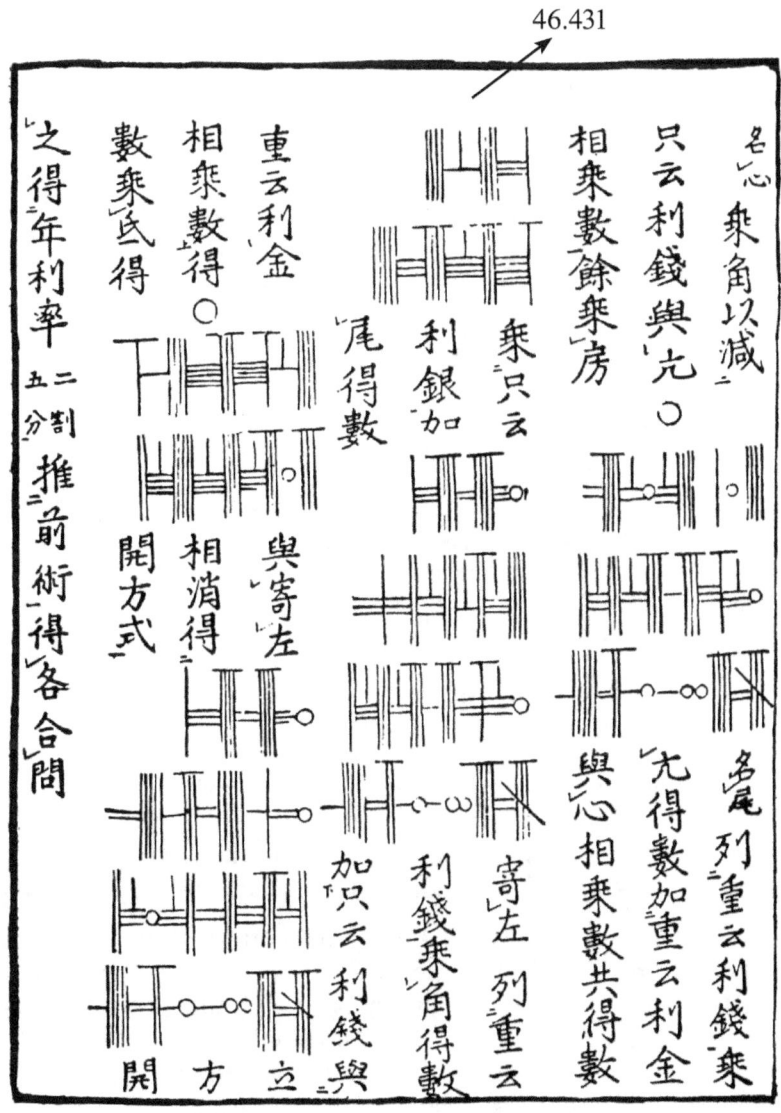

Picture 417. D. E. Smith: History of Mathematics, page 41 (1925, 1953, Dover Publications Inc. New York).

Finally, let's see the drawn variants of the numerals 1 – 10 followed by by 100 – 10,000.

We can see the variations of the numerals below. In the first row are the numerals used today, the standard in Japan as well. The change of the signs can't easily be followed in every case, but with a  graphically trained  eye one can follow the metamorphosis of most:

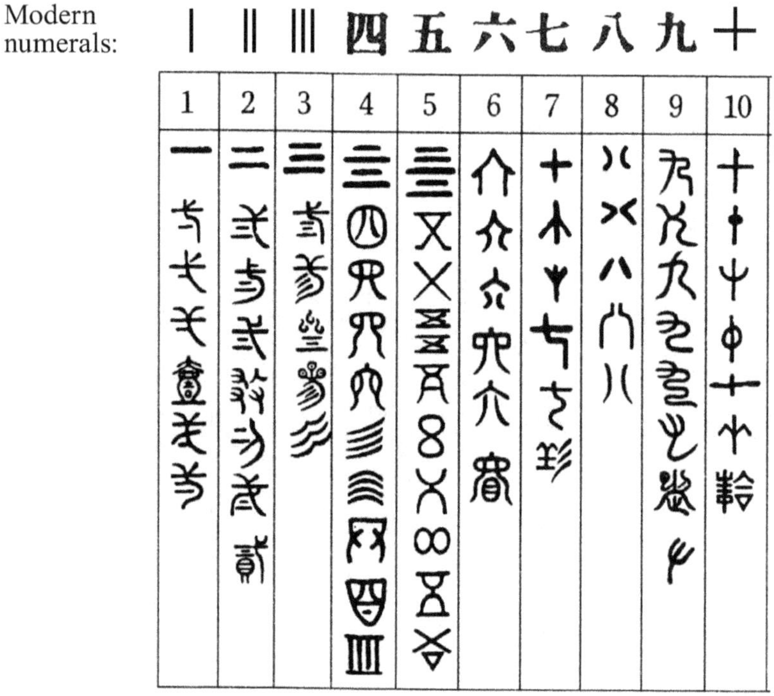

Modern
numerals:

Picture 418. It is easy to see the graphical inherence of the signs.

In the first three columns, the 1, 2 and 3 ancient lines remained in all variations: —  =  and  ≡       but mostly hidden in this sign:

The sign for four remained as 4 lines, but a drawn figure stepped in as well. The sign for 5 has been "infected" by the sign X and all following signs are derivatives of it (including the sign with the form of 8), but all these variants didn't endure (see the sign used today).

The variants of the signs 6, 7, 8, 9 and 10 do not need any explanation, their development under the hands of calligraphers can easily be followed.

The number-signs were often written as ligatures (contracted):

米 disintegrated 八十八 reading 8 x 10 + 8 = 88

8 10 8

茶 dis-integrated 十十八十八 reading 10 + 10 + 8 x 10 + 8) = 108

10 10 8 10 8

construction of this sign: 茶 ⟶ 十十 / 八 / 十 / 八 where 十 = 10 and 八 = 8.

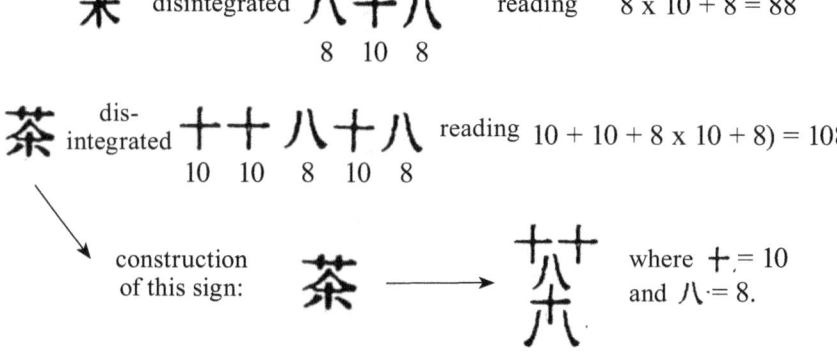

Picture 419. Two examples for ligatures.

The next example can easily be understood through an observation. The Sumerians and the Egyptians always wrote and certainly used to say it this way: "1" (one) hundred. This happens in our case as well, although occasionally we say just hundred. Therefore the one has always been included in the numeral 100. Accordingly, the sign for 100 is assembled from the signs for 1 and 100:

100 ⟵ 百 ⟶ 1

ONE hundred

*Leaving out the line for the 1 at the top,*
*then this really meant one.*

白

99

Picture 420. Chinese people always appreciated wittiness.

Chinese writing is characteristically perpendicular, and this applies to their number writing as well:

六 ⟶ 6
十 ⟶ 10
三 ⟶ 3

63

Picture 421.

In total, this is the complete sequence of Chinese numerals:

一 二 三 四 五 六 七 八 九
1   2   3   4   5   6   7   8   9

十   百   千   萬
10   100   1.000   10.000

Picture 422. The complete sequence of Chinese Numbers.

A large number:

五 萬 三 千 六 百 八 十 一

| 5 x 10.000 | 3 x 1.000 | 6 x 100 | 8 x 10 | 1 |
|---|---|---|---|---|
| 50.000 | 3.000 | 600 | 80 | 1 |

53.681

Picture 423.

The signs for the large place-values:

$10^4$   一萬   一萬萬   $10^8$

$10^5$   十萬   十萬萬   $10^9$

$10^6$   一百萬   一百萬萬   $10^{10}$

$10^7$   一千萬   一千萬萬   $10^{11}$

Picture 424. It is easy to read the numbers using the numeral-sequence top down.

# 5. ANCIENT GREEK NUMERALS

The original Ancient Greek number-sequences had round place-values as well (1, 10, 100, 1,000). It seems they started changing this method after the introduction of the half place-values (1, 5, 10, 50, 100, 500). They needed new signs for 5, 50, 500 and so on. These changes were not unified as their writing wasn't either. The different city-states didn't want to align themselves too easily in everything with the others.

For example, in the city of Epidaurus, the numbers for 1 – 90 drachmae were still written down in the most ancient way using exclusively round-place-values (1, 10, 100, 1000):

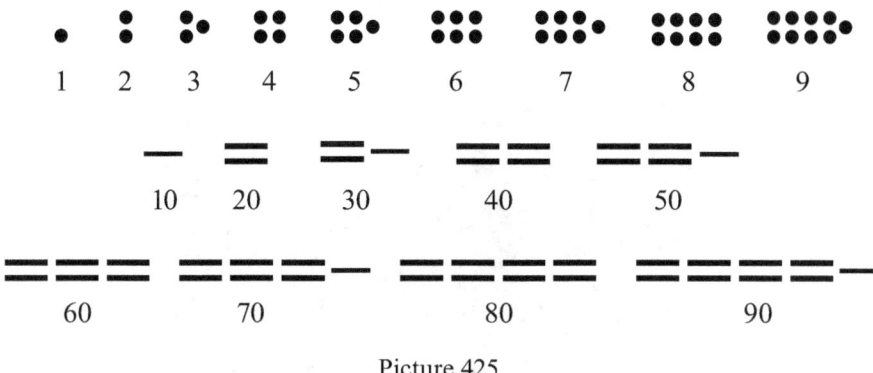

Picture 425.

However, the sign for 100 also became a letter here, although the half values had not yet appeared.

|  •  |  —  |  ⊟  |  X  |
|:---:|:---:|:---:|:---:|
|  1  | 10  | 100 | 1.000 |

426. kép.

The written sign is the first letter of the numeral's name. The sign ⊟ for 100 is the letter 'h', the first letter of hekaton, though today it is shortened to: hekto. Today we call 100 litres a hekto-litre. The sign x meant in Ancient

Greek the letter 'kh', the first letter of khilioi = thousand. Today we call that which contains 1,000 grams a kilo. The sign is our letter H, only we no longer draw the upper and lower horizontal lines. The Greeks however used it occasionally, even without the horizontal lines, as a numeral. It is interesting that the sign X (1,000) is identical with the X (10) of the number-rovás.

Introducing letters did not worsen number-writing in round place-values. According to the system, the place-value's sign must be written as many times as needed for the value of the number. Look again at the numerals 1-90 on the previous page, but the same is happening in the case of the numbers 100 and 1,000 when written with the new signs:

| 500 | 4.000 |

Picture 427

Here therefore, despite the changes, no half values appeared.

Look at the following picture however. We see Archimedes (287-212 BC), the Ancient Greek mathematician, physicist and astronomer:

Picture 428. Everybody knows the law of Archimedes: "Any solid lighter than a fluid will, if placed in the fluid, be so far immersed that the weight of the solid will be equal to the weight of the fluid displaced".

There is an "abacus" under Archimedes' hand (evidently, it was characteristic for him, but with what else could he count?). Well, it is clear even on this mosaic that it has 5 rows and not 10 for the beads. (This tells us that the maker of the mosaic knew the abacus and used it as well.) This however proves the use of the bisecting numeral system. The one takes the other for granted.

In other words, the half values had to be incorporated in between the round place-values of the Ancient Greek number writing. We see below a 2500-year-old example from Attica, in which a perpendicular line was used instead of a dot (this change happened often in ancient number-writing and we have already seen several examples of this). Let's see below the half values introduced into the round place-values:

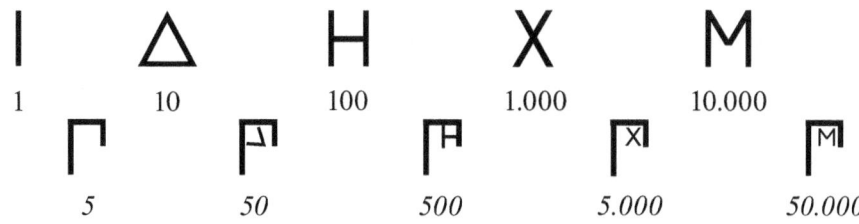

Picture 429. The bisecting Ancient Greek numeral set of Attica.

The picture tells us clearly that the half place-values have been secondarily introduced into the numeral system. The signs for the round place-values, except for the 1, have been marked by the first letter of the appropriate number-name. We mentioned previously the signs Δ, H (⊟) and X. The sign M is the first letter of the word müroioi meaning 1,000. The sign for 5 became the first letter of the Greek penta meaning 5.

A strong proof of the late introduction of the half values is the fact that all numbers are built from the sign for 5. The sign for the previously round value was uniformly written into it and the 5 had to be multiplied with it. Therefore, 50, 500, 5,000, 50,000 are respectively the signs for 5 x10, 5 x100, 5 x1,000 and 5 x 10,000. The method of number-writing however stayed the same:

Picture 430. Two numbers written with the Attica numerals.

One example from Athens:

Picture 431. An Athenian fragment with a number on it.

*This number is written in the emphasized middle line:*

# XXXᴦᴾHHHH△△ᴦ

Picture. 432 The number 3,935 written on an Ancient Greek Athenian table.
The table is broken at the number 5, thus, up to four signs may be missing.
The value of the number could be anything from 3,935 up to 3,939.

There are three 'T's at the beginning of the number : TTT meaning 3 talents. The whole script means: 3,935 drachmas and 3 talents.

The signs however lived a relatively free life. For example, the sign for 10 received the form of ☉, but this wasn't carried on into the sign for 50, which continued to be built as 5x10 using the old sign △. The introduction of ☉ therefore was a late modification. At the same time, no half values are found among the sequence of singles and hundreds:

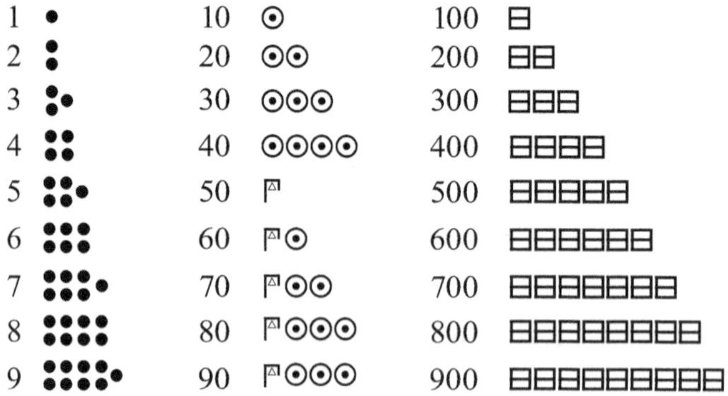

| | | | | | | |
|---|---|---|---|---|---|---|
| 1 | • | 10 | ☉ | 100 | ⊟ | |
| 2 | : | 20 | ☉☉ | 200 | ⊟⊟ | |
| 3 | :• | 30 | ☉☉☉ | 300 | ⊟⊟⊟ | |
| 4 | :: | 40 | ☉☉☉☉ | 400 | ⊟⊟⊟⊟ | |
| 5 | ::• | 50 | ᴦᴾ | 500 | ⊟⊟⊟⊟⊟ | |
| 6 | ::: | 60 | ᴦᴾ☉ | 600 | ⊟⊟⊟⊟⊟⊟ | |
| 7 | :::• | 70 | ᴦᴾ☉☉ | 700 | ⊟⊟⊟⊟⊟⊟⊟ | |
| 8 | :::: | 80 | ᴦᴾ☉☉☉ | 800 | ⊟⊟⊟⊟⊟⊟⊟⊟ | |
| 9 | ::::• | 90 | ᴦᴾ☉☉☉ | 900 | ⊟⊟⊟⊟⊟⊟⊟⊟⊟ | |

Picture 433. The ancient and the newer Greek number-writing became mixed here.

In Nemea therefore (around 2,400 years and earlier), people still used the number writing with round place values, but it was influenced by the "fashioned" number-writing method.

Changes were not restricted only to the Archipelago. The signs in Thespiae and Orchomenos for example were changed to the following:

| 1 | 5 | 10 | 50 | 100 | 300 | 500 | 1.000 | 5.000 | 10.000 |

Picture 434.

They tried to make it clear in the numerals, whether it was about drachmas, talentums or obols (obulus). They wrote the singles with the letter T and put little Ts under the rest of the signs, if the number handled talentums. These little Ts were occasionally quite wittily built into the number-signs. For instance:

3.263
talent

Picture 435. The little and large T-s meant only that we should read the number as talentums

The obulus was marked with perpendicular lines, therefore in the case of drachmas, the singles received a little stroke in the middle (⊢):

2.803 drachma és 5 obulus

Picture 436.

There were numerals for fractions as well:

X    Ɔ    C    O vagy I

1/8 obulus   1/4 obulus   1/2 obulus   1/6 drachma

Picture 437..

As I said earlier, the signs were never unified. Even major changes happened occasionally in the forms of the numerals. Georges Ifrah collected the most widely spread changes in the table below:

| Drachmas | Signs | Descriptions |
|---|---|---|
| 1 | 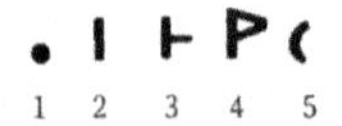 • I ⊢ P ( <br> 1 2 3 4 5 | 1 Epidaurus, Argos, Nemea <br> 2 Karystos, Orchomenos <br> 3 Attica, Cos, Naxos, Nesos, Imbros, Thespiae <br> 4 Corcyra (Corfu), Hermione (Kastri) <br> 5 Troezen, Chersonesus Taurica (Korsun), Chalcidice |
| 5 | 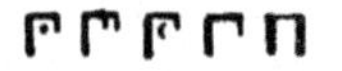 Γ Γ Γ Γ Π <br> 6 7 8 9 10 | 6 Epidaurus, 7 Thera <br> 8 Troezen <br> 9 Attica, Corcyra, Naxos, Karystos, Nesos, Thebes, <br>    Thespiae, Chersonesus Taurica <br> 10 Chalcidice, Imbros |
| 10 |  O ☉ – Ƨ <br> 11 12 13 14 <br> ↑ Λ Δ ▷ <br> 15 15 16 17 | 11 Argos <br> 12 Nemea <br> 13 Epidaurus, Karystos <br> 14 Troezen <br> 15 Corcyra, Hermione <br> 16 Attica, Cos, Naxos, Nesos, Mytilene, Imbros, <br>    Chersonesus Taurica, Chalcidice, Thespiae <br> 17 Orchomenos, Hermione |
| 50 |  Γ Γ Γᴵ Γᴵ Γᴬ <br> 18 19 20 21 <br> Γᴾ Ψ ΓΕ Γᴱ <br> 22 23 24 24 | 18 Argos <br> 19 Epidaurus, Troezen, Cos, Naxos, Karystos <br> 20 Nemea, Cos, Nesos, Attica, Thebes <br> 21 Imbros <br> 22 Troezen <br> 23 Chersonesus Taurica <br> 24 Thespiae, Orchomenos |
| 100 | 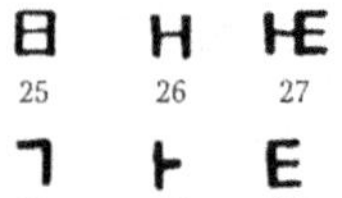 日 Η �muΕ <br> 25 26 27 <br> ⅂ ⊢ Ε <br> 28 29 30 | 25 Epidaurus, Argos, Nemea, Troezen <br> 26 Attica, Thebes, Cos, Epidaurus, Corcyra, Naxos, <br>    Chalcidice, Imbros <br> 27 Thespiae, Orchomenos     28 Karystos <br> 29 Chersonesus Taurica <br> 30 Chersonesus Taurica, Chios, Nesos, Mytilene |
| 500 |  ⅃Ⅼ ⅃· ⊢ᴵ ⊢ᴵ <br> 31 32 33 34 <br> ⋒ Γᴮ Γᴵ ΠΕ <br> 35 36 37 38 | 31 Troezen, 32 Epidaurus <br> 33 Karystos, 34 Cos <br> 35 Naxos <br> 36 Epidaurus <br> 37 Epidaurus, Troezen, Imbros, Thebes, Attica <br> 38 Thespiae, Orchomenos |
| 1,000 | X     Ψ <br> 39     40 | 39 Attica, Thebes, Epidaurus, Argos, Cos, Naxos, Troezen, <br>    Karystos, Nesos, Mytilene, Imbros, Chalcidice, <br>    Chersonesus Taurica <br> 40 Thespiae, Orchomenos |
| 5,000 | Γˣ     Γᵧ <br> 41     42 | 41 Attica, Cos, Thebes, Epidaurus, Troezen, Chalcidice, <br>    Imbros <br> 42 Thespiae, Orchomenos |
| 10,000 |  M   M   Ⅹ <br> 43   43   44 | 43 Attica, Epidaurus, Chalcidice, Imbros, Thespiae, <br>    Orchomenos <br> 44 Attica |
| 50,000 | Γᴹ     M <br> 45     46 | 45 Attica <br> 46 Imbros |

Picture 438 Georges Ifrah. Numbers, Page 184.

The Ancient Greek small letter (minuscule) script started in Byzantium probably after Carolingian times, around 675(?), with monk Alkuin. These newer letters were used later to mark numbers as well. I will present the letter-numbers first on page 389; however, they mixed these letters among the numerals, as seen on the previous pages. I must therefore present them here as well:

The small letter number-set:

| α | β | γ | δ | ε | ϛ | ζ | η | ϑ | ι | κ | λ |
|---|---|---|---|---|---|---|---|---|---|---|---|
| 1 | 2 | 3 | 4 | 5 | 6 | 7 | 8 | 9 | 10 | 20 | 30 |

| μ | ν | ξ | ο | π | ϛ | ρ | σ | τ | υ | φ | ·χ |
|---|---|---|---|---|---|---|---|---|---|---|---|
| 40 | 50 | 60 | 70 | 80 | 90 | 100 | 200 | 300 | 400 | 500 | 600 |

| ψ | ω | λ |
|---|---|---|
| 700 | 800 | 900 |

They started afresh from 1,000 on, but had to put a comma before the signs, for instance:

$$,α = 1.000$$
$$,β = 2.000$$
$$,γ = 3.000$$

and so forth up to 9,000. After 10,000 however, they used the capital letter-numerals. (A line was drawn above the letter-numerals to distinguish them from the letters):

$$\overline{ωμβ} = 842$$

Below is an example for the use of both kinds of numerals (from Attica and the small letters) in one number, where M = 10,000, β = 2:

$$\overset{\overline{β}}{M} = 20.000$$

Another more complex example:

$$,\overline{ϛροε}$$
$$M,\overline{εωοε} = 71.733.875$$

# 6. ETRUSCAN NUMERALS

We mentioned the Etruscan numerals previously in the introduction to number-rovás, since Hungarian and Etruscan number-writing and number-rovás originated from the same Scythian folk-group. The Etruscans took it with them when they left the Carpathian basin. You can see the detailed introduction and their concordance on page 170. Let's look again at the basic numerals:

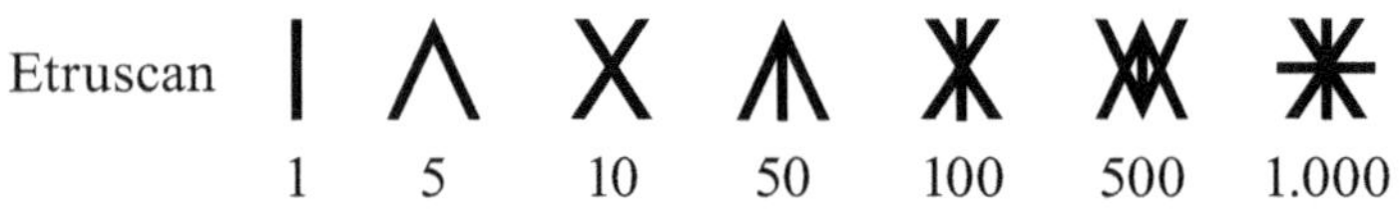

Etruscan

| I | Λ | X | Λ | X | X | ✳ |
|---|---|---|---|---|---|---|
| 1 | 5 | 10 | 50 | 100 | 500 | 1.000 |

Picture 439.

See now the inevitable variants:

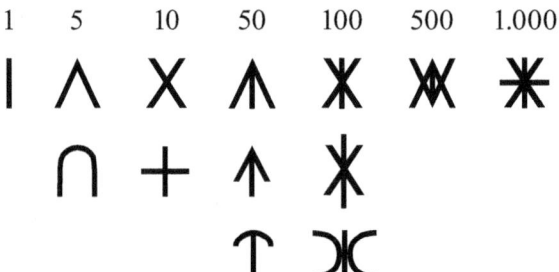

| 1 | 5 | 10 | 50 | 100 | 500 | 1.000 |

Picture 440. Variations of the same signs.

There was a different trend as well: to put the x (⊗) or a cross (⊕) inside the circle for (1,000), and this then led to the signs for 500 and larger round numbers:

| 500 | 1.000 | 5.000 | 10.000 |

Picture 441

# 7. ROMAN NUMERALS

According to the theory of Zsilinsky János, based on Livius and other Roman historians, Rome was founded by adventurers and outcasts from neighbouring city-states. Their first ruler (king?) founded the Roman state based on those people. This theory is supported by the fact that all the units of its social structures were created at once (this normally takes several generations to develop organically) and were in their form quite similar to those in the cities where the "outcasts" came from. Further proof of this theory is the regulation in the Laws of the Twelve Tables about the sequential inheritance and mutual guardianship of clan-members, which it wasn't necessary to regulate by law in the neighbouring city-states. They noted further in the Laws of the Twelve Tables that the clan-members were not agnate relatives, which means that even these relations were created artificially. (From Internet).

An independent Roman culture therefore did not exist. The absence of specifically Roman painting or sculpture also points to the lack of an intellectual base.

Probably due to all these factors, there is no "special" Roman number-writing, since there isn't a Roman past either. There are only numerals used by the Romans as well as by others. Their starting-base was the Etruscan-Hungarian number-rovás, which they modified over time, mainly by exchanging many numerals for letters, copying the Greek fashion. For this see the previous chapter.

This is how the numerals, known today as "Roman numerals" were developed:

| I | V | X | L | C | D | M |
|---|---|----|----|-----|-----|-------|
| 1 | 5 | 10 | 50 | 100 | 500 | 1.000 |

Picture 442. Roman numerals. The sign M marked in
Ancient Greek the number 10,000.

However, these are only those we still know today as Roman numbers. Let us first look at these in detail and then the different numbers.

The numbers were written with these signs as follows:

| | |
|---|---|
| X | 10 |
| L | 50 |
| C | 100 |
| CC | 200 |
| CCC | 300 |
| CD | 400 |
| D | 500 |
| DC | 600 |
| DCCC | 800 |
| CM | 900 |
| M | 1.000 |
| MCM | 1.900 |
| MM | 2.000 |
| MMM | 3.000 |
| MMMM | 4.000 |

There were signs, developed later, which a line above indicated multiplication by 1,000 and two perpendicular lines flanking the number laterally multiplied them by 100 again, thus in total by 100,000:

$$\bar{I} = 1000 \quad \bar{V} = 5000 \quad \bar{X} = 10.000 \quad |\bar{I}| = 100.000 \quad |\bar{V}| = 500.000$$

Example for an in-between number with the thousand-folder line (multiplier line):

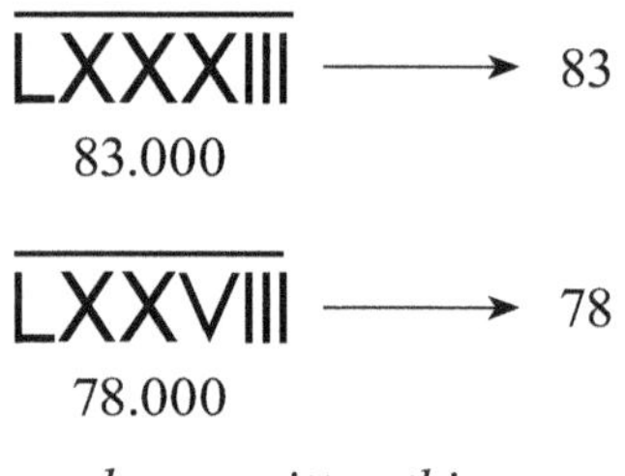

$\overline{\text{LXXXIII}} \longrightarrow$ 83

83.000

$\overline{\text{LXXVIII}} \longrightarrow$ 78

78.000

*In earlier times it might even been written this way:*

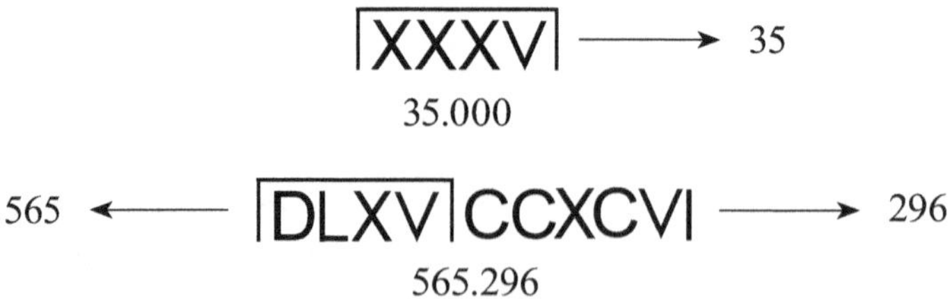

$|\overline{\text{XXXV}}| \longrightarrow$ 35

35.000

565 $\longleftarrow |\overline{\text{DLXV}}|\overline{\text{CCXCVI}} \longrightarrow$ 296

565.296

Picture 443. The writing of larger numbers.

The frame flanking the number became a 100,000-folder (multiplier) probably as far back as the time of Emperor Hadrian (second century BC):

⌈XIII⌉ ⟶ 13

1.300.000

Picture 444. The frame became 100,000-folder (multiplier) back in the time of the emperors.

The sign marking the higher place-value was a double-line above the basic sign:

$\overline{\overline{\mathrm{M}}}$ ⟶ 1.000

1.000.000.000
(1000 x 1.000.000)

$\overline{\overline{\mathrm{MMDCX}}}$ ⟶ 2.610

2.610.000.000
(2.610 x 1.000.000)

Picture 445. One could write even the milliards with this method.

There was no total uniformity. For instance: 99 was written as XCIX or IC and MDCCCCXXXXIIII (1944) could have been MCMXLIV as well.

The method of addition was similar to an abacus: identical signs were be grouped, arranged according to the place-values and added.

Let's add 1,114 to 1,123. The sign IV needs to be changed into only lines (IIII) and then we can group the identical signs:

MCXIV ⟶ MCXIIII

| | | | | | |
|---|---|---|---|---|---|
| 1123 | | M | C | XX | III |
| 1114 | + | M | C | X | IIII |
| grouping the identical signs | | MM | CC | XXX | IIIIIII |
| The number 7 written correctly | | MM | CC | XXX | VII |

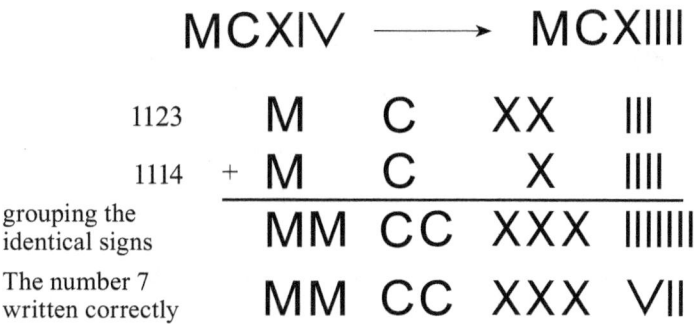

Picture 446. The method of addition.

Romans took on the original Etruscan-Hungarian number-rovás, they debased it. Using the abacus one didn't have this problem and the Romans used it in abundance as well for extended calculations. (See the chapter about the abacus.)

Don't let yourself delude when numbers only appear with roman numbers written on anything or carved into stone. The results of the moving "pebbles" of the abacus disappeared by touching them and had to be written down immediately. This again could only happen with Roman numbers. The written number therefore doesn't tell us anything about the method of calculation which led to the result.

There was somewhere in Italy, on a territory conquered by the Romans, a local culture which still kept a very ancient number-writing method. The Romans built their ancient place-value marking into their number-writing system as well. This is a number-writing method found already on the cave wall of Lascaux, written 17,000 years ago and not originating from the number-rovás. You can find a detailed comparison of the Roman signs with those found in Lascaux on pages 77-79.

The essence of these place-values' sign is a perpendicular line, which is bilaterally flanked by semicircles. The numbers of semicircles point to the value. The semicircles always tenfold and their number is unlimited.

1.000          10.000          100.000          etc.

*It is easier to draw this way:*

*In addition to these many other drawings were made.*

Picture 447.

See more about the variations at Lascaux on page 77-79.

Picture 448. Every one of the "splinters" on the left picture's straight line mean 10.

Italian variations on this kind of numerals:

| 1.000 | 10.000 | 100.000 |
|---|---|---|
| (\|) | (\|\|) | (\|\|)) |
| (\|) | (\|) | (\|) |
| (\|) | ((\|)) | ((\|))) |
| /\|\ | //\|\\\ | ///\|\\\\ |
| .\|. | .\|. | ..\|.. |
| ⌠\|⌡ | (\|) | (\|) |
| 人 | 人 | 人 |
| ⋃ | ⋃ | ⋃ |
| Ψ | Ψ | Ψ |

Picture 449. Based on page 197 of Georges Ifrah's book Numbers.

It is worth comparing a few signs from Italy with those of Lascaux:

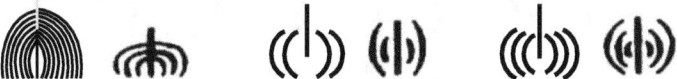

Picture 450. These pairs are only "variations on a theme".

It is easy to make half place-values from these signs. If we omit one side of a sign, the remaining other half will mean rationally half of the original value:

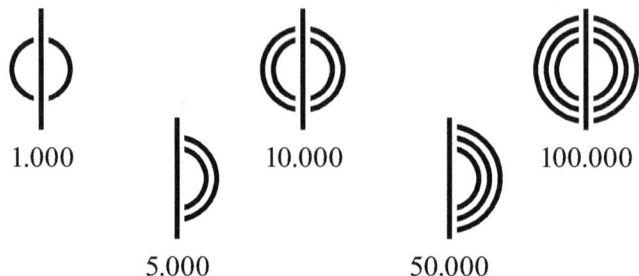

1.000    5.000    10.000    50.000    100.000

Picture 451 The half place-values are probably subsequent derivatives.

The half place-values were written in the same form as their round pairs. The sign for 500, drawn in the above form is missing. However, it didn't get lost. We know it in its constricted form.

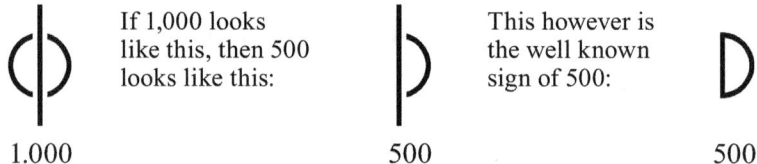

1.000   If 1,000 looks like this, then 500 looks like this:   500   This however is the well known sign of 500:   500

Picture 452. The origin of D, the Roman sign for 500.

Now we know why the Roman sign for 500 has a D form.

Otherwise, it is not easy to find one's way through Roman signs marking the place-values, because the signs of the number-rovás are mixed with the letters put in between, due to the Greek influence and the sign D has again a different origin as we have seen previously.

Let's repeat the basic Roman number-signs:

| I | V | X | L | C | D | M |
|---|---|---|---|---|---|---|
| 1 | 5 | 10 | 50 | 100 | 500 | 1.000 |

Picture 453.

Let's look at them individually:

1)

| I | V | X |
|---|---|---|
| 1 | 5 | 10 |

These three signs are the ancient signs from the number-rovás. They were already in use 15,000 years ago including the sign for 1,000.

2)

L

50

The L-form sign as a letter is the rotated form of the sign Λ. We can easily follow this modification on the old coins or by following the alphabets:

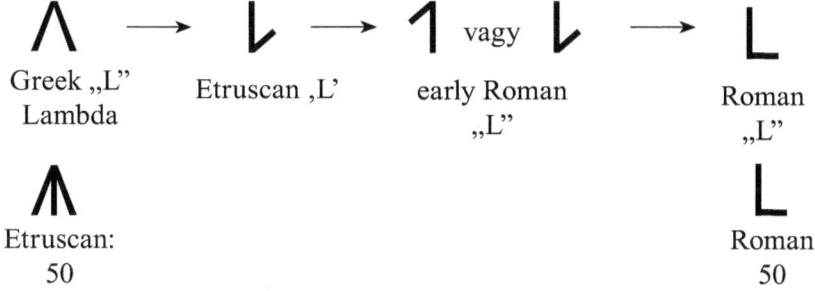

| Greek „L"  Lambda | Etruscan ,L' | early Roman „L" | Roman „L" |

| Etruscan:  50 | | | Roman  50 |

Picture 454 As we can see from this comparison, the relation is not merely superficial. We should further note here that the Greek lambda and the Etruscan 50 are much older than the Roman L.

3)

C

100

This is probably the first letter of Centum.

4)

M

1.000

The first letter of 1,000 (mille) is M. But we can't be sure of this connection, because this sign meant 10,000 in Greek.

# PART IIIIΛ

## SUBSEQUENT NUMBER WRITING METHODS

# 1. THE OBSERVATIONAL ASPECT

In this part, I will introduce some less well-known number-writing methods. We will see that these are as just as much continuations of the number-writing from ancient cultures as those of Egypt, China, etc. which we have discussed earlier in this book. Indeed, the further we look, the better we can prove that the ancient Palaeolithic dot-line (line-line) number-writing - using the place value method - was that which became widespread all over the world.

Not only does each of the number-writing methods we have seen so far bear witness to unbroken continuity, but on the other hand, not a single example of a different number system has turned up. Naturally, flourishing cultures often modified the simple signs; however, the starting point of the ancient number-writing method is always easily verifiable. In general, we can establish the fact that the signs first began to be modified around 3-4,000 years ago by people from a few outstanding cultures. But these modifications were more or less only small changes. In Egypt for instance, there was almost no change through several thousand years of history, while in India, the signs finally became "ciphers" "túl-cifrázva" (over-ornamented). Even this process can easily be followed back to the ancient number-writing. The Aztecs – very much prone to drawing - and the Arabs modified the numerals a lot, but even there the picture of the dot-line origin stayed clear. We shall see more of this shortly as well.

The conclusion of all the above is that once one almost eternally ancient culture spread throughout the world. We will see further proofs of that in the following.

Of course, there are exceptions from the rule of uniformity in the numerals designed for coding secret messages. They belong to the world of code writing. Furthermore, at one time in the past it was fashionable to use letters instead of numerals. We have examples of this too.

# 2. INUIT (ESKIMO) NUMERALS

Here we will look at the Inuit numerals:

Note the division into the sequences of 5, the numerals up to 19 and the sign for zero.

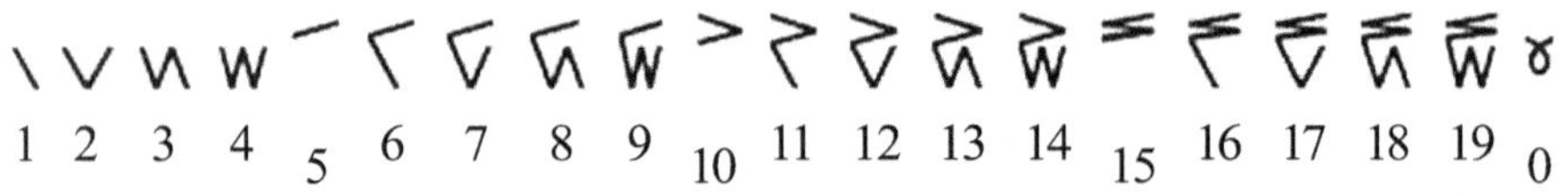

1 2 3 4 5 6 7 8 9 10 11 12 13 14 15 16 17 18 19 0

Picture 455 The numbers of the Eskimos

It is a clear base 20 number-system, divided into 4x5 parts with separate names for 19 numerals (similar to the Mayas and the Celts). The signs are identical with the ancient dot-line signs, *only the parallel lines are tilted towards each other:*

1 2 3 4 5 6 7 8 9 10
11 12 13 14 15 16 17 18 19

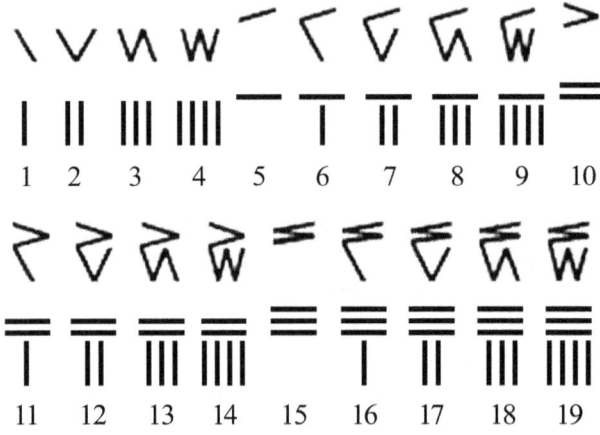

Picture 456. The lines might have looked better in tilted form.

The material that was used to write on (leather, bone) may have influenced the form. Let's do some time-travel in the following comparison.

Look at the first 9 numbers again, compared with the Chinese numerals, and notice that there is only an insignificant change from the ancient number-writing.

Picture 457. It's clear that both are in reality equal.

Both number rows are identical except for the sign for 5, and the differences from 6-9 are only that the perpendicular Eskimo lines are horizontal in the Chinese set.

The second comparison is based on the interchangeability of the dots and perpendicular lines. Now, let's mirror the Eskimo signs along a horizontal line to receive the following examples:

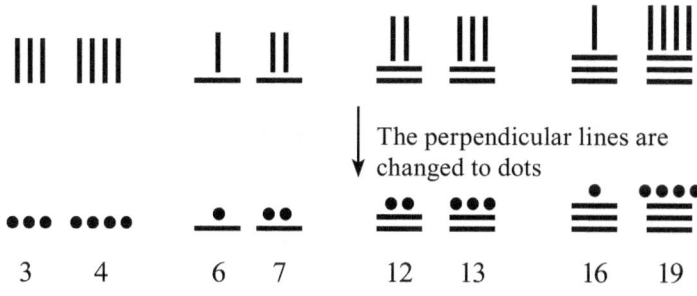

Picture 458.

In this case, what we see in Mayan number-writing (see page 87), for the simple reason that the Mayans and the Eskimos (the Celts too) were using the base 20 number-system. Therefore Mayan, Chinese and Eskimo number-writing originates from one common root, from the number-writing of the early Stone Age. Remember the 16-22,000-year-old number-sign found in La Passiega (page 84):

Picture 459. La Passiega, 16-22,000-year-old numeral

"Every path leads to the remote Stone Age"

# 3. MORDVIN NUMERALS

Two large folk groups make up the Mordvins: the Erzya and the Moksha. Most of them live along the rivers Insar and Moksha. They started to move apart in the 16th century. Today 2/3 of them are living in Mordvinia and the rest in little groups from Ukraine to the island of Sakhalin.

Ancient Mordvin numerals:

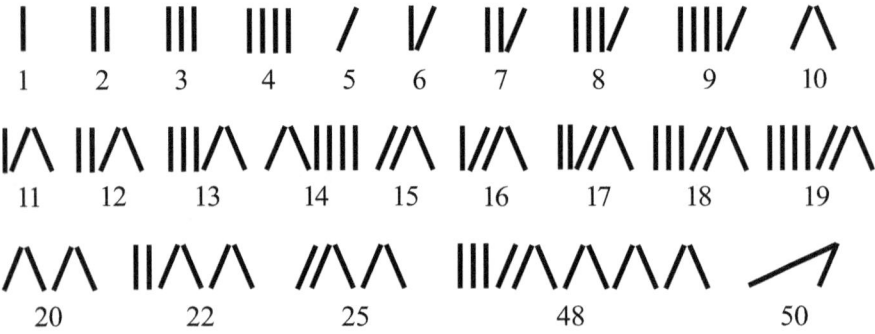

Picture 460. See the round numbers above 50 later.

This is a decimal system.

Let's use the same comparison method we used with the Eskimo numbers here as well. We turn the slanted lines to horizontal and put the perpendiculars above them or even make dots from them:

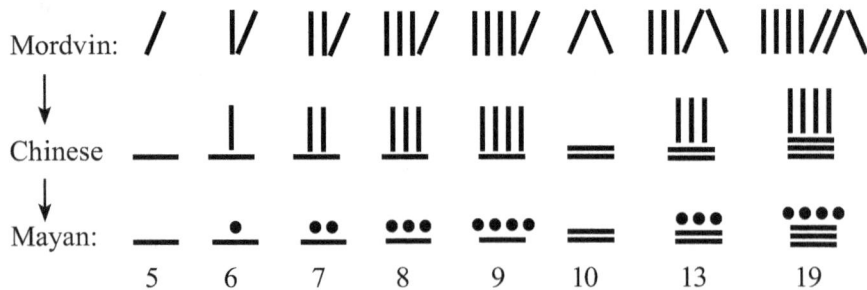

Picture 461 The principle stayed the same. The signs are merely "variations on a theme."

Well, the Mordvin and Eskimo numerals can be compared not only with Chinese and Mayan numerals. Having come so far in this book, we know that all number-writing (Egyptian, Brahmin and even the European of today) goes back to the same ancestor.

There are however some additional, special, inventions to the Mordvin numerals: The round numbers over 50 are special constructions:

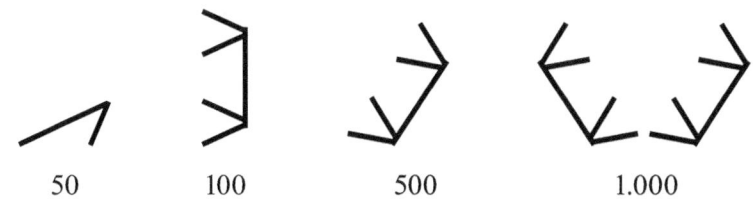

50       100       500       1.000

Picture 462. The Mordvin round numbers over 10.

The signs contain the long line representing 100, but for some reason, they are decorated with the sign for 10, disregarding the always clear logic of number-writing.

The signs are built from right to left, for instance:

59       519

Picture 463.

# 4. THE CHUVASH NUMERALS.

The Chuvash folk group belongs to the Turkish language family. Around three million Chuvash live in more or less smaller groups scattered from the River Volga to the Siberian territories. The Chuvash language belongs to the Bulgarian-Turkish type of the Ancient Turkish languages.

Their numerals are mixed. The numerals 1-9 are identical with the Mordvinians, but they continue above ten with the signs from the number-rovás:

| I | II | III | IIII | / | I/ | II/ | III/ | IIII/ |
|---|----|-----|------|---|----|-----|------|-------|
| 1 | 2  | 3   | 4    | 5 | 6  | 7   | 8    | 9     |

*from 10 on, however, follow the round numbers of the number-rovás*

| X | X | X | X | X |
|---|---|---|---|---|
| 10 | 50 | 100 | 500 | 1.000 |

Picture 464. Number-rovás signs in the Chuvash Number set.

The construction from the two different number sets is quite evident when we look at the numbers (the reading is from right to left):

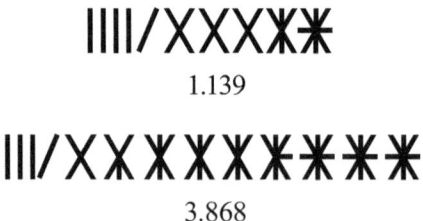

IIII/XXXX✳
1.139

III/XXXXXX✳✳✳
3.868

Picture 465. Two Chuvash numbers

# 5. HITTITE NUMERALS

The Hittites established a significant empire from 4-3,000 BC in Asia Minor, which had a large and long-lasting influence on their neighbours, among them the Greeks. *"Their cultural influence was immense, because they built up the contact between the 3,000 BC Middle-Eastern cultures and the Mediterranean cultures of the 1st millennium BC"* (Wikipedia)

Their number-writing was based on the ancient number signs as well, mixing in the sign X from the number-rovás. The sign for 1,000 reminds us of the Sumerian sign for 10 engraved with a stick, but it is possible that the long line became independently curled in their set.

| I | — | X | ⊂ |
|---|---|---|---|
| 1 | 10 | 100 | 1.000 |

Picture 466. The Hittite signs for the round numbers.

A few Hittite Numerals

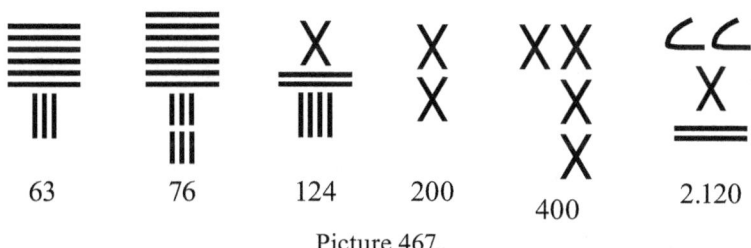

| 63 | 76 | 124 | 200 | 400 | 2.120 |
|---|---|---|---|---|---|

Picture 467..

They usually advanced from the top downwards, for instance, the X written 6 times one below the other meant 600. There are however, finds with numbers written horizontally. See below the 3,000-year-old number 4,400:

Picture 468. It is interesting that the reading goes, unusually, from left to right.

# 6. CRETAN NUMERALS

Cretan, in other words, Minoan culture flourished 4,000-3,400 years ago. The legends about the Minotaur, the Labyrinth and Ariadne are connected to this culture. Their beautiful frescos are well-known all over the world.

Naturally, their numerals are identical to the ancient dot-line number system. They designed one special sign for 1,000 and liked to tilt the perpendicular lines or bow them somewhat:

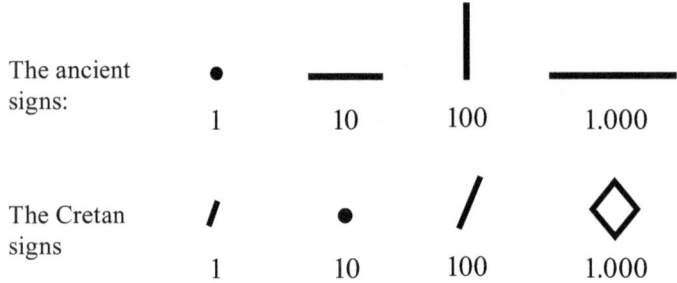

Picture 469. The comparison of Cretan with the ancient signs.

We note that they converted (interchanged) the original signs for 1 and 10. We see further their new invented sign for 1,000 (see the same sign in sr. Dán Péter's set too, on page 142, under Cretan connection). The tilting of the perpendiculars or their bowing is an insignificant modification. One example for this is an approximately 3,500-year-old number from Knossos (to be read from right to left):

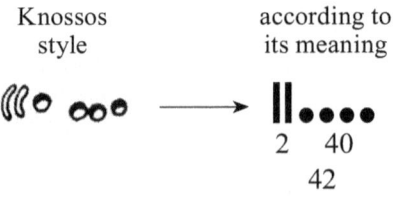

Picture 470.

But we find even these little changes merely in the "drawn" (hieroglyph-like) texts. Otherwise, in their usual scripts, the perpendicular lines were not tilted and the sign for 1 was not bowed; not even in their two different writing systems: in the "linear A" and "linear B" systems. The "linear A" was the earlier generally-used script, while system B came later and contains many signs used mostly by civil servants. Changes turned up later in the writing of the signs for 100, 1,000 and 10,000.

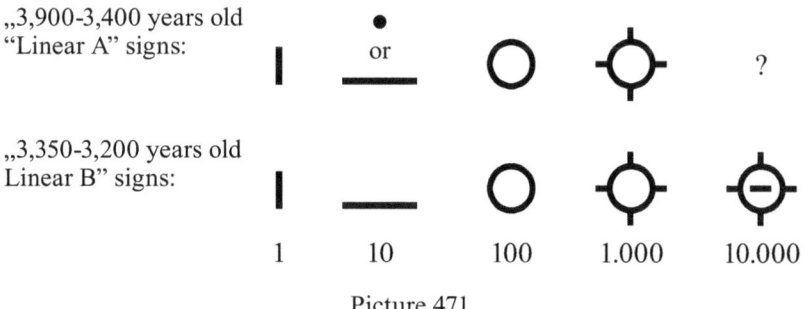

„3,900-3,400 years old "Linear A" signs:

„3,350-3,200 years old Linear B" signs:

1    10    100    1.000    10.000

Picture 471.

Let's look at an example for the use of "Linear B". The difference from "Linear A" is merely the occasional use of dots for 10

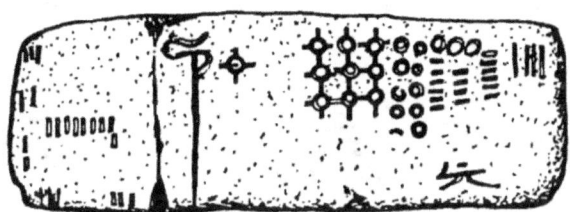

Picture 472. Source: Georges Ifrah: Numbers, page 179.

An example for "Linear A" number-writing, but only with signs for singles and 10:

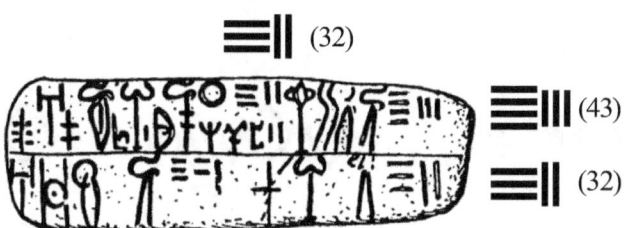

Picture 473.

It's remarkable that the Cretans wrote the numbers alternately from the right or from the left while using the "hieroglyph-like" script as presented at the beginning of this chapter. It seems that they changed from the left to the right direction if the numbers were over 1,000. In the "Linear A" script, however, they always wrote the numbers from left to right.

Here are some numerals written with "hieroglyphic" signs, where we can observe the slanted lines. Furthermore, the first three smaller signs are written from the right, but the greater than 1,000 signs start from the left:

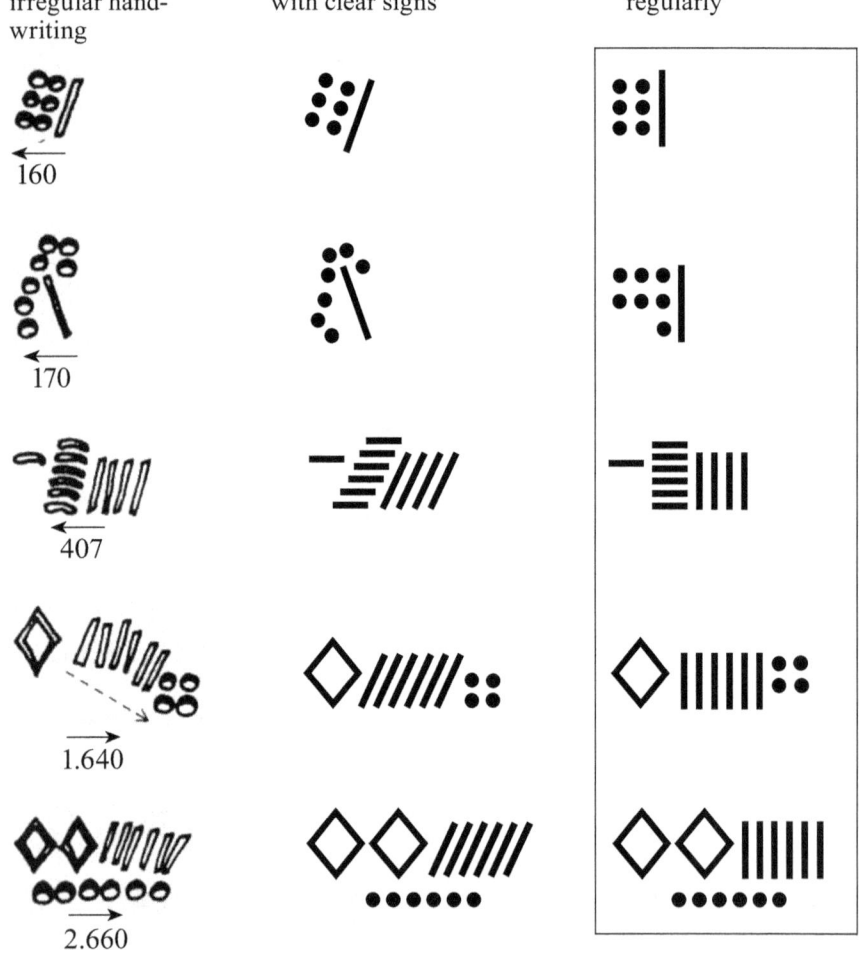

Picture 474. Humorously said: The Cretans liked to write their numbers a little carelessly.

Below, we see 3,6000-3,450-year-old "Linear A" numerals. Note that the tens are written alternately with dots or lines:

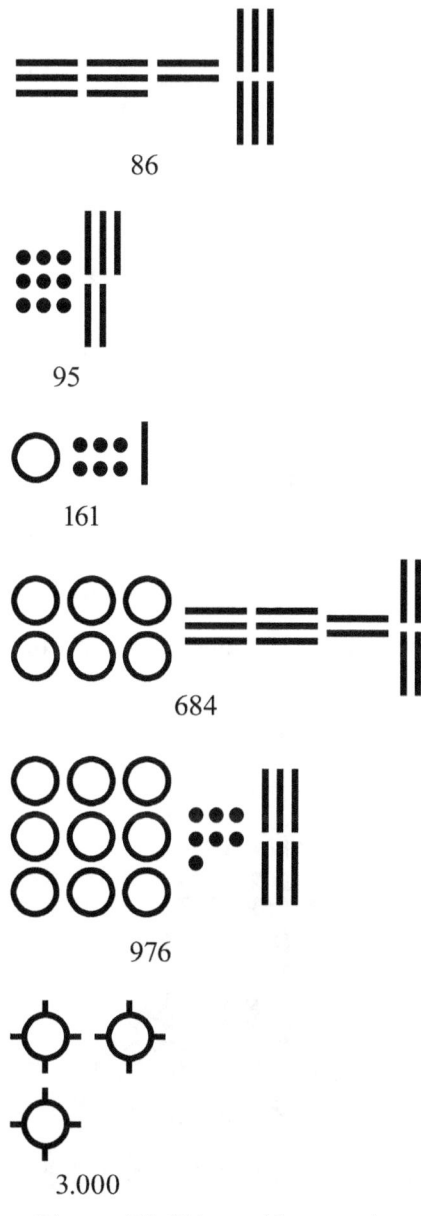

Picture 475. "Linear A" numerals.

# 7. NUMERALS FROM SHEBA

Sheba (Sabá) was a South-Arabic trading power around 3,000 years ago. It is noted in the Qur'an (Koran) and in the Bible as well (Queen of Sheba).

In this script only the signs of 1 (I) and 10 (the dot written as a little circle: O), and of course the method of writing, remained as in the original. It's remarkable that there were no half values written in their set, but we find the 5 and 50, which means that we see here a mixed system as well just as in the case of the Ancient Greek numerals.

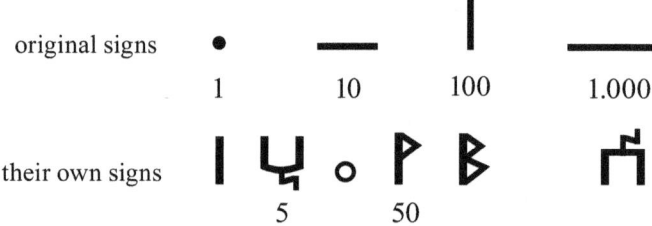

Picture 476: only the line (1) and the dot (10) remained.

The round numbers, the half values 5 and 50 in between, are shown here:

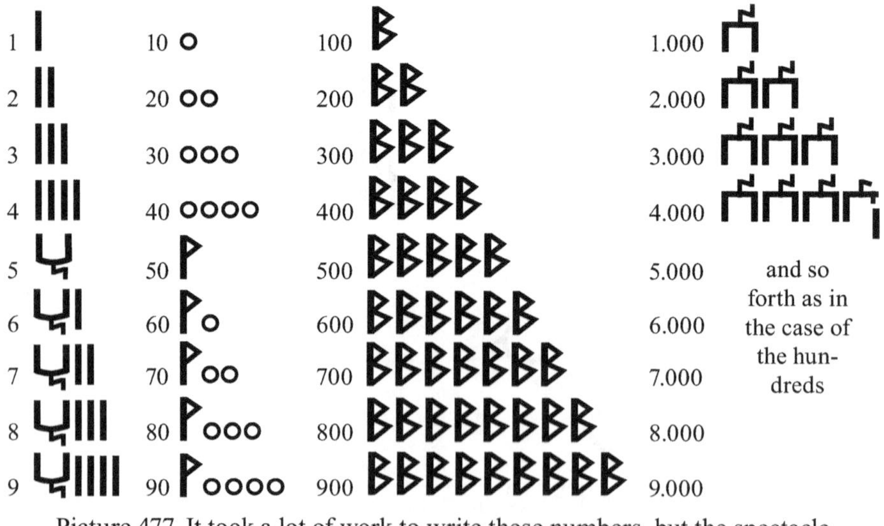

| | | | | | | | |
|---|---|---|---|---|---|---|---|
| 1 | I | 10 | O | 100 | ᵬ | 1.000 | ń |
| 2 | II | 20 | OO | 200 | ᏴᏴ | 2.000 | ńń |
| 3 | III | 30 | OOO | 300 | ᏴᏴᏴ | 3.000 | ńńń |
| 4 | IIII | 40 | OOOO | 400 | ᏴᏴᏴᏴ | 4.000 | ńńńń |
| 5 | Ӌ | 50 | Ꮲ | 500 | ᏴᏴᏴᏴᏴ | 5.000 | and so |
| 6 | ӋI | 60 | ᏢO | 600 | ᏴᏴᏴᏴᏴᏴ | 6.000 | forth as in the case of |
| 7 | ӋII | 70 | ᏢOO | 700 | ᏴᏴᏴᏴᏴᏴᏴ | 7.000 | the hun- |
| 8 | ӋIII | 80 | ᏢOOO | 800 | ᏴᏴᏴᏴᏴᏴᏴᏴ | 8.000 | dreds |
| 9 | ӋIIII | 90 | ᏢOOOO | 900 | ᏴᏴᏴᏴᏴᏴᏴᏴᏴ | 9.000 | |

Picture 477. It took a lot of work to write these numbers, but the spectacle was probably very important.

I think these very complicated signs were developed to be carved into stone. They certainly look quite spectacular on that surface.

The direction of writing changed for a particular purpose, which we will see later. The direction of reading can be seen from the direction of the signs for 50 and 100 ( ¶ and ᶾ or ᚹ and ᛒ ).

The direction change had a fundamental meaning. To understand the system, let's first look at a number written from right to left:

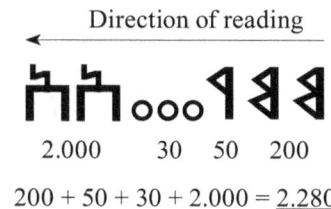

2.000    30   50   200

reading    200 + 50 + 30 + 2.000 = 2.280

*The value will change by reversing the direction of the writing*

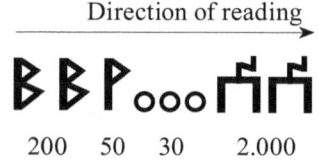

200   50   30   2.000

reading:
(200 + 50 + 30) x 1.000 + 2.000 = 282.000

Picture 478. This solution is seldom in the history of number-writing.

Similarly to the above:

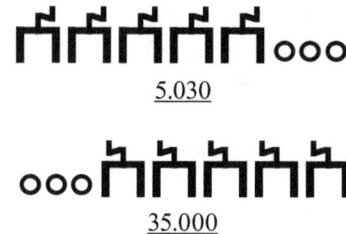
5.030

and reversed:
35.000

Picture 479.

# 8. SECRET TURKISH NUMERALS

We saw on page 75 that a certain group of the signs from Lascaux had this kind of form:

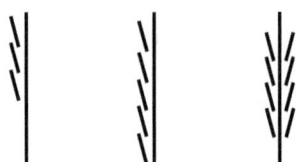

*This principle is connected to the number-writing method below and it returns later regularly. We discussed it previously, even the Aztecs used it. Now see the following:*

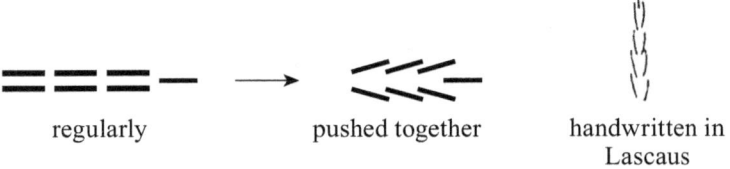

| regularly | pushed together | handwritten in Lascaus |

Picture 480. Every line means 10 on the later finds, the left numeral there-fore represents 70. The value of the line was often different, but always a round number. The sign from Lascaux can therefore only stand for 90 when the line's value was 10 as well, but we don't know that for sure.

The numerals of the Ottoman Empire's secret writing were based on the same principle. The system was cleverly extended to be able to write every number. This system has two known variants. One has only the 1, 10, 100, 1,000, and 10,000 round place-values. These were written with increasing numbers of lines put to the right of the perpendicular line.

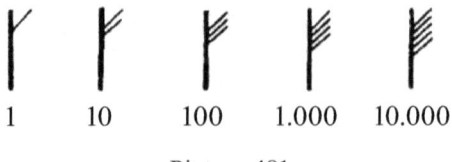

| 1 | 10 | 100 | 1.000 | 10.000 |

Picture 481..

There is however, an oddity in this system. Because the sign of 1 is empty on the left side ( ⌐ ), the number of the lines left is always one less, than the presented values.

The right side of the perpendicular line points to the place-value and the lines on the left tell us how many of that round number we mean:

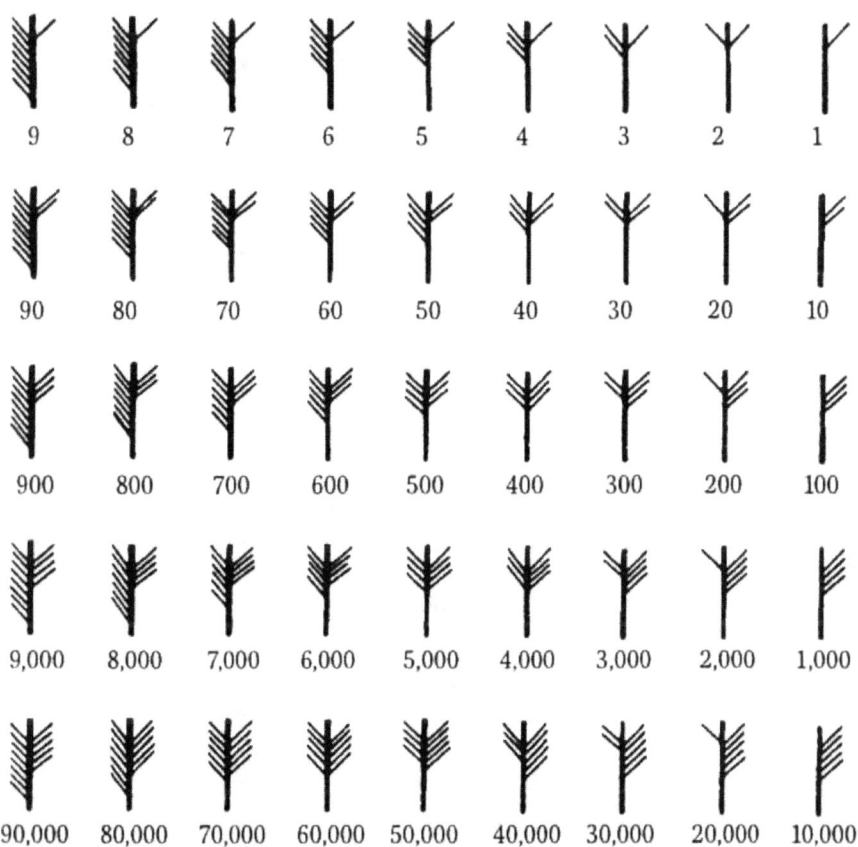

Picture 482.

The other secret number-writing method is however bisecting and we have seen many examples of it. Its sequence is 1, 5, 10, 50, 100, 500, 1,000 etc. According to its secrecy, there is a slip sideways and even this may have caused its mystery. A partially irregular system can only be understood and repaired after a longer investigation.

First take a look at this bisecting system:

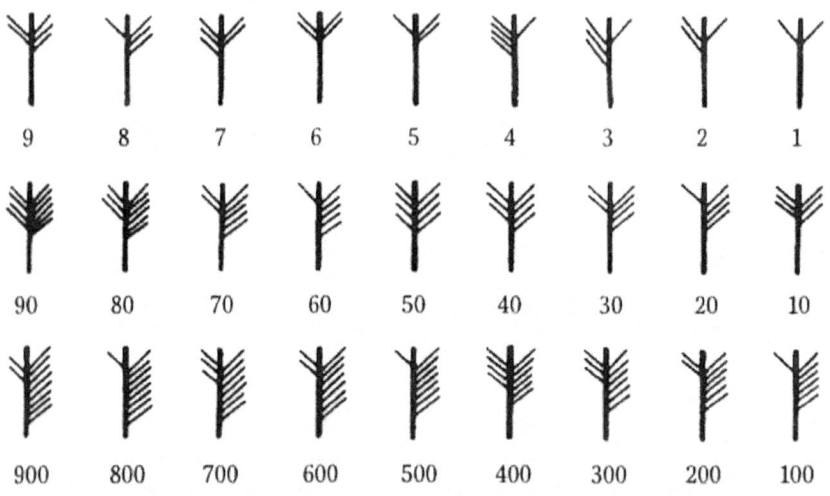

Picture 483. This is real secret number-writing. The numbers of the side branches are frequently not adjusted to the given number.

The place-value is marked by the line on the left side of the perpendicular line. This we can see in the case of the numbers 1, 5, 100 and 500. This should be the case by writing 10 and 50 as well, but here, due to the secrecy modification, the marking slipped one step to the right and we see first at 20 and 60 what should stand at 10 and 50:

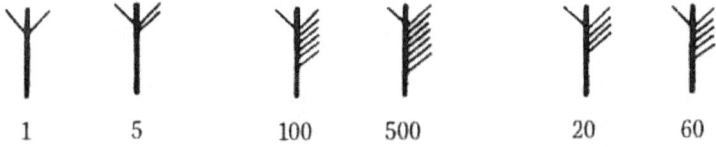

| 1 | 5 | 100 | 500 | 20 | 60 |

The marking of the round place-value is right here: one line on the left

Here shifted one step further due to the distortion of the system

Picture 484.

The one system is based on clear round place-values (1, 10, 100, 1,000, 10,000). The other however is based on the bisecting system (1, **5**, 10, **50**, 100, **500**). It is remarkable that these two, in principle very different worlds of number-writing, turn up even here.

# 9. AZTEC NUMERALS AND NUMBER WRITING

We have written data about the Aztecs going back to the year 200 AD. According to legend, their name points to their earlier homeland Aztlan (meaning the land of herons). They call themselves Mexican. Their empire was the largest during the 14-15th centuries, ruling over the territory between the Pacific Ocean and the Mexican coast. The leading people of the cities lived in spacious houses while ordinary people lived in small "apartments" separated by high walls. Small passageways and corridors ran between the houses.

The Aztecs were very aggressive, conquering their neighbours until finally they ruled over 4,000 folk groups. As conquering people usually do, the Aztecs, like the Romans, assimilated much of the conquered people's culture. This we can see looking at their number-writing, because it is a mixture of several different methods, quite inconsistent, and contrary to the Mayans' clear system.

They even used different signs for the same number:

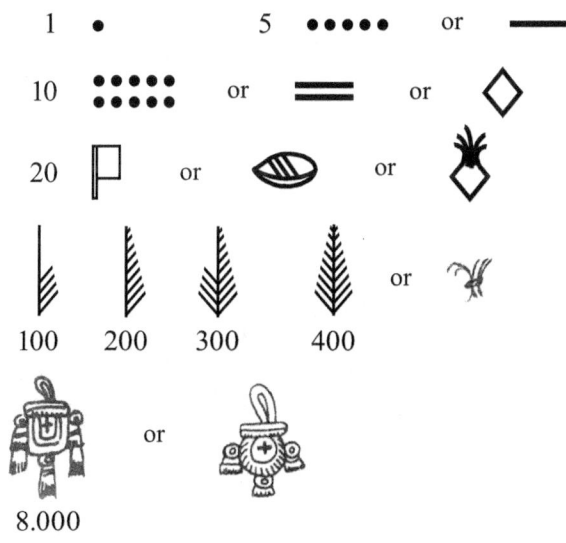

Picture 485. Aztec numerals.

It is unquestionable, looking at the numerals, that the Aztecs counted in the base 20 (vigesimal) numeric system. Their row of base-line was regular (1, 20, 400, 8,000, etc.) following the sequence of the power of 20, unlike that of the Mayans, whose sequence ran 1, 20, 360, 7,200, etc. Otherwise, there was a lot of disorder in Aztec number-writing.

1) Two signs for the number 5.

2) There were even 3 signs for the number 10, however 10 is not a base-number in the vigesimal system.

3) There were three signs also for the number 20: one was identical with the Mayan zero, another looks like a flag and the third is a modification of the sign for 10 (perhaps a heart with its arteries?).

They had separate numerals for 100, 200, 300 and 400, therefore they didn't follow the number writing with base-values. These numerals were the Lascaux-Italian "palm leaves" (seen on page 79). Furthermore, there was a second sign for 400 looking like a bound tuft of hairs or birch twigs. That is, they had along with the numerals needed for the vigesimal system (1, (5), 20, 400, 8,000), several other signs (100, 200, 300). It seems that they probably didn't practise much mathematics unlike the Mayans.

Because of the many different signs, they could write a number in different ways. Look at the number 8,375 written in four different ways:

Picture 486. The Aztecs could write the number 8,375 in four different ways

The Aztecs often didn't even care about the local values. You see below left a shield and the "flag" above it says that it stands for 20 of them. On the right, is a bag with cocoa-beans and the five "flags" above the bag mean 100 bags of it. On the other hand, they have an extra sign for 100, the mostly stripped "palm-branch".

 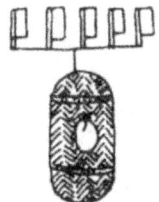

20 shields        100 bags cacao beans.

They had a sign for 200 as well, but in the honey depository below 10 signs for 20 marked this number. They drew as many honey-pots as 20 occurs in 200:

200 pots of honey

They occasionally used the sign for 400 as well:

400 decorated cloaks    1,600 cocoa husks .        800 deer hides

Pictures 487-488 Aztec book-keeping.

One more example of number-writing from one of the few remaining Aztec codices (Aztec Codex Telleriano Remensis)[72]:

We see a number in the lower right-hand corner of the picture above and the same enlarged below

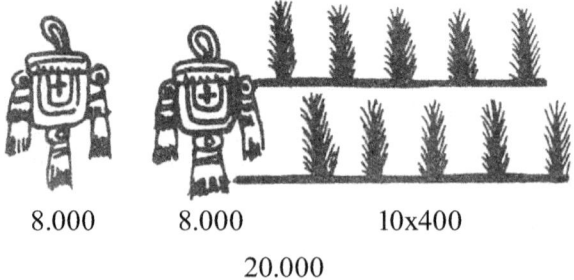

| 8.000 | 8.000 | 10x400 |
| --- | --- | --- |

20.000

Pictures 489-490. The number of a larger population-group written down in the Aztec way.

72  These drawings are from the book Numbers by Georges Ifrah, pages 303-307.

A few more Aztec numbers as examples:

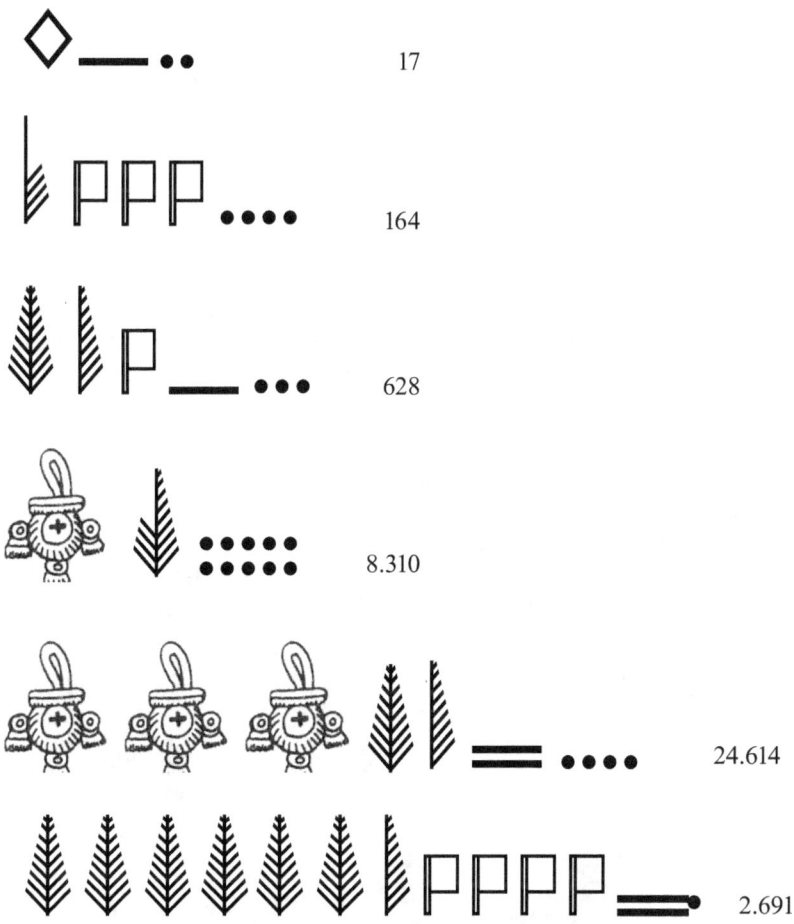

17

164

628

8.310

24.614

2.691

Picture 491. Many of the numbers are long-winded because they didn't use place-values

The numbers make it clear that it is the ancient dot-line number-writing, retaining the dots and lines, despite the fact that the Aztecs exchanged many of the numerals for partially funny drawings.

One can't imagine a large emporium without bookkeeping, inventory, tax-records, etc. I couldn't find any description for mathematical procedures except for multiplication. However, they must have used addition, division, subtraction and multiplication on a regular basis.

According to the description, multiplication was done in the following way:

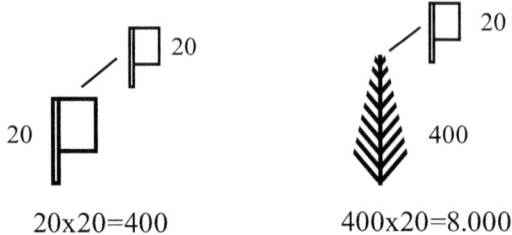

20x20=400          400x20=8.000

Picture 492. Marking multiplication by the Aztecs.

This, however, is quite modest. How did they get the results in more complicated cases? They didn't use the abacus; therefore they must have done the calculations with their fingers. The following pictures point to this possibility

Picture 493. The man standing on the right is counting with his fingers (Part of Diego Rivera's wall painting in the Mexican National Museum)

Looking at the picture above, at first it seems that the man on the right is showing 4 fingers. However this, is wrong. Looking more carefully, we can see his thumb pointing to the middle phalanx of his ring finger, which is the certain proof of counting with fingers. Try to put your thumb in this position. Counting with fingers is easy and fast. It was widely practised in Europe and Asia as well. We don't do it, but children still start counting this way.

Fingers are the best calculator, always with us, we can't leave them at home, but we have to learn to use them.

We can find an example for another ancient method of counting used by the Aztecs for counting several people at once, using their fingers and toes all together. Let's look at the numbers rendered to the fingers and toes on the picture below, where three people are doing the counting procedure together:

Picture 482. The Aztecs standing together to count something.

One man can count up to 20 (fingers + toes), the second from 21- 40 and a third man from 41-60. The number of counting men can be increased if needed.

Here we see where the number-names came from. Some characteristic examples in Aztec language: ce (1), ome (2), yey (3), naui (4), macuilli (5), chica-ce (6), chic-ome (7), chicu-ey (8), chic-naui (9), matlacti (10), matlacti-on-ce (11) ... caxtulli (15), caxtulli-on-ce (16), cem-poualli(20) , cem-poualli-om-matlacti (30), ome-pualli (40), ome-poualli-om-matlacti (50). (Flegg. 1983; Ifrah, 1987.)

Therefore:

**1   one**
2   two
3   three
......
17   seventeen
18   eighteen
19   nineteen
**20   one man**
21   one after the first man
22   two after the first man
23   three after the first man
.....
38   eighteen after the first man
39   nineteen after the fist man
**40   two men**
41   one after the second man
42   two after the second man
43   three after the second man
.......
58   eighteen after the second man
59   nineteen after the second man
**60   three men**
61   one after the third man
62   two after the third man
.....
77   seventeen after the third man
78   eighteen after the third man
79   nineteen after the third man
**80   four men**
81   one after the fourth man ......   and so forth

It is remarkable that the numbers are built using the same principle in Greenland among the Eskimos, in Japan among Ainus, in Senegal, Guinea, Benin and among Tasmanian people.

The Examples of Georges Ifrah:[73]

The number 53 is pronounced by the Eskimos in Greenland as "three on the first leg of the third man", yet people in Greenland don't live their life barefooted.

The Yoruba in Western Africa say: "ten and three before the third twenty" when they mean 53 while the Ainu people say: "ten and three from the third twenty".

This much agreement between distant areas of the world is only possible if this kind of counting had common knowledge before all these cultures moved apart from one ancient culture in an enormously distant epoch.

We could see on pictures 481 and 482 that the Aztecs performed their calculations using their fingers and toes, which is a timeless method of counting, spread all over the world.

---

73  Page 305 of his book.

# 10. ZAPOTEC NUMBER WRITING

The Zapotec live in Central America. Around 300,000 people speak this language today. Their culture was close to that of the Mayans.

We don't know much about their number writing. I found only two examples of it in the previously mentioned book of Georges Ifrah.Two numbers are shown. The sign for 20 is like the Aztec "flag" and the dots marking the singles became little squares on stems similar to the "flags". The positioning of the signs when writing numbers must have followed their wish to decorate rather than any preset rules:

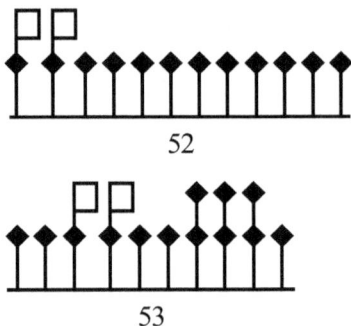

We can see on the map below the locations of the Mayan and Zapotec cultures:

Pictures 495-496..

# 11. ARABIAN NUMERALS

Arabian numerals built the base on which Ajtósy Dürer Albrecht drew our presently used numerals. The Arabians, however, took these signs over from India around 1,500 years ago. Thus, there are no original Arabic numerals. See the family tree on page 226.

The up-to-date forms of the Arabic numerals:

| | | | | | | | | | |
|---|---|---|---|---|---|---|---|---|---|
| ١ | ٢ | ٣ | ٤ | ٥ | ٦ | ٧ | ٨ | ٩ | ٠ |
| 1 | 2 | 3 | 4 | 5 | 6 | 7 | 8 | 9 | 0 |

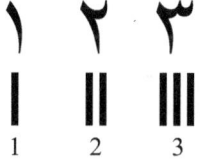

Picture 497. The Arabic numerals.

Time brought many changes to these signs as well, but we can see the connection to the ancient dot-line system best on the first three numerals:

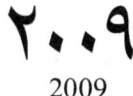

Picture 498. The ancient line-sign is visible in all three.

It is remarkable that the Arabs write from right to left, but put their numbers from left to right on paper, for instance:

٢٠٠٩

2009

Picture 499. The dots mark the "zero"

The Arabs also followed the fashion of writing the numbers with the letters of the alphabet, and even produced their own letter-numerals as we can see on page 391.

# 12. JAPANESE NUMERALS

The Japanese use Chinese numerals while writing their columns per-pendicularly, but they use Arabic numerals in their horizontal scripts. See the Chinese numerals in the chapter starting on page 332 and again in the following table

| | |
|---|---|
| 1 | 一 |
| 2 | 二 |
| 3 | 三 |
| 4 | 四 |
| 5 | 五 |
| 6 | 六 |
| 7 | 七 |
| 8 | 八 |
| 9 | 九 |
| 10 | 十 |
| 100 | 百 |
| 1000 | 千 |
| 10000 | 万 |
| 100.000.000 | 億 |
| 1.000.000.000.000 | 兆 |

Picture 500.

# PART X

## LETTER-NUMERALS

## 1. GEORGIAN LETTER-NUMERALS

The custom of writing numerals with letters turned up quite late in the history. Several folk-groups, among them Georgians, Greeks, Slavs, Hebrews and Germans built number-writing systems using letters of the Alphabets.

Here we see the Georgian letter-numeral set:

| | | | | | |
|---|---|---|---|---|---|
| 1 | ა (a) | | 9 | ∞ (t') | |
| 2 | ბ (b) | | 10 | ი (l) | |
| 3 | გ (g) | | 20 | კ (k) | |
| 4 | დ (d) | | 50 | ნ (n) | |
| 5 | ე (e) | | 100 | რ (r) | |
| 6 | ვ (v) | | 500 | ჳ (vi) | |
| 7 | ზ (z) | | 1.000 | მ (dz) | |
| 8 | ჱ, ჰ (ee, h) | | | | |

Picture 501. The Georgians started to write their numerals with letters during the 6th - 7th century.

The Georgians count in the vigesimal (base 20) system. They say for 73: "Three-twenty and thirteen".

# 2. ARMENIAN LETTER-NUMERALS

| | | | |
|---|---|---|---|
| Ա | 1 | Ճ | 100 |
| Բ | 2 | Մ | 200 |
| Գ | 3 | Յ | 300 |
| Դ | 4 | Ն | 400 |
| Ե | 5 | Շ | 500 |
| Զ | 6 | Ո | 600 |
| Է | 7 | Չ | 700 |
| Ը | 8 | Պ | 800 |
| Թ | 9 | Ջ | 900 |
| Ժ | 10 | Ռ | 1000 |
| Ի | 20 | Ս | 2000 |
| Լ | 30 | Վ | 3000 |
| Խ | 40 | Տ | 4000 |
| Ծ | 50 | Ր | 5000 |
| Կ | 60 | Ց | 6000 |
| Հ | 70 | Ւ | 7000 |
| Ձ | 80 | Փ | 8000 |
| Ղ | 90 | Ք | 9000 |

Picture 502.

# 3. GLAGOLIT LETTER-NUMERALS

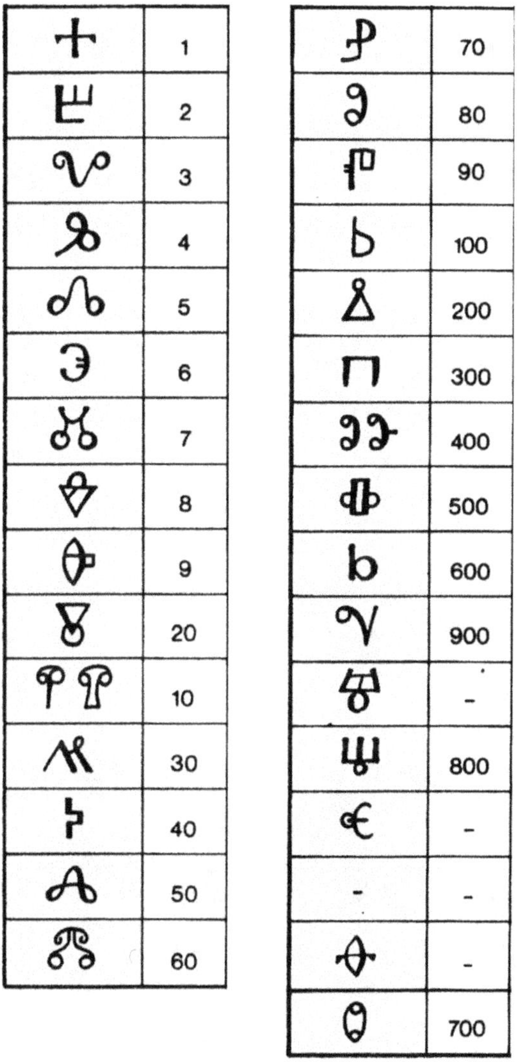

| Symbol | Value | Symbol | Value |
|--------|-------|--------|-------|
| † | 1 | ♪ | 70 |
| Ⴂ | 2 | Ә | 80 |
| ♈ | 3 | ⅌ | 90 |
| ♋ | 4 | Ь | 100 |
| ♏ | 5 | Δ | 200 |
| Э | 6 | Π | 300 |
| ♋ | 7 | ӘӘ | 400 |
| ♦ | 8 | Ⴔ | 500 |
| Ⴔ | 9 | Ь | 600 |
| Ⴒ | 20 | ♈ | 900 |
| ♈ ♈ | 10 | Ⴛ | - |
| ⋏ | 30 | Ш | 800 |
| Ⴆ | 40 | Є | - |
| Ⴀ | 50 | - | - |
| Ⴐ | 60 | Ⴀ | - |
|   |   | Ѳ | 700 |

Picture 503.

# 4. CYRIlLIC LETTER-NUMERALS

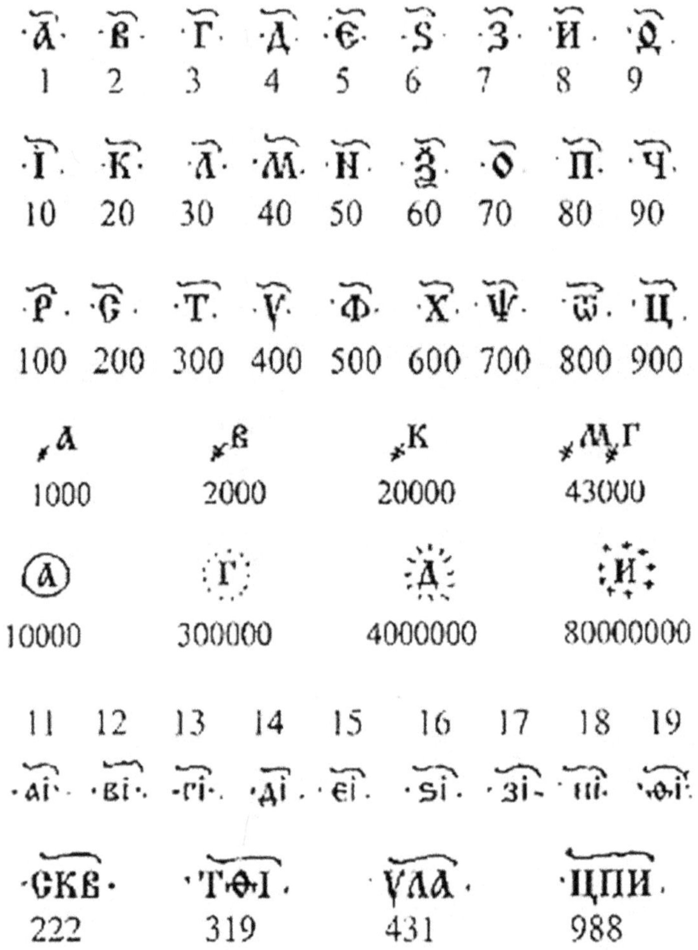

| а̃ | в̃ | г̃ | д̃ | є̃ | ѕ̃ | з̃ | и̃ | ѳ̃ |
|----|----|----|----|----|----|----|----|----|
| 1 | 2 | 3 | 4 | 5 | 6 | 7 | 8 | 9 |

| і̃ | к̃ | л̃ | м̃ | н̃ | ѯ̃ | о̃ | п̃ | ч̃ |
|----|----|----|----|----|----|----|----|----|
| 10 | 20 | 30 | 40 | 50 | 60 | 70 | 80 | 90 |

| р̃ | с̃ | т̃ | ѵ̃ | ф̃ | х̃ | ѱ̃ | ѡ̃ | ц̃ |
|-----|-----|-----|-----|-----|-----|-----|-----|-----|
| 100 | 200 | 300 | 400 | 500 | 600 | 700 | 800 | 900 |

| ⸯа | ⸯв | ⸯк | ⸯм̨г |
|------|------|-------|-------|
| 1000 | 2000 | 20000 | 43000 |

| Ⓐ | г | а | и |
|-------|--------|---------|----------|
| 10000 | 300000 | 4000000 | 80000000 |

| 11 | 12 | 13 | 14 | 15 | 16 | 17 | 18 | 19 |
|----|----|----|----|----|----|----|----|----|
| аі | ві | гі | ді | еі | ѕі | зі | иі | ѳі |

| ск҃в | т҃ѳі | у҃ла | ц҃пи |
|-----|-----|-----|-----|
| 222 | 319 | 431 | 988 |

Picture 504.

# 5. ARABIC LETTER-NUMERALS

| | | | | | | |
|---|---|---|---|---|---|---|
| 1 | ا | 10 | ى | 100 | ق |
| 2 | ب | 20 | ك | 200 | ر |
| 3 | ج | 30 | ل | 300 | ش |
| 4 | د | 40 | م | 400 | ت |
| 5 | ه | 50 | ن | 500 | ث |
| 6 | و | 60 | س | 600 | خ |
| 7 | ز | 70 | ع | 700 | ذ |
| 8 | ح | 80 | ف | 800 | ض |
| 9 | ط | 90 | ص | 900 | ظ |
| | | | | 1000 | غ |

Picture 505.

# 6. HEBREW LETTER-NUMERALS

| | | | | | |
|---|---|---|---|---|---|
| א | 1 | י | 10 | ק | 100 |
| ב | 2 | ד כ | 20 | ר | 200 |
| ג | 3 | ל | 30 | שׁ | 300 |
| ד | 4 | ם מ | 40 | ת | 400 |
| ה | 5 | ן נ | 50 | | |
| ו | 6 | ס | 60 | | |
| ז | 7 | ע | 70 | | |
| ח | 8 | ף פ | 80 | | |
| ט | 9 | ץ צ | 90 | | |

Picture 506.

# 7 COPTIC LETTER NUMERALS

| | | | |
|---|---|---|---|
| 1 | ⲁ | 60 | ⳅ |
| 2 | ⲃ | 70 | ⲟ |
| 3 | ⲅ | 80 | ⲡ |
| 4 | ⲇ | 90 | ϥ |
| 5 | ⲉ | 100 | ⲣ |
| 6 | ⲋ | 200 | ⲥ |
| 7 | ⳇ | 300 | ⲧ |
| 8 | ⲏ | 400 | ⲩ |
| 9 | ⲑ | 500 | ⲫ |
| 10 | ⲓ | 600 | ⲭ |
| 20 | ⲕ | 700 | ⲯ |
| 30 | ⲗ | 800 | ⲱ |
| 40 | ⲙ | 900 | ⳏ ⳓ |
| 50 | ⲛ | | |

Picture 507.

# 8. GOTHIC LETTER-NUMERALS

| | | | |
|---|---|---|---|
| 1 | A | 60 | G |
| 2 | B | 70 | Π |
| 3 | Γ | 80 | Π |
| 4 | d | 90 | Ч |
| 5 | Є | 100 | K |
| 6 | u | 200 | S |
| 7 | Z | 300 | T |
| 8 | h | 400 | Y |
| 9 | Ψ | 500 | Ⱶ |
| 10 | ι | 600 | X |
| 20 | K | 700 | ☉ |
| 30 | λ | 800 | Ꝗ |
| 40 | M | 900 | ↑ |
| 50 | Ν | | |

Picture 508.

# PART IX

## APPENDIXES

## 1. EPILOGUE

We might be convinced, reading this book, that the old dot-line (dot-rod) or its variant, the line-line number-writing system, from *at least* 30,000 years ago is still alive since it has been in use continuously since then. We must accept this continuity even if the distances of time between the very old finds are relatively large. The time distances between the finds get shorter the closer we get to our time, and prove that this is only due to the "ravages of time". Furthermore, independent of the time distances, the same number-writing turns up in the finds.[74]

This continuity has lasted even until today. The signs and the number writing-principle didn't change at all in China nor did its earlier version vary during the long period of Ancient Egyptian history. The numerals were still *being written in the oldest possible way* in the salt mines of Transylvania right up to the beginning of 20th century. Our "Arabic" numerals are no more than the distorted forms of the ancient signs.

Despite using the same signs and rules to write numbers over the last 30,000 years, people created several variants of number-writing. They did it because the system made it possible. It is the nature of the dot-line system that apart from the round numbers, all other numerals enclose more than one sign. Thus, in order to write a number, the dot, the line, the associated perpendicular line and the long-line – as numerals with round values – may be arranged differently compared to each other. Furthermore, the four signs' basic values can freely be substituted for each other and people amply used this possibility all the time. How much? I showed this in the chapter *"The complete sign-collection"* on page 48. The numeral forms created by us humans over the last 30,000 years are merely *"variations on a theme"*. This is supported by the fact that they can be mirrored criss-cross ("be transformed") into each other, as we have seen in the chapter *"Examples of alternating the values and positions of lines and dots"* on page 49.

All these prove the amazing persistence of traditions and peace of mind

---

74 If a cat disappeared behind a thick tree and turned up again on the other side, it would make no sense to argue that the cat is no longer the same cat.

in olden times. The changes were obviously due to some desire, as we could see in the discussion of the number-writing method of Sr. Dán Peter (page 145), but the framework and the traditional rules were earlier never broken. The guess that numbers and writing-signs were generally kept sacrosanct seems to be true, but certainly they were held in high respect.

We also proved in this book that ancient dot-line number-writing has necessarily always been a system using place values. Therefore the history of number-writing using position-values is at least 30,000 years old.[75]

One form-change still happened meanwhile. One branch broke away from this ancient number-writing-method – the group of the so-called number-rovás – at least 15,000 years ago.[76] The reason for the divergence is evident. The numbers had to be carved into wooden sticks (because the values could easily be collected and stored that way), therefore the perpendicular lines (bars) were modified to be suitable for carving. The other method of book-keeping, the 'quipu' (Japanese 'ketsujo') described on page 111, remained identical with the dot-line number-writing. I mention here again the interesting fact that the quipu and number-rovás shared the whole globe roughly equally. It appears that for a long time two ancient cultures ruled the whole earth in equal ratios, like brothers and sisters. Yet, one of the abacus' (soroban) numerous variants has been used for calculations in both cultures, all over the world. The most curious among these calculators was the Yupana used in Mexico. Its round numbers (place values) used the Fibonacci sequence.

A similar duality is seen with place values as well. The sequence of these is either round (1, 10, 100, 1,000) or bisecting (1, **5**, 10, *50*, 100, **500**). I was not able to clarify the history of these variants. Obviously, numbers can be written with fewer signs using bisecting place values.[77]

The very ancient clear forms of number writing started to become distorted - according to the finds - around 3-4,000 years ago, as if the spiritual order which had firmly subsisted for several 10,000 years, then started to deteriorate. Due to this or to some equally serious reason, even the writing sets and the way of writing started to *worsen* and become *needlessly complicated,* along with the numerals' distortion. [78] The reader can follow this

---

75 It is certainly older, since the 30,000 years is only the age of the oldest find yet, though there are the traces of a much older culture before it.

76 The earliest find is around 15-20,000 years old, but it doesn't mean that number-rovás started at this time. It is possible, but this find doesn't prove it.

77 Nearly the only exemption this kind is seen in the place-values in the base 60 Sumerian number system: 1, 10, 60, 600, 3,600, 36,000 etc., while the pre-Elamite system runs like this: 1, 10, 100, 300, 3,000 etc. (see page 331)

78 For example, to start syllable-writing, as in the Turkey or Crete.

in my book *"Signs Letters Alphabets".*[79]

Signs drawn with geometrical accuracy became more and more decorated, modified to the utmost on the Indian continent. Nothing remained here from the ancient forms. These distortions "infected" the Arabic world as well and from there it finally came to Europe. The inspiration to use letters for number-writing started around the same time. It seems to be an Ancient Greek invention and influenced mainly the Middle Eastern region. The Mayans invented, alongside the usual number set, a new sequence of numerals picturing gods' faces (see page 93). The Chinese even constructed a set of numerals looking like "labyrinths":

| | | | | | | | | | | | | |
|---|---|---|---|---|---|---|---|---|---|---|---|---|
| 1 | 2 | 3 | 4 | 5 | 6 | 7 | 8 | 9 | 10 | 100 | 1,000 | 10,000 |

Picture 509. These numerals never became popular probably due to their complexity.

We have seen innovations along with the discussed modifications in the history of writing as well, for instance: Rongo-Rongo or the alchemist's alphabet. Furthermore, need occasionally forced the creation of new signs as in the Morse alphabet or the signs of telegraphy.

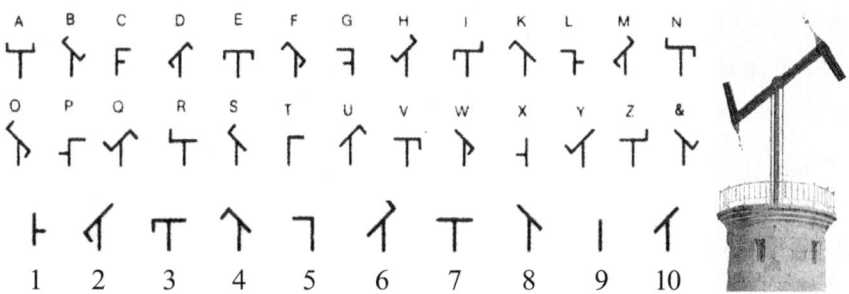

Picture 498. The letters and numerals of the telegrapher.

My presentation of the telegraph alphabet here shows that countless variants of sign-sets can be invented and this can be proven, almost single-handed, by the Indian "cifra" <czifra> (over-decorated) numeral set. (See page 223). This finally proves that the main trend of literacy has been continuous for 30,000 years. However, I also have very strong proofs for the

79 I mentioned dot-line numbers and number-writing in my book "Sign-Letters-Alphabets" (2001) on only a few pages (139-146). I didn't discover then - as several necessary finds were missing – the parallelism of the writing's and number-writing's history. These two are like Siamese twins.

continuity of language for the same length of time.[80]

*The history of writing-signs and numerals tells us that not only did the 30,000-year-old culture not die nor it was ever exchanged for something else, but it has been continuous until today. We must regard ourselves as equally continuous in the world.*

Finally, let's turn again to the 4 theses outlined on page 31. We can repeat them, but now in the form of statements:

1) Dot-line number-writing (and writing) can be attested for the last 30,000 years and before. Its continuity since then is undeniable.

2) Number-writing using place-value has existed as long as the dot-line number-writing system has been used. The oldest find attesting its use is 30,000 years old.

3) Marking zero is not a medieval invention. A sign for zero was a necessity. If the numeral's form marks the place-value of a number, then no zero is necessary. It is there, but no extra marking is needed.

4) The infinitely extendable numeral system has existed since people started recording numbers by any means, because reading the written numbers is only possible with a functioning numeral system. (Of course, largely extended numeral systems were used only in cultures where people liked to deal with large numbers.) The oldest finds – where we are able to prove the recording of numerals - are 35,000 years old (Lebombo on page 208) and 37-29,000 years old (Pekarna, see page 35), but there are numerous, around 30,000-year-old finds. Therefore, since at least this time numeral systems with place-values have existed, because there is a strict rule necessary in the numerals and a different system is unimaginable.

Finished: on 20th March 2012

---

80   See my book "Our words from the past", (page 97) (Frig Publisher, 2010)

# 2. WHAT DOES THE HUNGARIAN WORD 'SZÁM' (number) MEAN?

It is proper to speak about the meaning of the word 'szám' <saam> (number), since it is the most important word in this book. Furthermore, by finding out the meaning of this word, we may learn how our ancestors interpreted the process of counting.

We cannot follow just the set of numerals and their writing methods back to faraway prehistoric times, but should also follow a variety of cultural elements as words as well. For example, some European linguists, investigating the origin of nostratic and pre-nostratic languages, got as far back as 10-15,000 years looking for some word-roots. We will do likewise by following the path of the word "szám" (number).

I am analysing the word "szám" from the Hungarian point of view, not from the general viewpoint, since this book was originally written for Hungarian readers. We could it analyse from the German or even Ancient Greek point of view, but the method would be different. However, the word "szám" must be very old and was certainly part of the common "proto-nostratic" language, for it turns up in many other languages. It is easy to follow, because its form and pronunciation changed little in time.

The meaning of the word "szám" has shrunk in Hungarian to only a part of the original meaning's range. This narrowing down of meaning relates merely to the daily use of the word. The whole range of ancient meanings will appear, however, doing a deeper etymological examination of our vocabulary.

The root of "szám"[81] is the basic root **szöü, sző** <sœue, sœ> (weave like szövet [woven, textile]) and its *reverse* form **ösz** <œs> can be understood as öszvet [össze=öszve] (something put together, compound).

The word **ösz** = **szöü, sző** with "ad-glued" (agglutinated) -**m** (as in rém [horror] or gém [heron]): **szím, szám** [82].

---

81    Czuczor-Fogarasi: Magyar nyelv szótára (Dictionary of Hungarian Language), MTA, Budapest, 1862. Reversing word-roots is occurring quite often and no doubt, it had happened earlier as well. Examples of today: köp-pök <kœp-pœk> (spit), sivít-visít <shivit-vishit> (shricks), huśáng-suháng <hushaang-shuhaang> (bludgeon), ect. It is interesting in our case that szöv(et) (fabric) became after a newer reverse of root vász(on) (linen, canvas)

82   The word "szám" has nothing to do with "szem" (eye). The latter means "shining", its root is "szí" and is not connected to the words "szemcse" <sɛmchɛ> (granule) or "szemét" <sɛmét,> (garbage).

According to these, the original, clear meaning of "szám" is: something compound, put together, is together, and figuratively: similar, identical, suited to and timely, seen at once.

Therefore, "szám" means the result of an "add up process". Think of the Greek word "summa", pronounced Hungarian: "shumma", containing the word szám with a 'u' = szum.

Germans still use the word "szám" with its original meaning: together. The word "sammeln" (collect) is formally identical with the Hungarian számolni:

> samm|el|n (Ger. collect, accumulate, concentrate)
> szám|ol|ni (Hung. counting)

In Hungarian számolni (counting) means össze- = öszve-rakni, <œssɛrakni> (putting together), summing up. We mentioned earlier össze-olvasni (reading together), meg-olvasni (compute, count) until we get one szám (number). In Hungarian we call both the value and the numeral a "szám". These multiple meanings dim the original sense for us.

The pronunciation-variants of the word "szám" so far accumulated in Hungarian are as follows: s*zám, sim, cím*; changing 'm' to 'n': *szín, cin, csin*. We can prove with a little effort that they really only differ in their pronunciation.

We have seen above that **szám** is identical with the Ancient Greek **süm** = together. Now let's replace the variants of **szám** in the following Hungarian words with the Ancient Greek **süm**. It won't cause any confusion. On the contrary, this change lets us clearly understand why these words really have those meanings:

| | |
|---|---|
| **simul** <shimul> | **sümül: összül,** becoming one |
| **cim**bora <cimbora) | **süm**bora: "like grown together" |
| **cim**borálni <czimborázni> laborate | **süm**borálni: holding together, col- |
| **cim**balom <czimbalom> | **szüm**balom: (bal = bill) össze-bil- |
| lent (tips over) it is a striking instrument | |
| **cin**kos <czinkosh>**szüm**kosz: | plays together (accomplice) |
| **csin** <chin> | **szüm**: össze, (here collision), (for instance: the football-players got into the "csin") |

Ancient Greek examples:

**Süm**etros [συμμετρος] = **sim**ilar size, **sym**metrical.
**Süm**biosis [συμβιοσις] = „együtt biózás" = „együtt bujázás". Bio = buja (lush). Buja növényzet <buya nœvenjzɛt> (lush vegetation). The original meaning of bio, buja is reproducing, multiplying, in

reality living, therefore **süm**biosis = **sym**biosis = living together.

Several pronunciation-variants of the word **süm** (sym) came up, e.g. in Latin:

1) With the '**s**' changed to '**k**' (Latin '**c**'): com [this happened often e.g. Cesar, Szézár, Császár <chaasaar>, Kaiser or e.g. szomor <> komor (mournful, sadden)]. **Com**: together with something, at once. Latin cum: with something, e.g.: two examples:

Pag = fog [p-f] (hold). Putting s**züm** = **com** before it: **com**pages = össze-fogás [83]<œssɛfogaash> (collaboration).

Par = pár, putting **szüm** = **com** before it: **com**paro = **össze**párol, párosít = com|par, that is **szüm**|par = fellow, **cim**|bora.

In Hungarian, the root com, **kom** stands in the words **kem**ény <kɛmenj> (hard) and in kom|a (godfather) words.

The word **kom**, identical with **szüm**: together, but **koma** is just a part of the whole Latin word **com**|pater, meaning együtt-apa <ɛdjuett-apa> (together father), apatárs <apataarsh> (fellow-father).

The **kom** stands in the word **kem**|ény (hard) as well; people must have interpreted this word as holding together, pressed together. **Szüm**|eny does it and we call it **töm**|ény today, after [sz-t] change. **Kem**|ény and **töm**|ény <tœmenj> (concentrated) are pronunciation-variants as well as **töm**|ör <tœmœr> (solid). The etymology of the word **szám**talan (countless) is similar. It was earlier szám|éntalan and its variant is **töm**|éntelen (large amount). Töm|eny meant earlier a huge unit of a military troop: 10,000 soldiers.

2) **Kon**, com spread with 'n' as well: con e.g. con|kako. There are many examples in Latin: **con**tact, **con**fluo (flowing together), **con**suo = összeször <œssɛsœ> {suo = sző} (interweave) in Latin: (stitches up). The reverse of sző = ösz.

The meaning of the word **szüm** branched in two directions:

a) the main branch: together as in **summ**|a (shumma) = amount ($\Sigma$), but the word **szom**|széd (neighbours) contains the root **szüm** as well. They live beside each other having common borders, that is, they

---

83 With today's pronunciation compages = számfogás, but here, the original meaning „together" (szüm, summa) was used.

összesímulnak <œssɛsimulnak> (nestle close to one other).

b) the other branch: adequate, homophonous with something. Therefore is **szám** > széma = **séma** <shema> (scheme) and **szumma**, but **szüm** = **cím** <czíím> (title) or **cím**er <czíímɛr> (coat of arms), posta-cím (postal address). It numbers, figuratively labels. We use for this the word **szüm**ke = **cím**ke (lable).

**Cím** (cziim) is therefore a sign corresponding with something.
Cim, written with 'n' = cin. "**Cin**kelt lap": marked, labelled card.

If in Ancient Greek the sounds 'g, k or kh' [γ, κ, χ] followed the 'm' of szüm (sym), then 'm' became assimilated to 'g'. The word **süm**khoreo became **szüg**khoreo [συγχωρεω] = **öss**zekerül (e.g. meeting, coming together). **Szüg** = szüm = **össze** and the word khoreo(l) is pronounced today **kerül** in Hungarian (its root is ker, kör).

The word **szüg** (created by assimilation) is today identical with the words "**cég**" <czeg> (company) and "**cég**ér" <czeger> (trade mark) and mean: together.

The words **cím**er (coat of arms) and **cég**ér (trade mark) therefore mean the same, because **cím** and **cég** are both pronunciation-variants of the same word, which must have been **szüm** (sym). Therefore, szüm = com = szüg = cég is the cause of the identity of cég and company.

The pronunciation variant **szüg** stands behind the words "**szeg**ődik" <sɛgœdik> (accompanying, hires oneself out) and "**seg**éd" <shɛged> (apprentice, helper). Ellen-**szeg**ődni means opposing or refusing an összetartozást (a close relationship).

See a few international words out of the dialectical variants of **szüm** = **sim**, szim: **szim**ultán (simultaneous) shah game, **sim**ulator = doing **sim**ilarly, **sim**ulate, **sim**ulation (making it similar).

We can state that we found very many connections for the word "szám" going far back in time. It seems that, if we put the history of writing and number-writing along with these findings, then everything goes back to one point in the past. We see more and more uniformity going backward into the past. It seems there was only one single culture in the past if we go back sufficiently far.

Did the horses run apart?

# 3. BIBLIOGRAPHY

Andrew Robinson: *The Story of Writing*, Thames and Hudson, 1995.

Barabási László: *Az emberiség története* I-IV. kötet, Fríg Kiadó, 2007-2012. (The history of mankind)

Bartos Huba-Hamar Imre: *Kínai-magyar szótár*, (Hungarian-Chinese dictionary) Balassi Kiadó, Budapest, 1998.

Borbola János: *Királykörök*, (Royal circles) Írástörténeti Kutató Intézet, 2001.

C. B. F. Walker: *Cuinform*, British Muzeum Press, 1998.

Colin Renfrew-Paul Bahn: *Régészet – elmélet, módszer, gyakorlat*, (Archeology - theory, method, practice) Osiris Kiadó, 1999.

Cser Ferenc-Darai Lajos: *Kárpát-medence, vagy Szkítia?*, (Carpathian Basin or Scythia) Fríg Kiadó, 2008.

Daniel Pots: *The Potter's Marks of Tepe Yahya*, 1981, internet

David Eugene Smith: *History of Mathematics*, Dover Publications, Inc., New York, 1958.

Denis Schmandt-Besserat: *How Writing Came About*, University of Texas Press, 1996.

D. E. Smith: *History of Matematics*, Dover Publications, Inc, New York, 1958.

Egmont Colerus: *Pythagorastól Hubertig*, (From Pythagoras to Hubert) Franklin Társulat kiadása, A búvár könyvei XVII.

Filep László-Bereznai Gyula: *A számírás története*, (History of Number-writing) Filum, 1999.

Gáboriné Csánk Vera: *Az ősember Magyarországon*, (The prehistoric man in Hungary) Gondolat, 1980.

Geirges Ifrah: *Numbers*, Havrill Press, London, 1998.

Georges Jean: *The Story of Alphabets and Scripts*, Thames & Hudson, 2000.

Georges Jean: *Langage de signes*, Gallimard 1989.

Gönczi Ferenc: *Somogyi gyermekjátékok*, (Children-toys from Somogy county) Kaposvár, 1949.

H. L. Resnikoff-R. O. Wells, jr: *Mathematics in Civilization*, Dover Publications, Inc, New York, 1984.

John D. Barrow: *The Book of Nothing*, Vintage, 2001.

Juhász Zoltán: *A zene ősnyelve*, (The ancient language of music) Fríg Kiadó, 2006.

Kabay Lizett: *Kulcsképekhez kulcsszavak,* (Clue-words for important pictures) Főnix könyvek 24., Debrecen, 2000.

Károlyi Mária: *A korai rézkor emlékei Vas megyében,* (Finds of Early Cupper-Age in the county Vas) Vas Megyei Múzeumok Igazgatósága, Szombathely, 1992.

Makkay János: *A tartariai leletek,* (The finds of Tartaria) Akadémia Kiadó, 1990.

Mandics György: *Rejtélyes írások,* (Misterious writings) Akadémiai Kiadó, Budapest, 1987.

Marija Gimbutas: *The Living Goddesses,* University of California Press, 2001.

Mario Alinei: *Ősi kapocs,* (Etrusco: una forma arcaica di ungherese)( Allprint Kiadó, 2005.

Maurice Poppe: *The Story of Decipherment,* Thames and Hudson, 1976.

Mesterházy Zsolt: *A magyar ókor* (The Hungarian antiquity) I-II., Magyar Ház könyvek, 2002.

Michael A. Cremo-Richard L. Thompson: *Forbidden Archeology, The Hidden History of the Human Race,* Bhaktivedanta Book Publishing, Inc.,

Paul G. Bahn: *Prehistoric Art,* Cambridge University Press, 1998.

Peter T. Daniels-William Bright: *The World's Writing Systems,* Oxford University Press, 1996.

Sebestyén Gyula: *Rovás és rovásírás,* (Rovás and rovás-writing) 1909, Magyar Néprajzi Társaság. Utánnyomás: Püski Kiadó, 1999.

Szabó István Mihály: *A magyar nép eredete,* (The origin of the Hungarians) Mundus Magyar Egyetemi Kiadó, 2004.

Tóth Imre: *Magyar őstörténet,* (The Hungarians ancient history) Fríg Kiadó, 2000.

Varga Csaba: *JEL JEL JEL,* (Signs Letters Alphabets) Fríg Kiadó, 2001.

Varga Csaba: *Szavaink a múltból,* (Our words from the past) Fríg Kiadó, 2010.

Varga Csaba: *Az ősi írás könyve,* (The book of the ancient writing) Fríg Kiadó, 2002.

Várkonyi Nándor: *Az írás és a könyv története,* (History of writing and book) Széphalom könyvműhely, 2001.

Zolnay Vilmos: *A művészet eredete,* Holnap Kiadó, 2001.

See other books by Csaba Varga:

Signs Letters Alphabets
The  Living Language of the Stone Age
English from the Hungarian View
Our Words from the Past
Hungarian - English as well
The origin of mind

www.ingramcontent.com/pod-product-compliance
Lightning Source LLC
Chambersburg PA
CBHW051438170526
45166CB00001B/35